"The story begins on the banks of the ⟨...⟩ moves with the energy of an electric g⟨u⟩ ⟨...⟩ years of human spiritual decay. If you ⟨...⟩ destroyed the standing of women, po⟨l⟩ ⟨...⟩go, and gained the sheer insanity necessary for ⟨...⟩ ⟨b⟩e able to loot a hemisphere in the name of God and machines and greed, *Beyond Geography* will explain the answer to you."

—Charles Bowden

"This is no ordinary critique of modern civilization . . . *Beyond Geography* is clearly a work of brilliance and imagination, a compelling, disturbing, and uncommonly literate exploration of one of mankind's most basic dilemmas."

—*Atlantic Monthly*

"*Beyond Geography* takes us to a deeper significance of the American frontier than . . . ever dreamed possible. It is a step away from estrangement and thus a step toward all the Black Hills and toward accepting the long-standing invitation of Native Americans, 'the lands wait for those who can discern their rhythms.'"

—*The Nation*

"[A] hauntingly lyrical yet vitriolic indictment of Judeo-Christian civilization. . . . Will undoubtedly offend . . . those Americans who still fall back on the guilt-absolving cushion of Manifest Destiny."

—*Boston Sunday Globe Magazine*

"Belongs with Dee Brown's excellent *Bury My Heart at Wounded Knee*."

—*Denver Post*

"Five hundred words cannot do justice to Turner's vivid blow by blow account of the Pyrrhic victory of western man over wilderness and wastelands. Don't waste time on reviews, read the book! . . . A fresh, profound and dramatic perspective on American history."

—*Pacific Historian*

"Fascinating, provoking, and filled with fresh insights."

—Dee Brown

# Beyond Geography

## THE WESTERN SPIRIT AGAINST THE WILDERNESS

## Frederick Turner

RUTGERS UNIVERSITY PRESS
NEW BRUNSWICK, NEW JERSEY

*Fifth printing, 1994*

First edition published in 1983 by
Rutgers University Press under agreement with The Viking Press.
Reprinted by arrangement with Viking Penguin, a division of Penguin Books
USA Inc.

**Library of Congress Cataloging-in-Publication Data**

Turner, Frederick W., 1937–
  Beyond geography : the western spirit against the wilderness /
Frederick Turner.
    p.   cm.
  Originally published: New York : Viking Press, 1980. With new
pref. and introd.
  Includes bibliographical references and index.
  ISBN 0-8135-1909-8 (paper)
  1. America—Discovery and exploration. 2. Civilization, Western.
3. Indians—First contact with Europeans. 4. Indians, Treatment of.
I. Title.
E101.T87   1992                                            92-17569
970.01—dc20                                                CIP

For

ROY  HARVEY  PEARCE

# Extracts

(AFTER MELVILLE)

It is not down in any map; true places never are.

—Herman Melville, *Moby-Dick*

And I'm still not convinced that I didn't penetrate beyond geography.

—Saul Bellow, *Henderson the Rain King*

. . . there being a Geography of Religions as well as Lands, and every Clime distinguished not only by their Laws and Limits, but circumscribed by their Doctrines and Rules of Faith.

—Sir Thomas Browne, *Religio Medici*

The essential, as you know, is not of the political order, but of the spiritual.

—Pope John Paul II

> They soon forgat his works; they
> waited not for his counsel:
> But lusted exceedingly in the
> wilderness, and tempted God in the desert.
>
> —Psalms, 106:13,14

. . . the good olive, if neglected for a certain time, if left to grow wild and run to wood, does itself become a wild olive.

—Iranaeus, *Against Heresies*

. . . while a Christian is honestly serving God, he is a stranger even in his own state. We have been enjoined as strangers and sojourners to sojourn here but not to dwell here.

—St. Cyprian of Antioch

They were anchorites, i.e., withdrawers, because, being by no means satisfied with that victory whereby they had trodden underfoot the hidden snares of the devil (while still living among men), they were eager to fight with the devils in open conflict and straightforward battle, and so feared not to penetrate the vast recesses of the desert.

—John Cassian, *Colloquia*

If man is an explorer, the greatest quest of all is his search for the Creator. In this all lesser expeditions are contained. . . .

—C. J. W. Simpson, *North Ice*

The Church of God must no longer be wrapped up in Strabo's cloak; geography must now find work for a Christianography in regions far enough beyond the bounds wherein the Church of God had, through all former ages, been circumscribed.

—Cotton Mather, *Magnalia Christi Americana*

Genuine island 1500 miles to the West.

—Notation on the Bianco map of 1448

. . . islands . . . unknown to trade, to travel, almost to geography, the manner of life they harbor is an unsolved secret.

—Joseph Conrad, *The Secret Sharer*

It need not be supposed that a deep hereditary response to former habitats would necessarily be experienced with pleasure.

—Paul Shepard, *Man in the Landscape*

Hence it was that wherever the Admiral went the Indians came out on the paths to receive him with presents of food and some quantity of gold which they had gathered on learning that this was what he had come for.

—Ferdinand Columbus, *The Life of the Admiral*

I have come to the conclusion respecting the earth, that it is not round as they describe, but of the form of a pear, which is very round except where the stalk grows, at which point it is most prominent, or like a round ball, upon one part of which is a prominence like a woman's nipple, the protrusion being the highest and nearest the sky, situated under the equinoctial line, and at the eastern extremity of the sea. . . . I do not suppose that the earthly paradise is in the form of a rugged mountain, as the descriptions of it have made it appear, but that it is on the summit of the spot, which I have described. . . .

—Columbus writing to Ferdinand and
Isabella of his third voyage

. . . an Indian guided them through an immense pathless thicket of desert for seven days, where they slept continually in ponds and shallow puddles. Fish were so plentiful in them that they were killed with blows of cudgels; and as the Indians travelled in chains, they disturbed the mud at the bottom, by which the fish, becoming stupified, would swim to the surface, when as many were taken as were desired.

—Relación of the Fidalgo de Elvas in
*Narratives of De Soto*

. . . a man arrives as far as he can and not as far as he wishes. . . .
—Vasco Núñez de Balboa in a letter from Panama
to Ferdinand of Spain, January 20, 1513

. . . I shall set sail to the east with the fixed resolution to discover the termination of the Niger or perish in the attempt. . . . My dear friend Mr. Anderson and likewise Mr. Scott are both dead, but though all the Europeans who are with me should die, and though I were myself half dead, I would still persevere; and if I could not succeed in the object of my journey, I would at least die on the Niger.

—Mungo Park, *Travels of Mungo Park*

Be wise and learn my secrets, how disease is healed, how man and beast and plant may talk together and learn one another's missions. Go and live with the trees and birds and beasts and fish, and learn to honor them as your own brothers.

—Words of the last of the Stone Giants to an
Iroquois hunter lost in the woods

What does Africa—what does the West stand for? Is not our own interior white upon the chart? black though it may prove, like the coast, when discovered. Is it the source of the Nile, or the Niger or the Mississippi, or a Northwest Passage around this continent, that we would find?

. . .

. . . it is easier to sail many thousands of miles through cold and storm and cannibals, in a government ship, with five hundred men and boys to assist one, than it is to explore the private sea, the Atlantic and Pacific Ocean of one's being alone.

—Henry Thoreau, *Walden*

Was that what travel meant? An exploration of the deserts of memory, rather than those around me?

—Claude Lévi-Strauss, *Tristes Tropiques*

. . . I know that your deepest inclinations are impelling you to a study of the occult, and do not doubt that you will return home with a rich cargo.

Only don't stay too long from us in those lush tropical colonies; it is necessary to govern at home.

—Letter from Sigmund Freud to Carl Jung, Vienna 1911

A curious thing about the Spirit of Place is the fact that no place exerts its full influence upon a new-comer until the old inhabitant is dead or absorbed.

—D. H. Lawrence, *Studies in Classic American Literature*

The land! don't you feel it? Doesn't it make you want to go out and lift dead Indians tenderly from their graves, to steal from them—as if it must be clinging even to their corpses—some authenticity, that which—.

—William Carlos Williams, *In the American Grain*

# Preface to the 1992 Edition

An author's attitude toward a previous work is always a curious thing. For instance, if that work happened to be wildly successful, its author often appears to shun it and to discourage discussion of it, as if its very success cast a long and invidious shadow over all his subsequent works. The new work, the one in progress, he'll tell anyone who'll listen, is *the* work and the only one he's interested in talking about. If, on the other hand, a previous work was a failure, the author may exhibit toward it a melancholic tenderness, as if speaking about a lost child or a much-beloved dog who met an untimely death. And there are also writers who profess to be utterly indifferent to any and all of their previous works, whether successes or failures. A well-known American novelist and the first professional writer I knew well fit into this larger category. He once told me that as soon as he'd finished a novel, he lost interest in it and was consumed with ideas for a new one—rather like Balzac, who was said to have finished a novel at three in the morning and begun on the next at four.

*Beyond Geography* was neither a wild success when Viking published it in 1980, nor a failure. Instead, it pretty quickly disappeared from popular attention and became a kind of "cult" book, one of those singular works that has its life almost at the level of oral tradition. In the years since its first appearance I have often been surprised to meet up with devoted readers who wanted to speak to me only about *Beyond Geography*, and on these occasions I have known something of that desire to discourage such discussions and instead direct these devotees to more recent work. Had they read my *Such-and Such*? No, well I was sure they would find it interesting, etc. Usually these efforts were nugatory except to engender within the devotees of the book a mild disappointment with its author. Meanwhile, in its Rutgers University Press edition the book continued to sell modestly along, bobbing up on occasion into a wider attention—on the anniversary of Earth Day, for instance, when it was mentioned in various places and I was asked to read from it at an Earth Day rally.

In that same year, 1990, the audience for *Beyond Geography* became wider with its publication in Brazil as *O Espirito Ocidental Contra Natureza*, thanks to its devotees there, particularly the environmental activist José Augusto Drummond, who did the translation. And now there is this new English-language edition, timed in part to meet the rising tide of interest in the origins of New World history. My hope is that after the tide again recedes, *Beyond Geography* will continue to do the work for which it was designed: to provoke discussion of our mission and our obligations here in what remains of that new world Columbus discovered for a civilization that was about to visit upon it an unprecedented ecological destruction. Since the Admiral dropped anchor in the Caribbean islands on that fateful fall day in 1492, fourteen of the twenty-six species of parrots, parakeets, and macaws originally found there have become extinct, and the surviving species have all suffered massive declines in numbers. And this is only to introduce the smallest of illustrations.

Against the continuing extinguishing force of that civilization, the countervailing force of a book—paper, ink, glue—can seem terribly puny. Except for the words it contains. Finally, words, the power of language, will be what we find we are left with as we struggle to preserve what remains of the New World. It is with words that a new myth of the New World will be composed, a myth that celebrates the absolute virtues of all our fellow residents, from sequoias to spotted owls to snail darters.

Frederick Turner
Santa Fe, New Mexico
1992

# Preface

After I had read one of the chapters of this book to a small group at Amherst College, a young woman asked if I was aware that I was offering poetry as history. I gracefully accepted what I pretended was a compliment, but I knew at the time it was much more than that. In rereading the manuscript I knew even more surely that the question had identified the central problem I faced. And this was how to write on a subject whose scope was far beyond my competence—and likely to remain so.

What I wanted to write about was my vivid sense—let me risk calling it a vision—that the real story of the coming of European civilization to the wildernesses of the world is a spiritual story. To me it is the story of a civilization that had substituted history for myth as a way of understanding life. It was precisely this substitution that enabled Europeans to explore the most remote places of the globe, to colonize them, and to impose their values on the native populations.

The vision began, as the reader will see, on a day I spent roaming the hot and windy hills of the Pine Ridge Reservation in South Dakota. I saw myself there as both an inheritor of conquest and as an alien. I knew that both the Lakota and the Cheyenne had held sacred the Black Hills I could see in the westward distance, but I knew also that a belief in the sacredness of lands was not in my heritage. The distance I felt there was more than geographical. I could see the Black Hills. I was on a piece of aboriginal America. But I was estranged by history from them.

In subsequent months my vision became more coherent, less personal. A larger panorama began to disclose itself, and yet the thing was still a vision, not a thesis. The latter implies a logically constructed argument that moves by judicious use of evidence to a conclusion supported by the facts. What I had in mind was beyond that kind of proving, but I felt it to be truer than anything I knew of on the subject.

What I knew was (and is) relatively little compared with the mass of material that bore on the subject. Half a lifetime might be required before I could master enough information on the various constituent fields to "justify" the sense I already had of the phenomenon. Some years ago the historian Johan Huizinga was faced with a similar problem. "In treating of the general problems of culture," he wrote, "one is constantly obliged to undertake predatory incursions into provinces not sufficiently explored by the raider himself. To fill in all the gaps in my knowledge beforehand was out of the question for me. I had to write now or not at all. And I wanted to write."

The way I chose to write may be closer to that of the literary artist than to that of the historian. By this I mean only that I have been forced by the formidable scope of my subject to choose my materials with an eye for the symbolic, the revealing detail. For example, when I learned from Carlo Cipolla's *Guns, Sails, and Empires* that the same European bronze craftsmen who once were employed in the manufacture of church bells found themselves in the fifteenth century and thereafter employed as casters of huge siege guns and more portable field-artillery pieces, this fact told me much for my purpose, and I hunted through the various fields for similar luminous items that would tell me the truth I had already grasped.

The method does not always work as smoothly as it might. Occasionally the reader will see that I have felt compelled into so close an engagement with historical fact as to temporarily overshadow the vision of the whole. This was the insight of the young woman in that Amherst College audience. Yet I hope that by the time the final pages are turned, readers will find that the vision has been a consistent one and the invitation both clear and genuine.

Except for rare explanatory footnotes, I have dispensed with this overused and distracting device. But I have wished to acknowledge my many debts as well as to suggest to others works that might stimulate further thinking on the subjects discussed in these pages. Thus the brief essays and the bibliographies that are attached to them in the notes on sources. It should be understood that the citation of any particular work

does not mean that its author can be found in support of specific con-
tentions of mine. Rather, it means that I have found instruction in those
pages and have put it to my own uses. I trust this is notice of neither
subversion nor plagiarism.

# Acknowledgments

A number of individuals gave me ideas, suggestions for further reading, or lent me books, and I want to thank here Tamas Aczel, Normand Berlin, Peter Farb, and Louis Greenbaum for assistances. Malcolm Cowley, Paul Mariani, and Patrick Sullivan read portions of the manuscript and their comments gave me a good deal to think about. William Preston read the entire work and lent me his wisdom and the benefit of his long experience as a researcher in buried historical truths. Elise R. Turner also read the whole, bit by bit through many months, and heartened me in those moments when the range of my attempt seemed formidably vast. I also received spiritual sustenance and blunt prodding from Michael S. Harper and Ronald Sheffler. William and Juliana James helped me over a rough spot with the spontaneous gift of hospitality, friendship, and a place to work. The staff of Brush Ranch Camp, Terrero, New Mexico, were unfailingly generous with their office equipment when I was many miles from any town. In the last stages I was considerably aided by a leave granted me by the University of Massachusetts and by a grant from the National Endowment for the Arts. Through it all I had the benefit of knowing that my editor, Barbara Burn, and my agent, Carl Brandt, were at the other end, believing in me and confidently awaiting the finished work. Of course, all these friends are hereby given immunity from whatever errors may follow.

# The Travels of Turner

T. H. Watkins

A few miles up a little canyon in the Escalante drainage of southern Utah there is a huge "bandshell" of red slickrock, its sandy floor half in sun and half in shadow as it slopes down to a bright, cold, friendly little creek that chuckles through a bed of black shining rocks on its way to the Escalante River. There are a number of very large boulders scattered around on the sand here—remnants of the Jurassic age that frost and heat and unmeasured earth movements have pried from the ancient stone wall above them. On one of these is incised one of the most perfectly rendered petroglyphs I have ever seen, a depiction of the old Anasazi spirit whom the modern Navajo call Kokopeli—the flute player. Horned and hunchbacked, sporting an erection that is nearly as long as the flute in his mouth, the antic spirit seems to dance mockingly as he leans toward the coiled figure of a horned rattlesnake and plays an eternal piping whose melody can only be imagined, but is no less powerful for the fact.

To stand beneath the great curve of rock and behold this painstakingly graven image is to share in a spiritual experience older than history, older than conscious memory, older than anything we yet know how to measure—or at least older than anything *most* of us yet know how to measure. Frederick Turner might know how, I think; taking the measure of such numinous moments is what he has done throughout a book that is itself numinous with warps and layers and crannies of spiritual insight. Quite aside from the book's very real contributions to the philosophical history of the European experience in the New World, it is the quality of spiritual abundance that is the enduring virtue of *Beyond Geography: The Western Spirit Against the Wilderness*. Some of that value, I cannot help but believe, comes from Turner's protean instincts. With the publication of *The Viking Portable North American Indian Reader* and *Geronimo: His Own Story* more than twenty years ago, he established himself as an authority on the Native American experience, but over all the years I have worked with him as an editor and fellow field hand in the vineyards of the word, I have discovered very damned little that does not interest him—and, in interesting him, inspire him to find out everything he can about it, think about it, and then write it up and publish it, usually with twists of wisdom all his own.

My first encounter with his work came during my stint as an editor at *American Heritage* magazine in the mid-1970s, when Turner sent us a large portion of *Beyond Geography* (then a work-in-progress) with the offer to adapt it as an article for the magazine. The manuscript was passed on to me for first reading and, since it takes no particular talent to recognize genius when it is thrust in front of you, I soon urged my colleagues to take him up on his offer. They agreed. The article, which appeared as "The Terror of the Wilderness" in the February 1977 issue of the magazine, was an incisive and frequently lyrical exploration of the psychological, religious, and other cultural impedimenta that Europeans brought with them when they crossed the big pond and decided to conquer the North American continent. However well done, this was all more or less what one might have expected on such a subject; but then Turner started talking about the mythic character of the "captivity narrative" as both symbol and symptom of what was wrong with the ways in which our European ancestors thought about the continent and its inhabitants, and we were in spiritual territory never before explored in the classic (and sometimes staid) traditions of *American Heritage*. This made at least one of the editors uneasy, so in the author note that ended the piece we attached the slightly nervous observation that the article was "thoughtful" and "provocative," as if we were attempting to assure our readers that we knew we had just done something a little weird but that they shouldn't worry about it too much because we were not likely to do it again.

We didn't, either—but we did publish Turner again, and in formats that illustrate my point: his next piece was a definitively expert biographical essay on that Dust Bowl icon Woody Guthrie; this was swiftly followed by a historical and descriptive essay on Barataria, that marshy, islanded, faintly ghostly region that spills into the Gulf of Mexico south of New Orleans; and, finally, a wonderfully knowledgeable essay on New Orleans' Congo Square and the beginnings of American jazz—

with emphasis on the shadowy life of the legendary trumpeter Buddy Bolden—a piece whose language took on the richly inventive character of the music about which Turner was writing. In the years since, his work here at *Wilderness* magazine has included historical essays on the fate of irrigated civilizations and the development of the U.S. Forest Service, as well as works of conservation journalism that have illuminated the past, present, and problematic future of such essential American landscapes as the disappearing Louisiana coast, the arid Southwest, and the Southern Appalachian Highlands. For other publications he has written with equal facility and intelligence about baseball, literary movements, and food, and his three other books have continued his eclectic traditions—*Remembering Song: Encounters with the New Orleans Jazz Tradition*, a gathering on jazz; *Rediscovering America: John Muir in His Time and Ours*, an interpretive biography; and, most recently, *Spirit of Place*, a collection of essays discussing the influence of land and community on the work of several major American writers, from Willa Cather to Leslie Marmon Silko.

All of which is to say no more than that Frederick Turner hardly ever met a subject he didn't like. It is to say that his curiosity and its articulation in print are the symptoms of a writer who has spent most of his working life in search of connections among the varieties of experience on this planet. He is not a cultural anthropologist by training, but he shares the distinction of the breed when he seems to take as his specialty the entire tapestry of known and assumed human traditions. It was this erudite generalism that produced *Beyond Geography*, a poetic and complex book whose importance lies in the simple power of its message: that the most enduring strands that have characterized all cultures through all times have been those spiritual connections that bind us to the land—and that our greatest folly came when we were persuaded that our principal obligation to ourselves, our posterity, and our gods was to sever those bonds and pretend they had never existed.

We all share a common mythic past, one that goes back to that unimaginable moment when the fish that learned to walk (as Loren Eiseley liked to characterize us) began to blend cognitive thought with imagination to produce symbols. It was the wilderness that taught us this trick, Turner says. Forged in direct response to the demands of survival that the wilderness imposed, "in the deepest and truest sense myths are directions for the orchestration and recognition of life energies" and "refer to *all* that is to be found within Life's experiences as well as to the end of Life and what comes after. Thus living myth must include and speak of the interlocking cycles of animate and vegetable life, of water, sun, and even the stones, which have their own stories." This ritual confirmation of the interdependence of all life sustained countless human cultures through tens of thousands of years, at least until the first dibble stick was jammed into the first plot of cultivated earth to signal the beginning of civilization. As the structure of the human community became more centralized and intricately organized in the Mediterranean and then the European worlds, the spiritual character of our relationship to the land slowly evolved through a spiral of increasingly humanistic (and ultimately corrupt) religious systems until the anthropocentrism of orthodox Judaism and Christianity all but obliterated the traces of its earthly connections. By the time internal forces of empire and restlessness sent Europeans to exercise dominion over the lands and people of the "New World," reason had triumphed over myth, civilization over savagery.

So it has always been argued, and Turner narrates this venerable history better than anyone before him. But the true importance of *Beyond Geography*, I think, lies in the fact that at the same time he outlines the convolutions of this story with clarity and skill, he also takes it farther and deeper than anyone ever has, and in the reading of it in this context one is drawn to a conclusion that may enlarge our understanding of how and why our own civilization has developed such a complex and frequently contradictory relationship to the land and the wilderness.

At the center of this conclusion is the revelation that however profoundly European civilization attempted to distance itself both deliberately and subconsciously from the earth, it still could not fully escape its origins. Together with the baggage of dominion and conquest, of violent Christian certitudes that vindicated the pursuit of treasure and the wholesale destruction of resident cultures, Columbus and those who followed him to this uncharted world five centuries ago carried like an inarticulate dream the racial memory of paradise lost. Turner himself might even argue that the explorers had been driven into the Ocean Sea in an inchoate response to that loss: "Like . . . Desert Fathers repressing their bodily desires in mortification, the West turned to exploration as both a 'palliative remedy' (Freud) and a way of harmonizing the rest of the world with itself. If it could succeed in making the map of those spaces beyond itself match that of the Christian West, then certain torments of the spirit, certain cognitive disconfirmations (as psychologists would have it), might be better borne."

And what did the explorers find as solace for their cognitive disconfirmations? Nothing less than the wilderness their civilization had spent centuries denying and subduing—a dark, jungly, forested, and dangerous country, an immensity so vast and unknowably complicated that nothing in their memory could validate it, two continents that weighed on the European mind like the physical representations of a terrifying subconscious condition that could not even be acknowledged. What was even more unsettling was the fact that it was an *occupied* wilderness—and that those who occupied it could be recognized all too well. In the native residents of this New World, the European community suddenly was confronted by that common mythic past it had rejected. The waves from that disturbing shock of recognition still influence what we do here—for, as Turner makes clear, this meeting with the wilder half of the shared human experience presented the Europeans with choices they had almost forgotten how to make. The tragedy, of course, is that when given this opportunity, the choices they made were almost invariably those which continued the long alienation from the wilderness that had nurtured their beginnings.

It was largely a matter of refusing to listen—of allowing all the cultural noise of European tradition to drown out saner voices from that mythic past (or mythic self) they had so uncomfortably re-encountered. The voices were there to hear, had they been so inclined. The ancestors of the Native Americans began the long trek across the Bering Land Bridge from Siberia at least twenty-five thousand years ago, slowly wandering south into the midcontinent of America, spreading, multiplying, splitting into language groups and extended families, bands, tribes, societies, establishing communication routes and currency exchange, complex systems of religion, government, and community, engaging in wars of expansion and trade dominance, building technologically sophisticated irrigation societies in the Southwest, intricate mound-building cultures in the Mississippi River Valley. According to recent estimates, just in what is now the United States there was a population of anywhere from five to ten million waiting when European explorers discovered their world. These people did not know that their world was lost and in need of discovery, of course, though what some of them suspected early on was that it soon *would* be lost; Iroquois legends tell of ancestral dreams that "a great monster with white eyes would come from the east and devour the land." Among most of these resident and hugely variant cultures the power of connection remained at the heart of their traditions.

It could not have been more clear. "Our legends tell us," Luther

Standing Bear, a Lakota Sioux, related in 1933, "that it was hundreds and perhaps thousands of years ago since the first man sprang from the soil in the midst of the great plains. . . . In time the rays of the sun hardened the earth and strengthened the man and he bounded and leaped about, a free and joyous creature. From this man sprang the Lakota nation. . . . We are of the soil and the soil is of us." And, being of the soil, he might have added, it was necessary to listen to it and all else in that biotic mix collectively known as the land. What was there that could be heard? Let Brave Buffalo, another Sioux, tell some of it: "When I was ten years of age I looked at the land and the rivers, the sky above, and the animals around me and could not fail to realize that they were made by some great power. I was so anxious to understand this power that I questioned the trees and the bushes. It seemed as though the flowers were staring at me, and I wanted to ask them, 'Who made you?' I looked at the moss-covered stones . . . but they could not answer me. Then I had a dream, and in my dream one of these small round stones appeared to me and told me that the maker of all was Wakan tanka, and that in order to honor him I must honor his works in nature."

To honor the works in nature was the leitmotif of Native American cultures, however much they otherwise may have varied one from the other. A Hopi beginning the complex weave of ritual in a kiva to ensure a good corn harvest, California Yahi Indians chanting in their sweat huts to cleanse themselves of illness, a Cheyenne hunter giving thanks to the spirit of the buffalo he has just slain—all of these were acting out the protocols of interdependence, satisfying the human need to propitiate disaster and promote security in the only way known to them. To know the land, to listen to the land, to respect and even love the land, was to be nurtured and sustained by the land, to be given license finally to be "free and joyous." These resident peoples knew that the world and its creatures were not presented to us, we were joined to them, that our presence on this earth was not meant to be a conquest but a sharing. This was Original Knowledge—knowledge that we had lost and now could learn again because we had been blessed in a great turn of history by the re-encountering of our own past. It was the chance to embrace once again what Turner says "may yet prove to be our most successful response to life on this earth."

What we did then is another kind of history, sad and savage, a failure involving the destruction of both land and people. We not only did not listen to what we might have heard, we did our best to annihilate the messengers and then went on through most of our time here to view the land as a commodity—something to be used and, if necessary, used up in the pursuit of acquisition and power, blindly following the traditions that had already diminished us as a species. The consequent wreckage of land and wildlife is in evidence all around us as we fight with and among one another to discover salvation. It is both a good fight and a necessary one, and if the environmental movement succeeds in redeeming at least some of the damage our history has done, future generations may view it as the most important social movement of all time.

But the movement needs to remember where it truly comes from. It needs, to be precise about it, to remember the message of this book, which invests this new edition with an added significance. For however virtuous, the fight takes place in largely civilized venues—in the cloakrooms and cubbyholes of Congress, where legislation is crafted and argued; across large tables in agency offices, where administrative decisions are made; in the courts, where law is implemented and challenged; in the sundry forms of communication available to us, where the arguments for and against this and that are given the gross luminosity of print and television imagery. It is a terribly professional business, and it has gone on long enough by now to have acquired its own arid academic literature, as well as complex bureaucratic systems and restraints. This is not to denigrate professionalism. The most effective conservation organizations today are replete with professionals and are the better for their ecologists, biologists, economists, resource analysts, legislative specialists, journalists, environmental lawyers. Nor is it a criticism of bureaucracy. There are good bureaucrats and bad bureaucrats, and our civilization today is largely in the hands of bureaucrats.

But it is to say that our truest strength as a movement does not lie in our professionalism, but in acquiring and maintaining an understanding of those spiritual connections that are even more important to survival today than they were when human beings first learned to embrace the earth—because today we have the ability to so profoundly corrupt the rest of life that it cannot be reclaimed. Whether we know it or not, then, we are all of us engaged in a quest so demanding that we must seize wisdom where it is found—as it is found in these pages. It is not merely in the history that Frederick Turner gives us here that we will discover our way back to the dream of wilderness. It is in the spiritual landscape that he so carefully maps for us, the landscape of love and myth and human yearning that lives in the numinous world beyond geography.

# Contents

PART THREE  Haunts

PART ONE

# Loomings

# Estrangement

Once, a painter friend of mine interrupted his monologue on the New York art scene since de Kooning with a reminiscence of his discovery of the rest of America. Having spent all of his youth east of Newark, he had been genuinely astonished by the country he traveled through on his first motor trip to California. I recall especially his evocation, still charged with the primitive emotion of wonder, of the spaces that opened to him west of Chicago, the huge expanses of sky given dimension only by drifting, solitary clouds, the shadows these made on the waving fields, and the little hamlets, sighted miles off by the grain elevators bulking expectantly by railroad sidings. As he painted now with words, it called to mind other, earlier such discoveries, most immediately, I think, those of the Dutch—because of de Kooning and New York—and the unmistakable sense of amazement that somehow found its way between the lines and around the margins of their businesslike inventories of the New World. It was all so much grander, more various and plentiful, than anything the newcomers were prepared to accept. And so, for quite a spell, the Dutch huddled and built small.

My friend's provincialism and belated emancipation reminded me of my own narrow, youthful sense of America, growing up in a midwestern hub west of which a more expansive sense of place commences. In those days we lived with a vague and somehow exciting feeling that just west of the billboards of the vacant lots we played in there was something huge and powerful stretching endlessly toward cattle, cowboys, and sunsets. At night, the radio brought us unwanted stock and feed averages from out there while we waited for melodramas of the real West.

Summers spent in northern Wisconsin helped strengthen the impression that there was more to the country than south Chicago, overpowering though it was with its tense geography of wires, blocks, neighborhoods. But here in the north woods the sense was not of that expansiveness so intensely realized by the Bronx painter and guessed at by us amidst the weeds and broken glass of the vacant lots: here it was more the experience of density, the green and faintly damp groves of trees reaching up and down and blocking vision so that what was felt was more the weight, the impenetrable, immemorial mass of the country, than its reach. And once I remember my father taking my brother and me to a part of the woods we had never seen and remarking that this was a stand of virgin timber. The old and moss-clung trunks rose above us into foliage that seemed to be sighing something that had accidentally come to include us. In ancient times myth-bound peoples would spare certain magnificent trees in groves where they were cutting down all the rest; the stately survivors were left to house the spirits ejected from the fallen timber. Here, surrounded by second- and third-growth forests, these trees were perhaps mere oversight, but they made their statement.

Wisconsin gave us something else, too, and this was a brush with the continent's aboriginal presence. Traveling the county roads, dust boiling up over the fenders, we would pass the fenced boundaries of the Menominee reservation. While our parents busied themselves with roadmaps and sandwiches for kids, we would wonder why people chose to live in such squalor way out here in the country. At the turn where our road went into the woods an Indian family lived in a rocky field, their hut surrounded by junk, and it seems to me now that on several occasions my eyes locked with those of a dark-eyed, bang-shadowed boy of my age who stood amidst that alien litter. For many years that was the extent of my contact with the natives of my country. Only later did I come to connect my ignorance of them and their past with my terrifically incomplete sense of this New World I had been born into.

Thus, for example, I once lived in a Rhode Island cottage within easy walking distance of a monument marking the site of the "Great Swamp Fight" of King Philip's War. The monument, like the war itself, was meaningless to me. Nor was my sense of what had gone into the making of our land sufficiently alive to appreciate the fact that to get to the monument I had to pass through the woods ghetto of a band of mixed-blood Narraganset Indians. On sunny winter days their thinly clad, runny-nosed children could be seen squatting close to the roadbed, soaking up its meager warmth. In summer they played half-naked amidst the bottles, tires, and abandoned cars. And there was one man, chained or rooted to this spot, with eyes of pure ice in a furious face, who glared at my ignorant passages through this scene of historic misery.

Gradually, however, an implication emerged, and this was that my own ignorance of America, an ignorance amounting to an estrangement from the land, was in some real measure that of Americans generally. For me this was revealed both in the recondite records of our past and in travels through large stretches of North America. In these ways a feeling of American loneliness began to insist upon itself, a crucial, profound estrangement of the inhabitants from their habitat: a rootless, restless people with a culture of superhighways precluding rest and a furious penchant for tearing up last year's improvements in a ceaseless search for some gaudy ultimate.

This is an extraordinary phenomenon, and indications of it are to be found earlier than the political origins of the Republic. It was one of the first things traveling Europeans noted about those who had come to live in the New World. Here people seemed to have an itch, as if the living were uneasy, troubling, almost frantic. It was as if those who had inherited the fruits of exploration and conquest had been left a troubled bequest, as if there were some unplacated, unmet spirit of place dividing them from an authentic and comforting possession here.

Pondering this led inevitably to questions concerning the far larger phenomenon that Howard Mumford Jones once termed "the Europeanization of the globe." Why was it Western Civilization that had accomplished this gigantic, spectacular feat—especially since that civilization itself had either inherited or borrowed most of the requisite expertise and technology, had in fact been preceded to the Americas thousands of years before by Orientals, Africans, Celts, and Semitic peoples from the Mediterranean?

Of course, there is no single answer to this question unless one grasps that old and awfully handy one: that the "inherent superiority" of Western Civilization enabled it to push beyond its geography and into the unknown spaces of the globe, outstripping and encircling those older civilizations whose earlier speculations and probings had formed so much of the basis for the West's subsequent achievements. Indeed, this answer has been given often enough in the historiography of geographical discovery. Yet clearly it is internally contradictory since anyone who has studied the history of exploration knows that Western Civilization's contributions to the sciences of cartography and navigation were actually syntheses, applications, and extensions rather than true innovations. Paper, ink, and printing, without which neither maps nor navigational information would have been possible, originated with the Chinese, as did the magnetic needle. The map itself is probably a Babylonian invention, and the sea chart was a creation of Mediterranean precursors of the Greeks. Indeed, a great deal of Greek navigational lore, eventually transmitted to Europe via the Arabs, was probably inherited from these still more ancient and highly proficient sailors. The lateen sail, which made long, open sea voyages practical possibilities instead of suicidal risks, is a development of the Arab dhow. Even the machine itself, the distinguishing cultural artifact of the West, did not originate in the West but instead in the ancient Near East. And yet, as Lewis Mumford has shown, it remained for Westerners "to adapt the whole mode of life to the pace and capacities of the machine." It was Western Civilization alone, he remarks, that had the genius to exploit and bring to full force the ideas and inventions of others, to collect and resynthesize their "technical debris."

Above all, there is the brute, vivid fact that it was the West that went out in huge, rolling waves and made its incursions permanent facts. Sculpted portraits of Semitic merchant princes found in the Americas; scattered Celtic dolmens in New England; an ancient Chinese manuscript describing a coasting voyage in what was not yet called the New World; another from Egypt giving the outlines of a trip to the land of Punt (in 1492 B.C.); even the peopled islands of Polynesia that seemed so paradisal once to Spanish eyes and that may have been settled by Libyan sailors—all these tell us that between the aborigines and the Western whites others came, and went away, or else merged their blood and customs with native populations. None transformed vast expanses of the unknown into recognizable counterparts of their homelands.

No single part of what was once unknown to the West—or more than fugitively suspected—has been so radically transformed, so completely Westernized, as the land that opened out before my painter friend's eyes. As Jones once remarked, in the brief flash of three centuries Europeans transformed this "Stone Age landscape" into that of a modern European nation. North America is thus the epitome of whatever it was that moved the civilization of the West in its long push outward. *This* is what exploration and conquest has meant and has come to.

The full force of this transformation can most truly be felt in some portion of what remains of aboriginal America where the unplaced, unconciliated spirit of place is all but tangible. There it is possible to be aware of the burden of our history. So it came once to me, squatting amidst the high grass and under the wind-driven clouds of the Pine Ridge reservation in South Dakota. I sifted a few small but weighty stones in my hands, hearing their coarse scrapings, feeling their heft drag my hands earthward. Watching the wind that is almost constant out there wave the feathered lanceheads of the grass, I thought of the Teton Lakota proverb, "A people without a history is like wind on the buffalo grass," and I wondered how it came to be that neither in proverb nor in history have we been able to come to loving terms with what we must call home. The answer that came to me, not at that moment but only as I have thought about it since, came from the distance between myself and the gentle mountains off to the west—the Black Hills, sacred to the Lakota, whose tribal legends described them as the figure of a recumbent woman from whose breasts flowed Life itself and to whom both humans and animals went as children to a mother's arms. It was a distance longer than history, deeper than the mere accident of being born who and where I was. It was the distance between me—us—and human beings whose relationship to their lands was dictated by an oblique but strong recognition of human biology, by the particularities of those lands, and by a living mythology that celebrated all this.

And so it has seemed to me in thinking over that experience as a double alien on Lakota lands that the true story of Western exploration, and thus of America, is a spiritual one. It has its basis not in technology, not in the rise of nation-states and the consequent international rivalries that dominate the historiography of exploration, but rather in the history of that mythology that tied the West together into a quarrelsome, unloving, but nevertheless recognizable unit, sharing the same religious symbols, owning the same holy writ, deriving nominal spiritual identity from the same source, and drawing in its wilderness probings sustenance from the ancient desert analogs.

This, then, is an essay in spiritual history. As such it is necessarily tentative, but at last it is founded on that surest of realities: the human spirit and its dark necessity to realize itself through body and place.

# The Necessity of Myth

Just as the recorded history of the race begins with the artifacts of myth, so must an essay that seeks to trace the spiritual history of Western exploration begin with myth, that most basic expression of the human spirit which has nerved all civilizations and which, when destroyed, has left pathology and death in its stead.

But wherever we begin must also be *in medias res* since myth, as the basic expression of the human spirit, must be coeval with human culture itself—and may well be of an order even more ancient, as we shall see. But surely, as far back as artifacts can now take us, some sort of symbolic activity (of which myth is maybe the most gorgeous example) has been characteristic of human cultures. Almost half a million years ago humans were employing skillfully wrought tools. As Olivia Vlahos has remarked, the capacity to select and chip bone or stone into a tool is evidence of the capacity to symbolize, since the raw material must be envisioned as a finished implement before fashioning can begin. And when we reflect further that at this point stone tools had been developed to a finish and symmetry clearly beyond the merely functional, we begin to sense that powerful gravitational pull on our imaginations that caused Joseph Campbell at the outset of his great series on the history of myth to draw consolation from these words of Thomas Mann:

> Very deep is the well of the past. Should we not call it bottomless? The deeper we sound, the further down into the lower world of the past we probe and press, the more do we find that the earliest foundations of humanity, its history and culture, reveal themselves unfathomable.

Thus even to begin with the putatively ceremonial hand ax is to begin somewhere in the midst of process.

Yet if we confess to historical incompleteness, we are also compelled to assent to what the poet Robert Duncan has called the truth and life of myth. No matter how estranged we may consider ourselves from the tangled, improbable—indeed monstrous—world of myth, in various ways and under certain circumstances unwilling tributes to its signs, fictions, and symbols are wrung from us. As Jung would put it, for moderns the experience of these archaic visions is simultaneously strange and strangely familiar, as if, reading or listening to an unknown narrative, we should gradually become aware of a rhythm of events announcing itself in advance so that we foreknow the conclusion. "The myth," writes Duncan, "or pattern of elements in the story is a melody of events in which the imprint of knowledge—knowledge here in the sense of a thing undergone—enters the generative memory and the history of man takes on tenor."

As is so often the case with myth where apparent opposites turn out to be complementaries, resistance is also a sign of assent. When we meet these impossible fictions, we feel that the force of all our onward living ought to be at least equal to their obscure power. Thus young people, exposed for the first time to the myths of aboriginal America, so often refuse to enter these spiritual woods, fearing to lose the bearings our culture has given them, the rational and discriminating guides to a world already sufficiently bewildering. And yet, being young enough, they are tempted, and sometimes this is enough: they enter, encounter the alien, and at last find it deeply familiar.

There are, of course, a good many other ways in which this assent to myth may be elicited, ways that lie outside the translated texts in which the vestiges of our past lie embalmed. In some truly unimproved natural setting—one well removed from the reach, the sights, and maybe especially the sounds of our wonted culture—surrounded by the immemorial phenomenal world, whether trees, ocean, or the waves of prairie grasses, a change may overtake us, precisely to the extent that we are willing to remain where we are and resist what will be a gathering temptation to return to more certain comforts. It will not quite be fear, but it will be next to this: a kind of existential humility born of a sense of all the life that surrounds and includes us and that will go on without us. And this is the ground of myth—fear or humility and submission to the still unfathomed mystery of Life. That is why the most pervasive refrain in the myths of aboriginal America is an address to the Spirits, to the Master of Life, to the Great Mystery: "Pity me!" And that is why the heroes and heroines of myth, as Campbell has shown, are not conquerors but are instead those who submit, bending their energies into accord with greater forces.

Such an assent as that adumbrated here derives ultimately from the fact that after all we are animals, and myth should be regarded as a natural, adaptive response of the animal to its environment. And it may well be that myth—or at least its enactment in ritual—is not even an exclusively human activity. In recent years ethologists, psychologists, and mythographers have been suggesting that the search for the origins of myth and thus its ineluctable claims on us may take us well beyond the early "high" cultures of the near East; beyond the Paleolithic cultures whose artists rendered the round of myth in the animals and signs they placed in the sanctuaries of caves; beyond even the Neanderthalic evidences of myth in burial arrangements, bear cults, and ritual cannibalism—beyond all these into the organized life of other forms.

"Ritual" behavior has been observed in species as various as apes and fish, and elephants encountering the exposed skeletal remains of one of their own will redistribute these in a pattern that is to them internally satisfying. Even invertebrate forms such as ants and termites may have rituals, and the bees have their strange and marvelous rotating dance, symbolizing the distance and direction of a food source from the hive. All this suggests that we may be dealing here with a mode of behavior, a life response, far grander than ziggurats, deeper than caves.

In an essay that synthesizes much of this sort of evidence, Earl W. Count offers the view that myth-making is not only coeval with humanity but is "an aspect of our morphology." All vertebrates, he remarks, possess portions of the brain whose chief function is symbolic activity. This can only mean that myth is ultimately a product of "neurologic energies that are more ancient even than mammals, let alone the primates and the line of man." The symbolic process that finally manifests itself as myth is thus "both symptom and reflex of an edifice that has been compounded both phylogenetically and ontogenetically of primitive materials that represent a succession of elaborations." Seeing symbolic activity and therefore myth-making in this anciently functional way, we are prepared to understand that myths are not decorative and outmoded fictions but the instinctual and sure responses of the organism to Life. The symbol, according to Count, must be understood "as a means for contacting reality, and not as a device designed to shut out reality."

All of this should cause us to regard anew that much-remarked symbolic activity of the so-called primitive peoples of the wildernesses. Inference and the evidence of our wilderness documents confirm that these primitive peoples with all their rituals and "mumbo jumbo" have been acute students of their worlds and therefore of reality. Western whites, invading the foreign realities of the wildernesses, consistently noted with astonishment the nature-craft as well as the unaccountable prescience (pre-science) of the primitives, the dark others seeming to know and understand the totalities of their unfeatured worlds, both phenomenal and numinous, in ways whites could not but grudgingly admire. Indeed, the primitives' harmonious and precise knowledge of their habitats came in the process of the "Europeanization" of the globe to be the very mark of the primitive itself: feelings and expressions of kinship with the animals and even trees, stones, and water. Totem animals, sacred trees or ones with faces carved in them, soul stones all became the talismans of the savage, or, in the anthropological literature that followed conquest, evidences of the "childhood of the race." It seemed especially obvious, as whites watched the primitives imitating the dances of the animals, that these peoples had not developed even a rudimentary sense of human superiority over the lower forms of life. And yet, if dance is, as Martha Graham has asserted, the secret language of the body, in the mythic context it may be the expressive language of Life, and the primitive dances patterned on observations of buffalo, birds, and bees not evidence of a hopelessly fantastical view of reality but a beautiful way of remaining in nurturing contact with it.

Moreover, culture contact in the wildernesses made it clear that primitives believed that animals, as well as vegetation, stones, and stars, had soul life and languages appropriate to their kind. Primitive observations of animal behavior in particular fostered the conclusion that animals too had their myths and their ritual dances. So myths abound telling us of human visitors to animal kingdoms where they learn the language of a species and its ways, and learn thus to respect that particular form of life. And sometimes these myths tell us that the human visitors learn so well a wilder way that they do not wish to return to the human world but would rather roam forever with the herd or swim with the school. Such narratives express the interconnectedness and interdependence of all life. They tell us that it is an illusion to see true separations between the forms of being. Cumulatively they have the effect of sanctifying that life they describe.

As an example of what is meant here, let us briefly submit to the "melody of events" in this Zuñi Indian narrative of a young man's first deer hunt. His tutor is Coyote, that wise and cunning beast who so often plays the role of trickster-tutor in aboriginal American mythologies. Coyote gives the young man his hunting instructions, especially the sacred song of promise and offering, and he concludes his instructions with these words:

> Whenever you take a beast's body, give something in return. How can a man expect much without paying something? If you do not give creatures the wherewithal of changing being, how can you expect them to relish your arrows? So, whenever you slay a game creature, offer him and his like prayer-plumes—then they will feed you with their own flesh and clothe you with their own skins.

With this they set out on the hunt, Coyote showing the young man the proper tracking technics and outlining the strategy of attack. When the deer sights us, Coyote says, he will dart away, and I will run ahead of him to cut him off. You must do your part by singing the song of promise and offering:

| | |
|---|---|
| *Ná-a-le, ná-a-le!* | Deer, deer! |
| *Tom an te-á-nan,* | Thy footprints (I see), |
| *Ta-pan, ha í--a;* | I following, come; |
| *Há-lo-a-tí-nan* | Sacred favor (for thee) |
| *Ha thle-ai-é-ha* | I bring as I run. |
| *E-ha! e-ha!* | Yea! yea! |

All goes well as the tracks of a huge buck are discovered and followed through the light snow cover of a canyon floor, but as Coyote and his raw pupil close in on the prey, the young hunter's excitement overcomes his training. The deer, listening nearby, hears the young voice stammer and falter in the song.

> Deer, deer!
> Thy footprints (I see)
> I following, come;
> Sacred favor (for thee)
> I— I— (Oh yes!) bring as
>     I run.
>             Yes! yes—oh yes!

And the buck thinks:

> Ah! Young hunters sometimes forget their payments of sacrifices as they do their songs; I must be off!

And with this, he turns and races up the canyon, the chagrined youth
following but the wily Coyote out in front. Again the young hunter
sings the song, and this time the buck is better pleased, but still doubtful, he flees again before his tracker. He stops again farther on and hears
the song of offering drifting up to him. "Ha!" he thinks. "I should die
contented could I know that he would make payment—but who
knows?" Alarmed by his own doubt, he springs away once again. Now
for the final time the words of the song are borne to him, but this time,
turning to flee, he is cut off by Coyote. Caught between two tormen-
tors, the buck charges the trailing hunter whose fright-sent, ill-aimed
arrow shatters its shaft harmlessly against the boss of the antlers. Like
the first shot, the charge miscarries, and the buck wheels again against
the hunter. This time the young man is equal to his task, and his arrow
is buried to its very feathers in the massive breast of the charging beast.
The buck staggers, falls, struggles to rise again as Coyote shouts to the
young man to shoot again. But the novitiate can remember only that he
has not yet fulfilled the next step of the ritual, which is to embrace the
fallen prey, breathe his breath, and say, "Thanks, my father, this day
have I drunken your sacred wind of life." Now the young man dodges
between the frantically thrusting antlers and places his face against that
of the buck, breathing its life's breath, covered with its life's blood,
until finally the prey relaxes and relinquishes to him.

Coyote is joyous, for such fortune is a sign that the Beings of the
Game will befriend this hunter forever. After instructing him further in
the necessary sacrificial offerings to be made at the site of the killing,
Coyote reminds the hunter of their significance:

> Thou receivest flesh wherewith to add unto thy own flesh. For this thou
> shalt always confer in return that which giveth new life to the hearts of slain
> creatures.

And now the man can hunt alone.

The narrative's intentions are clear and large. It inculcates the lesson
of humility before nature and of gratitude for what nature may choose
to grant. It teaches also the lesson of trust in oneself, a trust founded on
both the proper observances of the hunting rituals and on a sense of
symbiosis as a law that embraces both hunter and hunted. And,
through the fiction of the language of the animals and the converse between them and the human being, it teaches the interconnectedness of
all things.

Such myths as this are to be found in numberless variations throughout the world. Wherever cultures have developed and retained a vital
and intimate contact with their habitats without the buffers of advanced
technology and its concomitants of removal, arrogance, and waste—
wherever, in sum, mythology is the ultimate technology—there we
find such artifacts of embrace and accommodation.

The art of the Paleolithic caves permits us another vision of this sanctified, unified world of the past—though, to be sure, that vision is partially obscured by our ignorance of the ancient specifics and by an enormous accretion of contemporary assumptions and counterassumptions.
Even so, for many in the West this parietal art may be their most direct
and vivid acquaintance with a living mythology, though it has been
sleeping in its caves for more than ten thousand years, like some of
those mythic heroes of later times who are said to have retired to similar sanctuaries to await in sleep and dream another time of need.

How symbolically appropriate for us to have to descend into these
caves of Europe and Africa, as if reentering the womb of time, in order
to reencounter myth's delight and mystery! And again, how apt that we
should realize with the discovery of more and more caves, the enormous variety and suppleness of this "caveman" art so that at last words
like "crude," "primitive," "credulous," and "uninformed" become
hopelessly nondescriptive, and our monolithic interpretations are resisted at every turn by the complexity of the evidence. When we enter
this world of myth we find to our astonishment that it is we who seem
primitive and comically serious surrounded by the esoteric and sophisticated play of the cave people. With torches, scaffoldings, flint tools,
and paints of ochre, charcoal, and animal substances, they celebrated a
theriomorphic world of almost incredible beauty, power, and harmony
in which man finds his place as just another being. The relative scarcity
of human figures on the cave walls and ceilings would not so surprise
us were we not the victims of a long-demythicized, anthropocentric
civilization. And whatever explanation we adopt for the few but startling animal/human figures—the bird-headed phallic figure of Lascaux,
the horned bison/man of Le Gabillou, the famous masked and horned
dancer of Trois Frères—whether we interpret them as shamans, sor-
cerers, or mere hunters in animal skins stalking their prey, we are
confronted by a vision of the glorious indivisibility of Life. The vision
was achieved over a long period of little technological elaboration perhaps, but one governed by the generative principles of archaic myth
that do not acknowledge such hierarchical divisions as those between

heaven and earth, animals and humans, body and soul. Indeed, the most significant recent work on Paleolithic art, that of André Leroi-Gourhan, emphasizes cave art's depiction and celebration of these generative principles. To Leroi-Gourhan, the bison and horses and the incised signs are symbolic representations of the creative act of coupling.

Despite our aesthetic delight in this Paleolithic art (which when first discovered filled contemporary observers with incredulity that shaggy, gut-sucking savages could ever have executed this), we have retained until quite recently a historically induced contempt for the world view that made the art itself possible. Our skill in discriminating and separating, a skill as overdeveloped as a weight lifter's biceps, permits us to appreciate the art but not its authorizing world view. Repeatedly over the last century and a half attention has been called to the primitives' "unwarranted extension of mind" into the external world, to myth-bound peoples' fallacious assumption that because they could think and desire, so could animals and trees. "Omnipotence of thought" this is usually called. The false attribution of soul life ("animism") to nonhuman aspects of the phenomenal world, it has been said, prevented primitives from seeing the world as it really is, prevented them from discerning the operation of process behind the apparently random singularity of events, and thus kept them *captives* of the world rather than allowing them to become *masters* of it. Believing they saw and felt life and spirit everywhere, primitives in this view were constantly in the craven act of genuflection before multitudinous power centers, their lives riddled with baseless anxieties, pocked by hideous acts of contrition and propitiation.

Spiritual possession—the "childish credulity" and "loss of control" endemic to this psychic state—is profoundly offensive to the post-mythic Western mind. In time the condition of letting go, of being entered into or taken over by the spirit of idol, sanctuary, or communal rite, became another of those convenient shorthands for all that distinguishes the childlike primitive world from the adult world of progress and civilization.

Such judgments levied against myth and its demands are, in their way, well founded, and no one attempting to understand any portion of the spiritual history of the West can afford to dismiss them with a crude antirationalistic counteroffensive. The world of myth-bound primitives *does* appear a fearful one with its multitude of unseen powers. Indeed, this is part of the power that speaks obscurely to us from the walls of the caves or in accounts of terrific, bloody initiation ceremonies. Melville, that not-quite-lapsed Christian, knew this from his sojourn in the mythy isles of the Pacific. "Though in many of its aspects this visible world seems formed in love," he wrote, "the invisible spheres were formed in fright." This primitive mythic world is a numinous one, and great power carries with it the possibility of great terror. As well as it expresses the joy and play of life processes, myth also expresses those darker tides of existence that no amount of electricity and reason may ever obliterate.

We must also admit that the attribution of soul life to various aspects of the natural world clearly does preclude manipulation of it much beyond a subsistence level (a favorable habitat may help here), and it has often been observed that civilized people's longings for a supposed lost paradise are only possible for those whose level of living allows the luxury of regret. The abandonment of that pervasive sense of the inviolability of large parts of nature is what permits their efficient uses, and this is the way to cultural elaboration and the accretion of comforts. But when handed the implements of modern agriculture which would have permitted them to move beyond their hunting-and-gathering economy, a tribe of Native Americans was spiritually horrified, and a chief asked the uncomprehending whites how he and his people might dare to violate their Mother's breast with cruel metal harrows.

We know also the truth of ghastly sacrifice and mutilation dictated by myth, stretching like a blood-red artery from prelithic cultures into the light of historical time. When we gaze, for example, into the tanned and sealed features of the human sacrificial offerings to the Great Mother, miraculously preserved for two thousand years in the acidic waters of Northern Europe's peat bogs, we are literally face to face with the monstrous teeth-and-bone-rending dimensions of archaic myth; and these are dimensions we must not avoid lest we fail to understand the horror that suffuses so many of our wilderness documents as they remark the experiences with the atavars of ways superseded.

But having said this much, we must remember that to concentrate on these darker aspects of the primitive mythic world is to distort it by attention to its parts and to the local and the temporal. For regardless of whatever local, temporal excesses may fairly be charged to myth's account, the gigantic and general purpose of archaic myth is the celebration of Life. After we have dissected and judged the follies and cruel demands of archaic thought and behavior, we have still to come to myth's core.

Witness: Late in the last century a team of distinguished Americanists set out to investigate a tradition of the Keres Indians which told that in mythic times they had lived in an impregnable sky city across the valley from their historic site at Acoma—where Coronado found them in 1540. According to this tradition, one day when the people were at work in their fields below, an awesome thunderstorm struck their rocky fortress of Katzimo, utterly destroying the scaffolding that had been their one mode of access. Only a lone woman and her child remained above. The tribe circled the fortress home for days, but they could not find even a niche for a fingerhold. Then, sadly, they abandoned Katzimo and traveled across the fields and low-shrubbed flats to a neighboring mesa where they built Acoma.

The Americanists immediately found at least one part of this tradition true: Katzimo *was* impregnable. Equipment had to be constructed and shipped, and with its aid the whites accomplished what legend says the aboriginal inhabitants could not. Atop Katzimo, the team set their trained eyes to search for archeological evidence of a mass dwelling site but turned up little except such random arrow points as might have been the casual litter of an adventurous hunter or two. Evidentially the legend was only that; the whites measured the mesa (2700 by 600 feet) and then left it.

Thus does ancient belief appear to succumb to historical investigation. Except that in conquering Katzimo and running the narrative to bare rock, the whites missed the core of both the narrative and the rock it was founded on. And this seems to be quite simply and wonderfully that the story of Katzimo is intended to give a particular people a strong, rooted sense of place. It seems to say: This is where we have always been; this is where the Great Spirit intends for us to remain. It is unreasoning but sure.

It becomes increasingly obvious in oblique ways that such supports are necessary to us too, and that whatever sacrifices they demand in "objective truth" are outweighed and justified by the comforts of old truths, old necessities, we carry within us. Over the long course, Life stripped of its numinous dimensions may be phylogenetically insupportable. Assessing modern civilization and its discontents, Freud was moved to a kindred observation—though he could not accept the notion that the major source of the discontent he observed around him was a spiritual one. Life as we find it, he writes, is too hard and makes too many severe demands upon us. Therefore man cannot do without "palliative remedies," of which there seem to be three major kinds: "powerful diversions of interest, which lead us to care little about our misery; substitutive gratifications, which lessen it; and intoxicating sub-stances which make us insensitive to it. Something of this kind is indis-pensable." But for Freud the "truth and life of myth" was far too costly a remedy; myth was instead a neurosis of the archaic past that man was struggling to live beyond.

Jung, on the other hand, came to understand the ancient mythic past modern man carries within him, not as a dump heap of childish neuro-ses, but as a continuing source of wisdom that poetically posits the ines-capable essentials of the human condition. This was ultimately why he disappointed his mentor, and why he may ultimately be of greater rele-vance. To Jung, the discontents of contemporary civilization were the inevitable consequences of Western Civilization's contemptuous and progressive disregard of the truths by which all men once had lived and which yet live within the individual. With this mythic past in mind, he writes that "whenever we give up, leave behind, and forget too much, there is always the danger that things we have neglected will return with added force." The fullest possibilities for human life, therefore, are to be realized in conformance with the grand, overarching shapes found in all mythologies, a conformance that is "neither a question of belief nor of knowledge, but of the agreement of our thinking with the pri-mordial images of the unconscious."

As a student of the world's mythologies Jung was hardly romantic about them: he knew too well the full spectrum of actions they sanc-tioned. But he knew also that the brain of modern man is essentially that of the Paleolithic cave artists, and that what moved them to their radically beautiful identification with their universe must also move us, though in obscure and thwarted ways. "We do not assume," he writes, "that each new-born animal creates its own instincts as an act of indi-vidual acquisition, and we must not suppose that human individuals in-vent their specific human ways with every birth. Like the instincts, the collective thought patterns of the human mind are innate and inher-ited."

The subterranean vitality of archaic myth, then, turns out to be something more than mere primitivism (which George Boas and A. O. Lovejoy succinctly defined as "the discontent of the civilized with civili-zation") or romanticism. It is instead attributable to the natural history of the race. For in the deepest and truest sense myths are directions for the orchestration and recognition of life energies. Stripped of their local manifestations, they are revealed as joyous and even playful celebrations of Life, celebrations in which those darker tides of existence lend the richness and tone of reality itself. And the Life so celebrated cannot be the narrow, unsexed, and anthropocentric version that Western Civilization has become uncomfortably familiar with through our attenuated successors to the myths of the primitive past. It must, instead, refer to *all* that is to be found within Life's experiences as well as to the end of Life and what comes after. Thus living myth must include and speak of the interlocking cycles of animate and vegetable life, of water, sun, and even the stones, which have their own stories. It must embrace without distinction the phenomenal and the numinous. In such ways these vital fictions turn us toward the unchanged realities we must live amidst. They may yet prove to be our most successful response to life on this earth.

# Bearings from
# the Ancient Near East

It may well be objected that myth, even in its most primitive state, is evidence that man is already estranged from the rest of creation, that the capacity to symbolize and conceptualize animals, other forms of life (though these are curiously absent in Paleolithic parietal art), and indeed the wholeness of life is an indication that the primal unity so ardently sought has been lost. Maybe this is the real significance of those ancient ghostly hands found on the cave walls, for it is the human hand with its marvelous dexterity guaranteed by opposable thumb and forefinger, together with our bulging cerebral cortex, that makes man what he must be—the animal that can imagine itself as well as other things, that can make distinctions and separations, that can envision a future and long for a past.

It might further be observed here that one of the better ways of defining myth might be as a strategy of *control* as well as of propitiation and cooperation. In this context many may think of the theory proposed by nineteenth-century anthropologists that to represent something linguistically or graphically is to attempt to establish some power over it. Sir James G. Frazer styled this "sympathetic magic" and thought it one of the two fundamental principles of all magic practice. Thus the Zuñi myth of the young man's first deer hunt could be thought of as an attempt to control nature through employment of the tool of language. This view logically leads to the conclusion that *all* men desire to control the natural world, but that only some have developed efficient, sophisticated means for doing so. Then what separates the savage from the civilized is that whereas the former tried to master the world by talking to it and imitating it in graphics and dance, the latter moved beyond such pathetic attempts to the invention and employment of extra-somatic means that actually achieved the desired results. In terms of technology, the difference could be stated as that between control-as-worship and control-as-use. Or, as many have noted, it is the difference between the cosmos considered as "Thou" and the cosmos considered as "It."

No doubt there is truth in this line of thinking. No culture desires to be wholly at the mercy of nature. And no doubt myths do try to coerce the powers of existence to act generously toward puny man, his desires and needs. But the technological impulse is not the difference between the primitive and the civilized, for, as Mumford has pointed out, it is as native to the species as symbolizing. The difference is in the degree and strength of that impulse, its relationship to other fundamental aspects of humanity, and the cultural and spiritual results of what must be seen in a late-twentieth-century light as its hypertrophic activity in the history of the West.

What we must ask here is what it has meant for us that Western Civilization had its foundations in the soil of the ancient Near East, where humans began to enact the dream of mastering the natural world. For not only are we the inheritors of the technics of agriculture and animal husbandry and the designs of polity worked out there, but also, and more significantly, we are the inheritors of certain attitudes toward the natural world, certain "civilized" strategies of survival and control. The main conduit for these attitudes into the civilization of the West has been the sacred history of the ancient Israelites, the spiritual matrix out of which Christianity arose and that eventually formed the first half of Western Civilization's holy scripture. The Israelites shared much of the same landscape and its challenges with the Sumerians, Babylonians, Canaanites, and Hittites, and thus also certain attitudes toward it. As these hardened semi-nomads viewed from their tents the luxuries of the cities of the valleys, the city itself and the settled life it nurtured became a reference point in many of their traditions, standing in ever starker contrast to that dry, marginal existence they themselves led. Later, as sedentary agriculturalists in lands given them by their god, the ex-wanderers looked out in fear upon that wild and chaotic world through which they had once moved. Through them the wilderness assumed the forbidding shape that Christianity would emphasize even more strikingly.

What I seek to show here is that by the fifteenth century, when the West's exploratory probes beyond its geographical confines took on the earnestness of design, the civilization was possessed of deep-set, long-established attitudes toward the wilderness and indeed all unimproved nature, toward those who lived in the wilderness, and toward the relationship of "civilization" to these. These attitudes, codified and embedded within scripture, are in a real measure traceable to the struggles of the ancient Near Eastern peoples with their marginal and tricky environment.

These peoples were among the very first to act out one of mankind's most alluring dreams, that of a sedentary existence. Such a mode beckons with a cluster of securities, the most basic of which may be psychological: "perceptual expectancy" it has been called, the security of a known and fixed abode invested with names, stable reference points, and the proximate bones of the ancestors. There is also the economic security of a known and dependable supply of water, pasture, and shelter. In brief, a fixed home place seems to offer a defense against some of the more random and rigorous of life's challenges.

It is not known whether this is what the early settlers of the Near East were seeking when they wandered into the region. We do know, however, that the region in which they planted themselves presented them with very difficult challenges that they only gradually and painfully learned to meet. We commonly think of the Near East as the "Fertile Crescent," a place furnishing ready-to-hand conditions for the reaping of agricultural surpluses that make sedentary existence possible for large populations. The impression is a false one. In comparison with Europe and the Americas, the conditions for large-scale agriculture were severely limited. The area of alluvial soil between the Tigris and Euphrates rivers was but a fraction of the region and was surrounded by dry stony ground, steppelands, desert, and marginal lands where only adequate rainfall could hold back the advances of the desert. Even in the land between the rivers there were serious obstacles to settled living, for this was a place of great extremes, of floods and burning dry spells. The former turned large areas into reedy swamps so that the early settlers had to make and remake their fields by means of drainage ditches. The latter necessitated the development of technics for the cooperative management of water resources through extensions of the ditches that could also serve for irrigation. They had to learn, perhaps from the dwellers on the Nile, technics for raising river waters to the level of the fields, and they had to learn gradually the skills of seeding, cutting, crop rotation, and animal domestication. Slowly over centuries, the reed-hut villages surrounded by fields in which the people toiled with stone hoes were succeeded by larger towns of mud houses situated farther from the fields.

Through all this the people were developing a markedly new habit of mind, that of foresight, of thinking constantly in terms of the economic future. And they were inventing the requisite technology to conserve for that future. Mumford says the major technological invention of the "Neolithic Revolution" was the perfection of the container: bins, granaries, jars, cisterns. So a Canaanite myth describes man himself as a clay vessel, and in the Israelite creation myth the verb used to describe the god's shaping of the human form is that denoting the work of a potter.

The cumulative result of their labors and ingenuity was the creation and insurance of *margin,* of surplus, without which neither city, nor town, nor even stable village is possible. But this struggle had enormous spiritual and cultural consequences.

Every environment encourages a special mythology. Into the sacred narratives, into their ritual enactments, into the personalities of the deities, are filtered and fibered the areal weather patterns, the size of the sky and what it brings, the shapes of clouds, the contours of the terrain and its dominant colors, the flora and fauna, the natural rhythms of movement, mating, molting, and perhaps above all human adaptive responses to all this as they develop over time. Here in the Near East the hard-won evolution from encampments to villages to the towns that eventually grew into large cities—achieved over thousands of years in a difficult environment—nurtured the belief that "civilization" meant the walled, blocked, and grain-stocked city and that civilization could only be achieved and perilously maintained by unremitting hand-to-hand combat with a nature that would of itself grant little. *Margin,* that which protected civilized man from nature's caprices, that which separated culture from the wild and uncultivated, was the work of human hands, assisted not by the earth, which was hostile, but by the gods of the sky—far removed from earth. Whereas the mythologies of the earlier settlements seem to have been based at least partly on the earth, with the development of towns and cities the locus of divinity shifted to the sky and the irrational, violent gods who dwelt there.

Perhaps the most significant spiritual result of this Neolithic Revolution was the revolution in the sexual bias of the areal mythology and culture. Though the evidence is partly negative in nature and in any case is hard to assess accurately, it seems that the mythology and cultural character of the earliest village agriculturalists were feminine-oriented. Those of most agricultural peoples are, for the mystery of the solitary and lifeless seed deposited into the darkness of the nurturing earth and magically reappearing as new life naturally suggests the enormous, singular power of the feminine principle. Many anthropologists now think the first farmers were women, and we know that in many cultures it was they who were entrusted with the rituals of the fields.

By contrast, the mythology and cultural character of pastoralists are usually aggressively masculine, pastoralists being, as Paul Shepard says, essentially domesticated hunters. Partly this is because pastoralism is a tough way to live, seeming to call for the "masculine" traits of willpower, fortitude, and forcefulness against the continuous challenge of nature. Animal *husbandry,* it is called, and the vigilance and strength of the shepherd is what sustains life and "produces" new life among the flocks. Nothing comes easy for pastoralists, and in their cultures and in their mythologies the status of women is low and patriarchal authority supreme.

It is clear, however, that the rise of the ancient Near Eastern citystates would not have been possible without the intimate cooperation of agriculturalists and pastoralists, producing at last a mixed barnyard economy. But, cooperative as these two types were, the life style and the spiritual orientation of the pastoralists triumphed over what seems to have been the more pacific, feminine culture of the agriculturalists. Why this was so is not altogether clear, but some guesses can be made.

Mumford says that with the growth of settlements the role of the shepherds and hunters altered correspondingly: they became semisedentary themselves, serving the settlements as protectors of the fields and stock. (Indeed, this pattern persists to this day in the Near East.) Thus they might still exercise their aggressiveness in a constructive way. But in time, he suggests, this aggressiveness became turned against those they were supposed to serve: in an ancient version of our contemporary "protection" racket, protection became not a service but a threat, and the settlements found it more prudent to accept the domination of the pastoralist/hunter than to attempt to dispense with him.

Psychologists offer another explanation that in no way contradicts Mumford's speculation and may even amplify it. Bruno Bettelheim and others have suggested that male assertiveness and aggressiveness and male-instituted contempt for the feminine principle have their genesis in the male's knowledge that in the most basic of ways he is not the life equal of the female. No matter how he might try, he cannot bring forth life. Thus to Bettelheim and Karen Horney the history of the elaboration of culture might be read in terms of the male's unwearied attempts to manipulate the external world and in so doing become a creator, the mark of his generative equality with women being his tireless assertions of superiority. Bettelheim makes the case that male puberty rites were magical attempts to become genitally equal to women, and that only after the demonstrated failure of these rites to achieve that parity did men turn to the external world as the arena in which their baffled creative desires might find some expression. To him the "larger forms of society" are male-dominated compensatory formations. And in a study of the role of sexual rivalry in history, Amaury de Riencourt quotes Karen Horney to the same effect:

> "Is not the tremendous strength in men of the impulse to creative work in every field precisely due to the feeling of playing a relatively small part in the creation of new living beings, which constantly impels them to an over-compensation in achievement?"

So in the Near East the triumph of masculine mythology, of masculine-oriented culture, and the revolutionary elaboration of culture might be seen as the compensatory consequences of the displaced pastoralists' feelings of inferiority and secondariness. Thus man's "rise to civilization" might be seen not so much as the triumph of a progressive portion of the race over its lowly, nature-bound origins as a severe, aggressive *volte-face* against all unimproved nature, the echoes of which would still be sounding millennia later when civilized men once again encountered the challenges of the wilderness beyond their city walls. In this light the distinction between the savage and the civilized looks different. Bettelheim writes that "perhaps certain people remained 'uncivilized' because they did not wish or feel a psychological need to progress beyond the 'passive giving oneself up'; *i.e.*, living dependent on what nature by itself provides."

Bettelheim's speculation bears on another entailment that must be mentioned in considering the influence of these early civilizations on our own, and this is the cumulative spiritual effect of the steadily increasing margins that the restless ingenuity of men were able to create. Every added protection against the natural world contributes its bit to the steadily building illusion of independence from nature, so that in time that greatest of illusions is erected: the omnipotence of man. The growth and incremental sophistication of technics of animal husbandry, hydraulic agriculture, storage, distribution, and shelter made possible the emergence of the city, of civilization, and this phenomenon occurred both on the landscape of the Near East and in the geography of the mind. The sheer visual stimulation of numbers of people living together thanks to human inventiveness must have fostered a burgeoning sense of the efficacy of human willpower—and this is the progenitor of the will to power, of the urge to dominate the land, and of the belief that *all* of nature may ultimately be dominated.

Evidence of this aggressive, domineering attitude toward the natural world is to be found everywhere in the ancient Near East: in the structure of the cities, in the brutal history of warfare these early civilizations carved out of each other's hides, and most especially and truly in their mythological records.

It is hard to think of another group of kindred mythologies as unrelievedly brutal, violent, and male-dominated as that of these early high civilizations. Originally the deities appear to have been thought of as indwelling spirits that resided in various natural phenomena. Later they are represented in animal form, and still later they emerge in anthropomorphic guise as rulers, parents, tyrants. These sky beings are wholly monstrous in their rages, in their unappeasable appetite for vengeance, in their capricious turnings against one another and against the cities of men. Parricide, infanticide, and rebellion are the lifeblood of their narratives, and their concern with human life is mainly manifested in the forms of floods, storms, droughts, famines, and the destruction of cities. Thorkild Jacobsen translates a Sumerian lament that speaks of such unmerited visitations of divine hatred. In it the god Enlil has called down a devastating storm on the city of Ur:

> Good winds he took away from Sumer.
> > The people mourn.
> Deputed evil winds.
> > The people mourn.
> Entrusted them to Kingaluda, tender of storms.
>
> .          .          .
>
> The Storm ordered by Enlil in hate,
> > the storm which wears away the country,
> covered Ur like a cloth,
> > veiled it like a linen sheet.

Another Jacobsen translation powerfully illustrates the remoteness and arbitrary nature of these beings, now so removed from the earth. In this one, a man has been ordered to build the god Ningirsu a temple, but the god has given the man no more specific instructions. So this prayer, like so many of the old clay tablets and cylinder seals, is stamped not only with piety but with terror also:

> O my master Ningirsu, lord,
>     seminal waters reddened in the deflowering;
> able lord, seminal waters
>     emitted by the "great mountain,"
> hero without a challenger,
> Ningirsu, I am to build you your house,
> but I have nothing to go by!
> Warrior, you have called for the "proper thing,"
> but, son of Enlil, lord Ningirsu,
> the heart of the matter I cannot know.
> Your heart, rising as [rise the waves in] mid-ocean,
> crashing down as [crash] the breakers,
> roaring like waters pouring [through a breach in a dike]
>     not to be stemmed,
> warrior, your heart, remote [and unapproachable]
>     like the far-off heavens,
> how can I know it?

Creation is a work of cosmic destruction in these texts, often taking the form of a violent victory of the male gods over their female adversaries, who here exhibit that terrible face that is but one of the aspects of the Great Mother. Thus in a myth widely spread through the region, the firmament of heaven and the foundation of earth are formed of the dismembered carcass of Tiamat, the primal goddess, defeated in battle by Lord Marduk: he smashes her skull, splits her body like an oyster, and the obedient winds whisk her blood away. Little wonder that the earth was eventually perceived as hostile with such a murderous conception of it.

The triumph of the male gods guarantees the relegation of the formerly dominant goddesses to the roles of thwarted adversaries, marplots, or supernumerary helpers. We find that the mother goddess of the herd animals, Ninhursaga, becomes the demoness of the stony grounds that ring the arable soil of civilization. In a lament her own daughter asks, ". . . to whom should I compare her?/ To the bitch that has no motherly compassion. . . ." Even in agricultural myths where originally the goddesses had been preeminent, they are now debased, as, for instance, in Sumerian mythology where the male god, Enlil, is credited with the gift of the primal tool (the pickax) for

field work and construction. Thus he is made responsible for both agriculture and the culture of the city. Significantly, perhaps, Enlil's pastoralist origin is revealed by his epithetical title: the Shepherd. Sumerian
mythology, so influential for the traditions of the Israelites, also shows
the male god, Enki, as directly responsible for the fertility of field and
farm, and it is he who guards the implements of agriculture. Only after
he has called the cultivated fields into being does Enki assign the
goddess Ashnan charge of them.

Buried in these awesome texts like evidence of archaic encampments
beneath city walls are signs of that earlier, more harmonious agricultural way alluded to above. And within these faint vestiges, which form
the deepest substratum, is to be found evidence of longings for that still
older (oldest?) presedentary freedom, of that radically integrated spiritual existence of Paleolithic cultures. Thus in the greatest epic of the
Near East, the epic of Gilgamesh, which Theodore Gaster has called the
area's *Iliad* and *Odyssey,* we find the presedentary Paleolithic substratum
in the figure of Enkidu; the fall of this man from a state of natural harmony; the debased woman as agent of the fall; and the rise of the hero
of consciousness—the fully aware doer of deeds—quester, explorer,
and at last tragic exemplar of mortal limitations.

The episode recounts how a huntsman out from the city accidentally
discovers the natural man, Enkidu, roaming with the beasts of the field
and releasing them from the huntsman's snares. Like his fellow creatures, Enkidu feeds on grasses, drinks from the stream, and beds down
with the herd at night. The huntsman, terrified at the sight of this
shaggy, skin-clad atavism, rushes back to the city to inform its mighty
ruler/tyrant, Gilgamesh. The ruler then devises a scheme of capture, instructing the huntsman to procure a girl of the streets and take her to
the place where Enkidu was seen drinking with the beasts. When the
creature comes to water, the girl is to strip off her garment and entice
him with her nakedness. Once he has embraced her, reasons Gilgamesh, the animals will recognize him as a human and not a fellow creature and will forsake him. Then in his loneliness the creature will be
drawn to the world of men and forced to abandon his savage ways.

The plan works perfectly and Enkidu, who once released the captured beasts, is himself snared in the slim arms of the city girl. Gaster
translates the result:

> For a whole week he dallied with her, until at last, sated with her charms, he
> arose to rejoin the herd. But the hinds and gazelles knew him no more for

one of their own, and when he approached them they shied away and scampered off. Enkidu tried to run after them, but even as he ran he felt his legs begin to drag and his limbs grow taut, and all of a sudden he became aware that he was no longer a beast but had become a man.

In this figure is a microcosmic record of the succession of one stage of mythology and culture by another, for his fate now lies in the city of men and not in the fields of animal freedom. From it he will go forth with Gilgamesh into what has become for him the wilderness to test himself against the decrees of the gods. And in the wilderness he will die, estranged from his old home.

In sum then, the rise in the Near East of civilization as the West would one day come to think of it witnessed as a significant concomitant the supplanting of the older, organically derived feelings of gratitude toward nature and of the vital interdependence of all things by the masculine notions of force meeting force, of the enduring opposition of man and nature. The old notion of the fecund and mothering earth was transformed into the symbology of a settled struggle, the iron phallus/ plow driving its pregnant seed into the subtly resistant womb/earth; the cities erect, stark and bristling on the land, incised lions and bulls guarding their gates against all that lay beyond them. Civilization as it emerged here was consciously walled off from organic harmonies and defined in terms of oppositions. As Joseph Campbell writes, this point of view "is distinguished from the earlier archaic view by its setting apart all pairs-of-opposites—male and female, life and death, true and false, good and evil—as though they were absolutes in themselves and not merely aspects of the larger entity of life." Mumford calls the new mythology of this Neolithic Revolution a "mythology of power" and characterizes its cultures as possessed of "armored personalities." Not only did they now wield power over the wild animals that had so long harassed herdsman and agriculturalist alike, but, flexing their newly evident muscles, they now sought power over all of a nature that seemed to resist with its own force the force of civilization. And inevitably they sought power over one another. Mumford makes a strong case that nothing in the known earlier stages of human culture equals the disciplined force and scale of the warfare between the Near Eastern city-states, and he concludes that the great innovation here was the mechanization of men into fighting units of such destructiveness as that boasted of by the Assyrian tyrant Ashurbanipal in his sack of the cities of Babylonia:

[as for] the rest of the people: guided by the guardian spirits . . . I now . . .
ploughed those people under alive. Their flesh I fed to the dogs, pigs, vultures, eagles—the birds of heaven and the fishes of the deep.

. . . I took the corpses of the people . . . who had laid down their lives
through hunger and famine and the remains of the dog and pig feed, which
blocked the streets and filled the broad avenues; those bones [I took] out of
Babylon, Kutha, and Sippar and threw them on heaps.

This mythology of power fed upon itself, its achievements, and the
margins thus created between men and their environment. What we
witness then is an early and crucial instance of the technological impulse
becoming an affective substitute for the mythological, supplying some
of the questions and some of the answers that had once been referred to
the authority of myth. Significantly, Mumford and V. Gordon Childe
see the explosion of technics at this stage as the greatest until our own
time. The walled city is itself the cumulative artifact, lying, as an ancient tablet tells us, like a storm cloud on the horizon. The wall, as
Mumford so rightly observes, is not merely a physical entity but a
"spiritual boundary of even greater significance, for it preserved those
within from the chaos and formless evil that encompassed them." What
lay beyond the walls became by emerging definition something other
and less than civilization; peoples who lived outside the walls became
by that placement less than civilized, objects of hatred, fear, and deri-
sion. And so a fragment of a Sumerian myth describing in contemptuous terms some nomadic Semites who wandered beyond the city of
Sumer portends future civilized attitudes toward other wilderness peoples. Martu, a god of these nomads, wishes to marry a goddess of the
city, but her handmaidens attempt to dissuade her thus:

> He lives in tents, buffeted by wind and rain,
> Eats uncooked meat,
> Has no house while he lives,
> Is not brought to burial when he dies.

# The People of
# the Book

I t has by now become an axiom of historical study that peoples who
have gone through a more or less voluntary transition of culture look
back with a mixture of nostalgia and fear on those earlier stages out of
which they have translated themselves. Usually those earlier days seem
ones of comparative hardship and deprivation, and almost always for
this reason they are equated with a greater moral rigor and purity of
purpose. And yet, just as surely the people fear to go back; they do not
truly wish to somehow reverse the course of their culture and become
"primitives" once again. Indeed, it seems that a certain level of cultural
security and stability is a necessary condition for nostalgia, which at-
tains the status of a leisure activity. One might go further and guess
that to the extent that the truths of the old days are remembered, and
the conditions, artifacts, and most especially the gods and rituals are
taken seriously, they will be invested with a strange negative power,
having a sort of chthonic authority to summon up feelings of loathing
and dread. It is as if one dwelling in the common light of modern times
should suddenly come upon an unaccountable kind of movement
among the fond, familiar megaliths that litter his plowed fields and feel
his hair and skin bristle.

Freud and Jung, both interested in how the deeds of the ancestors
might be communicated to and affect succeeding generations, suggested
that there were ways entirely independent of conscious transmission by
which people might carry a kind of blood knowledge of what had gone
before them. Freud thought it a general law of cultural history that
what has been outgrown and displaced becomes feared and despised.
Seeking to treat a whole culture on the model of the individual psyche,
he speculated that what gave the old, outgrown modes and artifacts

their peculiar resonance was the culture's retention of the impulses and preferences that had once engendered the old ways and been proper to them. This would explain on a psychological basis why one people encountering another who live in the ancestral way might feel a strange commingling of attraction and hostility; and why people regard their own past in such a tangled fashion. There are more immediate considerations involved here too.

The Sumerians, founders of the high civilizations of the Near East, had once been wanderers themselves, and though archeological evidence indicates they displaced no one when they entered the lower Mesopotamian Valley sometime after 4000 B.C., once they became entrenched, they had ample reasons, both psychological and military, to dread the appearances of successive groups of wanderers. For these others were in search of precisely that which had driven the Sumerians to the river valley from the east: a place to call their own, one where the natural advantages might lessen the pressures of existence. The Sumerians had found this in the lower valley and, in the process of establishing themselves there, developed what we must honor as natural fears of and antagonisms toward atavisms of their own past. So too with *their* successors and competitors, until all the civilizations of the Near East found themselves in the same defensive posture against all that lay beyond their small spheres of cultivation and influence. The city as it has come down to us from these peoples is thus a kind of barricaded, fortified oasis. What lay beyond was unimproved and probably unimprovable nature, which the Sumerians and their neighbors considered the abode of frightful *jinns*, shaggy satyrs, and wanton and fatal demonesses. This territory was the more frightening because it was so obviously vast in comparison with the tiny, precarious niches where "civilization" had established itself. And out of this great Other, from its deserts, mountains, and steppes, came group after group, looking thirstily toward the cities, their watered and waving grain fields, their shade trees, livestock, fruits of the vine, their dependent towns and villages. "Civilization" now had to entail the contrasted concepts of "wilderness" and wilderness peoples. In marking out boundaries to their worlds—walls, man-made shelters, granaries, cisterns, reservoirs—civilized peoples took the first long strides in the ongoing process of insulation from the rest of creation. Surveying this phenomenon, Henri Frankfort was moved to conclude that from their inception these cities were at variance with the natural order of things. If in many ways our world is richer for the advances made here, in other no less significant ways it has been narrowed. What was within the walls, within the reaches of the irrigation ditches, was nature subdued, controlled, put to proper use; what was beyond, whether land, people, or spirits, was savage, unpredictable, malevolent.

The terms of man's stark strife with nature here were profoundly affected by the environmental extremes of the region and the comparatively small amount of rich, alluvial soil available for large-scale agriculture. Indeed even in areas developed by modern agricultural methods the marginality of the region is suggested by the fact that in many places it is possible to stand with one foot planted in productive land and the other in semi-desert. In Palestine, for example, Salo Wittmayer Baron writes that

> the desert and its winds seem eternally to be warring against the sea and its breezes. When the latter finds an accomplice in man, in a period of progress, the desert recedes, yielding several miles of its territory to human habitation. When man's hold weakens, however, the winds sweep over the border line and soon vast stretches of land are again covered by desert sands and lost to civilized life.

Yet, of course, civilizations did arise in this environment and on a scale and possessed of cultural achievements unprecedented in the known previous history of mankind.

This paradox is resolvable only through attention to the character traits of these people, their ingenuity, aggressiveness, and initiative that mastered the challenges of the lands they chose. The boasts of the neolithic rulers incised on steles, recording how much they had extended the irrigation systems and built up the cities, testify to areal traits as real as weather.

But if Mesopotamia, a watered valley, and Palestine, where sea breezes bring adequate seasonal rains, present hard challenges to sedentary agriculturalists, how much more difficult the lot of those ancient semi-nomadic pastoralists of the steppes and desert edges who haunted the fringes of the civilizations. Though the Renaissance was to create an idyll of the shepherd life as an antidote to its own urbanized realities, the fact is that there is nothing idyllic about this mode of existence. Fredrik Barth, the distinguished authority on nomadism, has called it a "precarious specialization."

Its origins are those of agriculture in the Near East, and it developed in symbiotic relationship to farming. But with the economic specialization that is always a feature of cultural elaboration, full-time herders appeared and began to exploit pastures farther and farther removed from the expanding centers of cultivation. Pastoralism in the Near East had to become semi-nomadic since plant life in the arid stretches beyond cultivation is sufficient for pasturage only with seasonal migrations. Sedentary pastoralism out there is impossible. This condition prompted Donald P. Cole to call semi-nomadic pastoralism a cunning exploitation of areas otherwise unfit for cultivation or habitation.

The cultural traits of these pastoralists must be a match for their environment. If the steppes and desert fringes are tough, the soil barren of all but the most hardy plants that can send up spikey blooms with the barest blessings of rain; if the wind and wind-driven sand are constant irritants; if grazing itself is, as Paul Shepard says, like sandpaper on the skin of the earth, so there must be a hardness, even a fierceness, in a people locked in struggle with this marginal environment and an ability to make values of adversities.

Such an occupation in such an environment seems to encourage other traits besides that militancy for which the pastoralists have been historically famous—and feared. Long periods of waiting under wide, burnished skies fostered introspection, philosophic abstraction, and narrative inventiveness. Sharp cleavages between day and night, waiting and the rush of migration, desert and oasis, fostered dualistic habits of mind. Finally, and for us most significantly, these Near Eastern herdsmen, like their agriculturalist neighbors, found divinity in the sky, in the mountains, in the wind. Even more than those who built the towns and cities, transferred their devotions from the earth to the sky, and built ziggurats to reproduce in the polis the sidereal conformations, the pastoralists had reason to expect little from the earth. If the city dwellers with their hydraulic agriculture, crop rotation, and storage devices seemed to forcibly take more from a resistant nature than it wished to yield, those dwellers in the stony, sparse lands engaged in a struggle that was even more emphatic.

There was much besides an adversary relationship with nature that tied together the sedentary peoples and the mobile husbandmen of this region, although they are popularly supposed to have been diametric opposites. They developed in symbiosis and continued this relationship even after the pastoralists had been forced into ever more marginal territories. From ancient times to the present, pastoralists have traded their military proficiency for the pastures and water of sedentary peoples, coming with their flocks and tents to the harvested stubble fields (which their herds would then fertilize) and returning in winter to the steppes and desert edges. There is also a long history (as Genesis shows us) of sedentary peoples providing temporary asylum for pastoralists driven by drought from their home territories. In such instances the pastoralists might settle on the verges of the towns or cities and adopt for a time a semi-sedentary mode.

Yet for obvious reasons there exists an equally long record of distrust and hostility between these groups, and indeed the history of the Near East is replete with incursions of the nomads into the domains of the settled, whose wealth must have seemed fabulous. Beginning c. 3000 B.C., wave upon wave of nomadic invaders molded the history of the region in the rise and fall of empires. For despite their attachment to the desert and steppelands, the pastoralists seemed to be irresistibly attracted to the settled areas. The best evidence for this is that though there are known instances of *forced* reversion to a wandering existence by those who were dislocated by military defeat, political shifts, or persistent famine, there are no known instances of *voluntary* reversion, whereas there are numerous instances of voluntary conversion of nomads and semi-nomads to sedentary living. This tells us much. Nor do certain well-known examples from other parts of the world qualify the essential pattern here. In North America, for example, both the Lakota and the Cheyenne tribes reverted from a semi-sedentary existence to the life of hunters on the Great Plains—and they loved the new life so much that farming and a settled existence came to seem repugnant and restrictive to them. But it is important to know that these tribes had been forced onto the Plains by the gun-wielding tribes of the Upper Midwest and that their new habitat was a grassland dark with huge herds of bison, deer, and antelope. What the evidence of the Near East indicates is that whatever the rewards and windy freedoms of the semi-nomads, they were outweighed by the apparent comforts of stable supplies of water, grain, pasturage, shelter, places to rest, to settle, to be buried in—"all that's kind to our mortalities," as Melville wrote of the vision of home port to the tempest-tossed sailor.

And so the typical evolution of a people in the Near East was from a nomadic existence in the deserts or mountains, to a semi-nomadic mode on the fringes of the territories of the agriculturalists, to invasion or infiltration of those territories, and finally to a sedentary life in which they retained only dim memories and ambivalent feelings about the old

primitive freedoms.* And when a new horde appeared on their horizons they met it with a suspicion and hostility born of recent history
and old memories. So in Numbers 20–22 we find recorded the terror of
the Edomites and the Moabites, themselves ex-nomads, when the tribal
leader Moses asks right-of-way for his people through the cultivated
lands, promising not to go through the fields or the vineyards, nor
drink of the water of the wells, nor turn to the right or the left until
they have passed the borders. These are the Israelites, forty years in the
wilderness, pitched on the plains before the settlements, a dark and familiar menace:

> And Moab was sore afraid of the people, because they were many: and
> Moab was distressed because of the children of Israel.
>
> And Moab said unto the elders of Midian, Now shall this company lick up
> all that are round about us, as the ox licketh up the grass of the field.

From such did Abram come, and through the history his people were
to enact and record, the whole spiritual/ecological past of the Near East
would become part of the heritage of the West, part of the equipment
Westerners shipped with as they thrust out into the undiscovered
beyond.

Not much more is known of the ultimate origins of Abram's people
than that they were Semitic tent dwellers who came up out of the wilderness into the area of the established civilizations. Their distinguished
historian Salo Wittmayer Baron refers to them as one of "those stray
groups of *condottieri*" who raided the sedentary populations and performed as mercenaries in small wars between neighboring city-states.
This contradicts the weight of Jewish tradition, which gives us a predominantly pacific picture of the patriarchs, though it is very doubtful
this part of the tradition can be relied on. Indeed, beneath the layers of
appliqué that make up Old Testament tradition it is possible to catch
glimpses of what looks like the oldest substratum, and this shows us
Abram as a military leader (Genesis 14:1–21) and Jacob as having taken
his territory by force from the Amorites (Genesis 48:22). The first cen-
sus of the tribes under Moses is a military head count (Numbers), and
Max Weber, in his study of ancient Judaism, goes so far as to claim that

---

*Marco Polo reported of the Emperor Kublai Khan that this nomad-turned-
philosopher caused some grass seed from his native Mongolian steppes to be planted in
the courtyard of his Peking palace to remind himself of the source of his people's greatness.

the only times such groups were ever pacific were when the balance of military power had so shifted to the cities and their dependencies that it was prudent for the pastoralists to adopt temporarily the strategy of peace.

Something further is revealed of origins by the cosmogonic myth that begins the scriptures, for this is a memory and a reflection of the desires of a nomadic group of the Near Eastern wilderness, recorded by a scribe after the people had become sedentary. It is not hard to imagine that such a people would have seen the gardens of the Mesopotamian cities as Paradise. And that is precisely how Eden is imagined: as one of those Mesopotamian walled gardens wherein various birds and animals were kept for scenic and sporting purposes. Here the great god himself, like an Oriental potentate, seeks refreshing shade in the heat of the day (Genesis 3:8). These gardens, as Paul Shepard says, were for the city dwellers a formal recognition of the delights of nature, not in the raw or of itself, but a nature tamed, humanized, and walled about like the cities.

In keeping with the Near Eastern shift to male gods and with the sexual bias of pastoralist peoples, the divinity who orders this paradisal garden is male. Some scholars think that in the account of his creation of the cosmos there is an echo of the great myth of the slaying of the female chaos-dragon Tiamat. No attempt at theogony is felt necessary: he simply is; or, as he will later say, I am that I am. In creating the garden, he acts as man subsequently must: he makes order out of natural chaos, creating a civilized clearing at the center of the cosmos, pushing the old chaos back to the edges where it remains a persistent threat—as the desert is to cultivated land (Psalm 104:6–9; Jeremiah 5:22). The divine meddler who creates out of primal oneness separations and divisions and an anthropocentric hierarchy gives such behavior divine sanction. All cosmogonic myths describe separations, for this is what creation means. Yet it has been the West that has taken this most to heart:

> Be fruitful and multiply, and replenish the earth and subdue it: and have dominion over the fish of the sea, and over the fowl of the air, and over every living thing that moveth upon the earth.

This vision of life is spiritually light-years removed from that mythic community of the living etched on the walls of caves that infused the tribal life in the forests and along the river courses of the aboriginal Americas, and that lies buried within us as a phylogenetic memory.

Like the conception of Paradise with man as the lord of all earthly creation, the scriptural circumstances that begin the story of humankind come out of Near Eastern tradition and landscape. The false advice of Satan by means of which the god's favored creature forfeits immortality is a motif found in other mythologies of the area, as for example in a Babylonian myth in which a man loses the chance for immortality through the bad advice of his divine tutor. As for the god's curse of toilsome tillage, this looks like a reflection of the semi-nomadic pastoralist's historic contempt for agriculturalists, a contempt mingled with and mastered by deep envy. This ambivalence explains the apparent contradiction of the curse: man was originally an agriculturalist in the paradisal garden (Genesis 2:15), yet his punishment is to remain one in his exile from the garden (Genesis 3:17–19). Under the curse, man enters the wilderness of the world, estranged forever from nature, which becomes a cursed adversary, eternally hostile to man's efforts at survival. He is also now fated to be an enemy of the animals (Genesis 9:2), and so existence in this world takes on the character of an unremitting contest with nature wherein man toils to fulfill the god's command that he subdue the thorny, thistled earth and establish dominion over it. Paradise is closed and the cherubim set to guard its gates are the very creatures that stand sentinel at the gates of the Near Eastern cities.

Within such gates at an indeterminately later period are found a remnant of the original exiled people. Genesis 11 shows us Abram and his clan living in Ur of the Chaldeans, the city now thought to be the original center of Sumero-Accadian civilization. The evidence of the text suggests that Abram's people came to Ur from Haran in northwest Mesopotamia; in a larger sense, they have been connected to a wave of nomadic incursions of the second millennium. Carleton Coon speculates that a succession of excessively dry years set off this movement and so brought these nomads into the light of history (the city), out of the unrecorded darkness of the wilderness. Perhaps Abram's clan was one that adopted a sedentary mode of existence for a time before again moving on. Consensus chronology has settled for a date of c. 2000 B.C. for the momentous departure from Ur.

Purely rational-historical interpretations of any people's mythology or sacred history illuminate little of real significance. To take seriously the spirits that spoke in aboriginal America and those that may yet speak in the transformed New World—and my essay shall ask that the reader do this—is also to take seriously the promptings of the as-yet-unnamed deity who set Abram's face toward a far and unknown land. Which means that the god's call to Abram, his covenant with him and his descendants, cannot be dismissed as merely a later "fixing up" of tribal history to justify the conquest of Canaan by sanctifying it as the Promised Land. We resign nothing of intellect or learning in taking seriously the voices and spirits of places, the spiritual dimensions of a people's history. On the contrary, to dismiss the dictates of the gods and the spirits of places as either transparent rationalizations for aggression or childish credulities is to recommit old mistakes—both of conquest and of historiography. There are tragic consequences for both. That which is essential is invisible, as Voltaire says, and those matters that have moved peoples and conferred form on history have not been solely (or maybe even predominantly) those that have been documentable and quantifiable and thus susceptible to rational interpretation: economics, politics, or the conscious designs of men.

So here it is necessary to accept as affective fact that Abram (whether an individual or the personification of a generation) felt a spiritual urging to leave the city, travel up the Euphrates to Haran, and then, still seeking his spot, leave that well-watered haven for the alien land of Canaan. Nor does it conflict with this to observe of the original covenant delivered in Canaan (Genesis 12:7) that it is probably impossible to overemphasize that the god was promising Abram and his descendants arable land (*sadeh*, in contrast to the arid lands out of which they had come) that could be settled, seasoned with the bones of the generations, titled, and defended. Moreover, as is subsequently made clear to the Israelites, this gift, this fruitful land, bears its plenty for them by the grace of the god and not by the cunning devices of men, for it is watered with the rain of heaven and not by irrigation as in Mesopotamia and Egypt (Deuteronomy 11:10–12,17). Abram and his people are thus to be miraculously delivered from wilderness wandering to settled civilization, which amounts to a kind of earthly salvation.

That this promise makes the Israelites fit the pattern of invading pastoralists of the Near East invalidates none of the tradition. Neither does the current scholarly opinion that the portion of Genesis in which the tradition is recounted is probably the work of a scribe writing well over a thousand years after the events are supposed to have taken place. Any chronology of the writings of the Old Testament must reckon with that bedrock of oral tradition that vastly predates even the earliest known written texts; and our knowledge of the sacred literatures of the world tells us that when myths and legends are at last written down, they are largely faithful to their oral bases. Sacred narratives, while they are still felt to be sacred, are not often the playthings of revisionist historians.

What seems to have happened is that in the interval between the emergence out of the wilderness and the sojourn into Egypt the Israelites felt themselves guided to the outskirts of Canaan where they existed as metics, engaged still in pastoralism but now also doing some farming—as Genesis 26:12–13 would seem to indicate in the Jacob and Esau story, in which it is recounted that "Isaac sowed in that land, and received in the same year an hundred-fold: and the Lord blessed him. And the man waxed great. . . ."

It was at this time that the god, now identified by the common Semitic designation "El" (God), as "El Shaddai" (God, the One of the Mountains), reaffirmed the covenant with the man henceforth known as Abraham through the ritual mutilation of circumcision. And the outward sign by which the god manifests his power and so coerces belief, his first theophany, is the visitation of terrific destruction on the cities of Sodom and Gomorrah for sexual practices the god now designates as abominable. This act announces three of the grand themes of the Old Testament: that the Israelites are not to do as other peoples do; that both as fact and as metaphor what most distinguishes this holy people from those who are henceforth to be thought of as pagans is sexual conduct; and that the god will speak to all humans in the terrible syllables of natural disasters.

Yet significant as the Abrahamic covenant is in establishing the framework for all future expectations, this first Canaanite period and the Egyptian bondage that follows it are both parts of an interlude. For the real locale of the Pentateuch (and perhaps of the Old Testament as a whole) is neither Canaan, the Promised Land, nor Egypt, but the wilderness. It was the wilderness out of which these people had come and which formed their culture and desires. It was the wilderness to which they were sent out of Egypt by the voice of their god. It was there that he ratified for all time the covenant and specified its conditions. There, too, they murmured against the god and his intermediary and were severely punished. And from the wilderness they were at last delivered into the comforts of civilization. Once in the Promised Land again, they were, as George Williams has said, obliged over and again to rethink that vivid locale in the light of serious problems of adjustment to sedentary existence. Leviticus 18:3–4 places the emphasis securely on that no-man's-land between settlements:

After the doings of the land of Egypt, wherein ye dwelt, shall ye not do: and after the doings of the land of Canaan whither I bring you, shall ye not do: neither shall ye walk in their ordinances.

Ye shall do my judgments, and keep mine ordinances, to walk therein: I *am* the LORD your God.

In this perspective it seems inevitable that the god should call the people out of civilization into the wilderness to reveal the ultimate content and direction of their special historical destiny. However much Judaism (and its descendant Christianity) has come to seem a religion of the city, its psycho-spiritual camp is in the steppelands and the desert: its god is a wilderness god, and he is thunderously insistent that his people never forget this or that they too are of the wilderness. So he sends them out from the Egyptian fleshpots, the Nile and its blessings, to immense skies, burning sands, the stony upthrusts of a world unmade by men. It is, as Moses says, "a great and terrible" place (Deuteronomy 1:19), and like the paraphernalia of initiation rites it is meant to scar the people forever. "Circumcise your hearts," the god tells the tribes, and the wilderness is to aid them in this.

In his appropriate surroundings the god chooses to make known his essential character, emerging out there from the shadowy voice that prompted the patriarchs, and now revealed as a capricious war god (despite unending efforts at theodicy) whose weapons are those of natural disasters. He has already sent a flood that all but blotted out his entire creation, plagues of locusts and snakes, a famine, and an epidemic, and his hot breath has brought the dreaded desert wind, the *khamsin*, that blackened Egyptian skies with clouds of sand and dust. His words are volcanic eruptions, thunderstorms, and earthquakes. The cumulative effect of all this is to emphasize the destructive aspects of nature and to reinforce the anthropocentric, adversary attitude toward the natural world announced in the paradise myth. This is what we should expect, given the origin of the tribes and the temper of the lands through which they moved. Behind it lies the entire experience of the Near Eastern peoples in their long struggle with a marginal environment.

This antinature bias, as Baron, Weber, and Johannes Pedersen have pointed out, is reflected in the grand covenantal experience at Sinai, both in the specifically *historical* character of the religion there spelled out and in the monotheism that sets it apart from the nature-based polytheisms of all other peoples—indeed, not only apart but *against* them in a war to the death.

Already the god had prepared his priestly interpreters for this crucial break with nature by telling Moses and Aaron that the old organic festivals of the year would now be reinterpreted in the light of the history the god was causing the tribes to enact:

> This month *shall be* unto you the beginning of months: it *shall be* the first month of the year to you.
>
> .   .   .
>
> And this day shall be unto you for a memorial; and ye shall keep it a feast to the LORD throughout your generations; ye shall keep it a feast by an ordinance forever.
>
> (Exodus 12:2,14)

The feasts referred to here are those of the Passover and Unleavened Bread, originally thought to have originated as pastoral festivals founded on seasonal changes. Here they are converted to historical remembrances of the god's intervention into human history:

> And it shall come to pass, when ye be come to the land which the LORD will give you, according as he hath promised, that ye shall keep this service.
>
> And it shall come to pass, when your children shall say unto you, What mean ye by this service?
>
> That ye shall say, It *is* the sacrifice of the LORD's passover, who passed over the houses of the children of Israel in Egypt, when he smote the Egyptians and delivered our houses.
>
> (Exodus 12:25, 27)

At Sinai this historical articulation of the religion continues in the first words uttered by the god to the tribes encamped at the base of his earthly seat:

> And Moses went up unto God, and the LORD called unto him out of the mountain, saying, Thus shalt thou say to the house of Jacob, and tell the children of Israel;
>
> Ye have seen what I did unto the Egyptians, and *how* I bare you on eagles wings, and brought you unto myself.
>
> Now therefore, if ye will obey my voice indeed, and keep my covenant then ye shall be a peculiar treasure unto me above all people: for all the earth *is* mine:

And ye shall be unto me a kingdom of priests, and a holy nation. These *are* the words which thou shalt speak unto the children of Israel.

(Exodus 19:3–6)

Distinct historical events are referred to here, processional and consecutive, looking ahead to a kind of salvation. They narrate what Cornelius Loew has called "sacred history" as distinguished from cyclical and therefore timeless myth, and it seems no accident that the Israelites, as is widely acknowledged, were the first people to create and inscribe a consecutive historical record. "Write this for a memorial in a book," the god commands Moses. Of this unique departure, Baron concludes:

History is the all-pervading dominant sanction for the most fundamental ideas, including the concepts of messianism, the chosen people, the covenant with God, and the Torah. God created the world at a certain time; later He created man; still later He selected Israel as His nation of priests; led them out of Egypt; gave them their law; commanded them to observe that law for their inner sanctification—and all this in the interest of an ideal goal in the messianic future. In that age, "history" will finally vanquish "nature." . . .

The religion, then, from its formal announcement here is to be historically rather than mythologically oriented. It describes human existence as moving relentlessly forward to the achievement of a special destiny, and the supplications for renewal that lie close to the heart of archaic myth are now no longer necessary. For, as Pedersen observes:

Yahweh was outside and above ordinary life, separated from nature, and it was not necessary for him to be radically renewed. Thus the creation in primeval ages does not become the mythical expression of what is annually repeated in the cult, it becomes an event in time, which once took place at its beginning. Herein we find the germ of a change in the old view of time.

True, events repeat themselves, but the repetition is incremental in character and serves mainly to reinforce the sense of historical continuity. In this fundamental way the religion must stand in contrast and antagonism to all natural religions, which, to the extent that they are truly bound to the natural order and rhythm of things, must be deeply repetitive and ahistorical. And so in the course of their history the Israelites were to turn a razor-edged scorn and indeed a sharp sword against all nature-bound cults that could neither look back to events that had happened once in historical time nor forward to prophesy those to come. One feels here again the impress of environment and occupation on this antinature, prohistorical bias, since for this people nature was not a power with which they could establish a celebratory, reverential relationship. It was a power, all right, but one they sought deliverance from rather than surrender to. Nature might not be exactly evil, but it did exercise a cruel power over these wanderers, and they sought emancipation from it. The people of the New Testament, far more divorced from myth and nature and lacking even the vital concept of a promised land, went further and sought to suppress the world of nature.

The prohibitions announced at Sinai against imagery and idolatry (Deuteronomy 4:15–40) are a necessary part of such a governing bias. There are two reasons for these prohibitions. First, images are fated to be representational to some extent, and so, whether bull, sun ray, stone, star, or ear of corn, connected to that cyclical nature the god himself was so infinitely removed from and was now commanding his people to live beyond. As Pedersen writes:

> . . . when the God was detached from the life of nature, and his relation to it consisted only in the creator's display of power, then the psychic strength was removed from nature, it became merely an instrument for the creator, a means for him to display his power. Then it would be absurd to seek divine life and holy strength in the things of this world. And if idols were formed in the shape of animals or men, it could only be understood as a ridiculous attempt to degrade the creator by ascribing to the limitations of creation that power which He alone possessed.

All images, he rightly concludes, are the fit objects of destruction because they are aimed at dishonoring Yahweh.

The second reason for the antiimage prohibitions is that traditionally images had been associated with shrines, which are in turn attached to specific localities. Here were people on the march; no attachment then was possible to the land, such as it was, and none was tolerated. Even Sinai (Horeb), the place of the grand theophany, was forgotten as a specific place, and the Promised Land toward which the god was turning the people's wilderness-weary eyes, was not to be revered for itself but only as a constant reminder of one portion of the bargain here sworn to.

The separation from nature and myth and the commitment to history is emphasized more dramatically and with greater political and cultural results in the new religion's monotheistic character. Though possibly there had been parallel conceptions among other peoples—and Freud based an entire theory of Jewish history on the short-lived monotheism of Amenhotep/Ikhnaton of Egypt's Eighteenth Dynasty, whence he claimed the captive nation derived the idea—it was the Israelites who established monotheism in the spiritual geography of humankind. And with it came the terrible concomitants of intolerance and commandments to destroy the sacred items of others (Exodus 23:23–24; 34:13–16) and to "utterly destroy" polytheistic peoples wherever encountered. Deuteronomy 7:16 commands the holy nation to "consume all the people which the LORD thy God shall deliver thee; thine eye shall have no pity upon them: neither shalt thou serve their gods. . . ." And Deuteronomy 13:16 goes so far as to specify that entire pagan cities must be offered up as burnt sacrifices to the one god, as odors pleasing to him. For polytheism is like imagery connected to nature in its concrete particulars and in its numina. It is for this reason that whatever savageries primitive peoples have visited upon one another, they have usually feared to desecrate idols and altars: there was felt to be too much power in these things, and besides, the gods of one people were quite often recognizable to their adversaries. This goes far to explain why the conception of genocide is foreign to polytheistic cultures. But the distinctions raised in the covenant between religion and idolatry are like some visitation of the *khamsin* to wilderness peoples as yet unsuspected, dark clouds over Africa, the Americas, the Far East, until finally even the remotest islands and jungle enclaves are struck by fire and sword and by the subtler weapon of conversion-by-ridicule (Deuteronomy 2:34; 7:2; 20:16–18, Joshua 6:17–21).

Moreover, in a curious way the very oneness, the *singularity*, of this god emphasizes the separation from nature, for though he created the earth and claims all of it as his, yet he is not to be found everywhere in it: not in the primal chaos at its edges, nor in the cities of the idolators, nor in the deserts given over to demons. Light, truth, and holiness are to be enjoyed only where the god dwells, and he chooses to tabernacle exclusively in the camp of his people, thereby establishing a center of civilization beyond the boundaries of which lie darkness and death, a wilderness peopled with beasts, bestial pagans, and their theriomorphic deities. If the city in the ancient Near East is an oasis, the camp of these semi-nomads serves the same function and defines in the charged terms of religion the chasm between what lies within and what beyond. Thus in Leviticus 16:7–10, 20–22 are found the ritual prescriptions for sending the scapegoat bearing the tribal sins out of the god's camp into the wilderness, the territory of his adversary, Azazel. Thus too the charred corpses of the sons of Aaron, Nadab and Abihu, who unlawfully offered a fire of incense to the god and perished for it; they must be dumped outside the boundaries of the camp where all ritually polluted things are to be thrown (Leviticus 10:1,2,4,5; Numbers 5:1–4; Deuteronomy 23:10–14). And so, though the tribes are now traveling through the heart of the wilderness, they are not really *in* it but are instead insulated by the god against it. Under no circumstances are they to surrender to it or to its temptations. This, of course, is what they continually threaten to do, for the wilderness tempts to disobedience, to riot and rebellion with its hardships, its disorderliness, its radical naturalness.

Here then is the tangled but necessary truth of the Israelites' wilderness experience: though they were originally a desert people, though their god is a desert god, though their covenant with him is ratified for all time in the Sinai wilderness, yet they must *resist* the wilderness and its dark temptations. They are not to remain here, but to move on. They have been promised civilization with its fruits and comforts if they are equal to the challenges of the wilderness experience. For the god wanted "to know what *was* in thine heart, whether thou wouldest keep his commandments or no" (Deuteronomy 8:2).

They fail this ordeal, of course, and Moses tells them that the forty years' wandering is divine retribution for their surrender to various temptations. And surely a long list might be drawn up of their transgressions, most especially the incident of the golden calf. For at the *very moment* that their heroic intercessor is receiving on the mountain the tablets of law from the hand of the god, they are abandoning themselves below in orgiastic fashion to the worship of a fertility image. The placement here of this greatest of all apostasies seems intended to establish forever the tone and temper of wilderness perils. Forever after for peoples of this book the wilderness will be thus stigmatized: as a great and terrible place where terrible things occurred, as at the oasis of Kadesh where the god slew fifteen thousand rebels; where the tribes were tried by the god and found faithless; where the deprivations imposed by the landscape found out the weaknesses of the flesh.

In the subsequent history of the people they are never allowed to forget this experience. The journey from Sinai to Kadesh, the conquest of Canaan, the attacks of the desert and mountain nomads on their lands, the decadent nature cults of their neighbors, and finally the very nearness of the wilderness itself continuously reminded the Israelites what had happened out there. Transgressions against the god's ordinances would result either in the encroachment of the wilderness into the fields of civilization or in the tribes' expulsion from civilization once again into that savage landscape among savage peoples (Leviticus 26; Deuteronomy 11:17; 28:21–24; 29:23; Psalm 107:33–35; Isaiah 5:5–6). The god will again make you dwell in tents, his prophet Hosea threatens. And this is a great fear.

# A Crisis Cult

Throughout the late months of A.D. 246 and on into 247 preparations went forward in Rome for a giant celebration of the first thousand years of Roman glory. But then the games, pageants, and sacrificial observances had to be postponed for a familiar and disquieting reason: the emperor who was to have opened them was out on another urgent campaign against the barbarians. Ironically this emperor was not an old-style, true-born Roman but one who in the good old days would have himself been called a barbarian. Before Trajan (98–117) a non-Roman emperor was unthinkable, but now only a few arch-conservatives thought anything of the fact that a non-European wore the purple.

Nor had this Philip the Arab come to power through the legitimate will of the Senate. That body, though still accorded nominal respect, had almost wholly surrendered its role in deciding Augustan succession, and it was now the army that determined the brief, fitful reigns of emperors. So it had been with Philip, whose troops apparently lynched the teen-aged emperor, Gordian III, and so thrust Philip into power. Over the next twenty-three years thirty emperors would be proclaimed, would reign, and would die, often violently, and in all this the Senate would have no say. From the ascension of Marcus Aurelius (161) to the death of Constantine (337) there were eighty emperors, and in the year 238 alone there were seven, six of whom had been assassinated, leaving Gordian III the sole survivor. Now he too was gone, and the stern-visaged Philip in his place pursued the latest threats to Roman hegemony.

Philip's campaign was successful after the short-term fashion of the time, but few could have been bemused by such temporary settlements. The good old days were going, had been for a long time, and none

knew what to expect in their stead. From a century before the birth of
Jesus to the reign of Constantine what bestrides the empire like a colos-
sus is not the emperors nor even the empire's vast power, extent, and
wealth but instead a gathering mood of uncertainty, pessimism, even
suicidal despair. The pervasive theme of the literature, social patterns,
and political acts is that the Romans felt themselves in the grip of a
strange, indefinable crisis, a widening, deepening gyre of calamity that
could only lead to some kind of death.

To begin at the outside, there were, of course, the barbarians. In con-
vulsive chain reactions stretching through these centuries one horse-,
cart-, or ship-borne group after another harassed and thrust into the
confines of the empire like some giant archer of steadily improving aim
until even the interior cities, those *loci* of civilization, were regularly
vandalized by Goths, Alemanni, Borani, Juthungi. The countryside that
fed the cities was regularly plundered and burned by the barbarians and
the pursuing, requisitioning legions, whose ranks were increasingly
filled with superficially acculturated outlanders. Morale among the
soldiers was not high, and in one of Philip the Arab's campaigns there
were major defections to the putative enemy. With efforts that must be
called at least militarily heroic, Rome was able to stave off defeat,
though it had to give up much, had to wall itself off as best it might
from the threatening wildness beyond with a system of encircling fortresses. But the cost was ruinous. In order to hold fast to its crumbling
empire, Rome had to maintain an army of almost insane proportions.
In consequence the economy became so hopelessly deformed that the
emperor Caracalla (211–217) is reported to have remarked that no one
but himself should have money, and he must have it only to give to his
soldiers. But as more money was coined for this purpose, it was worth
less, and eventually the government met its army payroll payments
with goods extorted from the populace with a system of spies and
secret police enforcing the desperate decrees. Michael Grant estimates
that inflation drove prices up nearly 1000 percent between 258 and 275
and that at the end of the century prices rose even more steeply.

Partly because of this economic deterioration, great faults were opening in the political landscape of the empire. Gaul, Spain, and Britain instituted separatist movements. Palmyra, first under the resourceful
Odenath and then under his even more ambitious wife, Zenobia, tried
to break away and form a rival power with stolen bits of Egypt. Sleeping Persia was roused by King Shapur to become a constant, strength-
sapping threat. Like the ruler in the Russian folktale "The Golden

Cock" who could not ride against his enemies on the west without hearing alarums from the east, the harried emperors tried to face their legions in all directions at once—all the while uneasy about the loyalties of the troops.

Closer in, other disturbing changes cast shadows. The birth rate dropped steadily throughout this period as if, under these threats, the Romans wished to end things as nearly as they could on their own terms. There is evidence of widely practiced birth control by emancipated Roman matrons, declining numbers of marriages, and increasing rates of infanticide. Extra-legal sexual liaisons, on the other hand, increased considerably, especially short-term, unsanctified marriages and homosexuality. We have record of some extraordinarily lavish homosexual weddings, complete with brides, dowries, and large audiences. Riencourt goes so far as to state that the Romans gradually committed ethnic suicide, and while this may claim too much, it might be ventured that to many in these times the conventional sexual roles, the family, and child rearing seemed unappealing or nugatory.

One aspect of the declining birth rate that has wider implications for the state is the various visitations of plague, for the furious spread of communicable diseases is an index to the fundamental problem of the empire. Here as elsewhere it is now clear that Rome had simply lived too far beyond itself: the plagues that swept the empire with increasing frequency and severity from A.D. 165 on were largely attributable to the gradual changeover from a primarily agricultural state to an urbanized one and to the movements through Italy of large numbers of troops and captives from the most distant areas of the known world. Both conditions are favorable to the spread of disease, both existed in the swollen days of empire, and neither could be accommodated by the existing knowledge of sanitation. The designs of empire like those of polity had outrun civilized man's understanding.

The first great plague of which much is known began around A.D. 165 and it lasted, with interruptions, until nearly the end of the century. Hans Zinsser tentatively identifies it as smallpox, and at its worst, in 189, there were two thousand deaths a day in Rome alone. Marcus Aurelius died in this plague, which, with fatal irony, is thought to have begun when soldiers looted a temple somewhere in the Near East. Another great plague spread fifteen or sixteen years after a Gothic victory over Roman forces at Trebonii in 250. Of this affair, Zinsser quotes the medical historian Haeser:

"Men crowded into the larger cities; only the nearest fields were cultivated; the more distant ones became overgrown, and were used as hunting preserves; farm land had no value, because the population had so diminished that enough grain to feed them could be grown on the limited cultivated areas."

Contemporary reports suggest it was feared that civilization itself was endangered, that the whole cultivated world was slowly settling back into bogs, swamps, and wastelands, and that the race of men was becoming extinct. Of all the forms of disruption a society can suffer perhaps none is so deeply demoralizing as mysterious disease, for in its ungovernable career it not only destroys the social fabric, obliterating generations and the intricate ties that bind humans to one another, but it also threatens the philosophical and spiritual assumptions of the populace. The much more heavily documented Black Death of the Middle Ages shows this clearly. When this level of stress is reached, disintegration of the society is a distinct possibility, and the plagues in intolerable combination with the other factors touched on above had clearly brought about such a state in the Roman empire by the beginning of the fourth century.

Some evidence of a crisis mentality can be traced back far before A.D. 300. As early as the first century B.C. there are telling images of decadence and the corruption of older republican virtues. In the following century the frequency of these images increases and primitivistic longings are a generally acknowledged way of commenting on Roman cultural ills, as, for instance, in Tacitus's account of the barbarian tribes of Germany.

But the level of stress and anxiety rose markedly as the second century drew to a close, and it continued to do so throughout the third century. The celebrated melancholy musings of the fated Marcus Aurelius seem to epitomize this trend, although E. R. Dodds, among others, has shown that it would be a mistake to believe such attitudes were confined to those with sufficient leisure to wax melancholy. Many were asking with Aurelius the kinds of doom-gripped, ultimate questions that characterize a people possessed by a sense of apocalypse. Happy people, ventures Dodds, do not perpetually wonder about the reasons for living; they do not ask, Why are humans on earth at all? For happy people the joys and intensity of living are both justification and answer. But in this period historians have discerned a steadily increasing weariness with living and an increasing tendency to disbelieve that this world and its sorry existence were all that was meant to be. Perhaps the world was growing old and like the the empire was fated soon to die: its colors seemed faded, its harvests of grain and ore more meager, and life in general seemed meaner than in the past. And there was an open, savage contempt for the human body, variously styled a betrayer, a bag of filth and gore, clay, even excrement. Marcus Aurelius characterized human life in the mood of his time:

> Yesterday a drop of semen, tomorrow a handful of spice and ashes. In the life of man, his time is but a moment, his being an incessant flux, his senses a dim rushlight, his body a prey of worms, his soul an unquiet eddy, his fortune dark, his fame doubtful. In short, all that is of the body is as coursing waters, all that is of the soul as dreams and vapours. . . . An empty pageant; a stage play; flocks of sheep, herds of cattle; a tussle of spearmen; a bone flung among a pack of curs; a crumb tossed into a pond of fish; ants, loaded and labouring; mice, scared and scampering; puppets jerking on their strings—that is life.

Beneath all this was the immeasurable factor of the decadence of Roman religion, which by the middle of the third century was a welter of conflicting and competing cults. In ancient times the Italians had lived in a numinous world in which forests, hills, rocks, and waters were alive and sacred. They had listened then to the voices of birds, noted their numbers and positionings, made expiations for the clearing of groves, and lustrations for the marking out of fields. Though such practices and the feelings that prompted them lingered long in the countryside, the locus of religious activity and concepts of divinity gradually withdrew into the cities where temples were built to the anthropomorphic guardians of the urban communities.

In the great days of empire down to the end of the second century B.C. religion served well the needs of the state. But there were tendencies in the direction of decay, not least of which was a kind of legalistic attitude toward ritual, as if Romans were saying, We do these things for you, and now you must do these things for us. This hope is always implied in worship, yet when it reaches the level of a dry contract, something of belief in the magic and mystery of divinity has been lost. Frederick Grant believes the legalistic tendency was endemic to Italian religion, but he also suggests that the increasing influence of Greek rationalism and skepticism was a contributing factor. Indeed, the evolution of worship in Greece shows this same movement from the numinous to the anthropomorphic to decay, the last stages marked by fatal attempts to interpret the ancient myths historically and the gods as historical personages distorted by ignorance and poetic license, as in the work of Euhemerus (end of the third century B.C.). At last only blasphemy was offered to the silent oracles and the dead gods.

So it was now with the Romans. In the first century B.C. Varro had apparently followed the euhemeristic tendency of considering the myths as poetic fictions that were too often contrary to the nature of divinity. By the middle of the next century Seneca described the images of myth as monstrous, the rituals disgusting. His contemporary Petronius remarks that "no one any longer believes that the gods are gods." In the old days, he writes,

> the mothers, wearing their finest gowns, climbed the hill with bare feet, loosened hair, and pure minds, and prayed Jupiter to send them rain. Then it used to rain instantly, in buckets!—for it was either then or never. And they all came home, soaking wet, like drowned rats. But as it is [now], the gods come shuffling along with wool-clad feet, because we have no religion. And so our fields lie [parched and baking]!

There is evidence in this period of the neglect of shrines and temples, of offerings of trash to the divinities, of widespread cynicism and bitter joking about them and their stewardship of Roman fortunes. It was alternatively felt that this very neglect and cynicism were responsible for their desertion of the empire (thus its present condition); or that they had lost their powers (the senescence theme again); or that they had not been the right divinities to begin with. A similar desperate cynicism had gripped the Greeks in the declining days of their hegemony when it was often said that the great gods were either far away and indifferent to Greek concerns or else had never existed. As with the Greeks, so now with the Romans, whose oracles had lapsed into grim silence, and despite efforts to revive them, by the end of the second century a disbeliever could crow without fear over their demise.

But neither neglect, nor cynicism, nor the cool influence of Greek rationalism was sufficient for these times. Some other form of religious conviction was desperately needed. To fill the gaping spiritual chasm dozens of "mystery cults" were imported, most of them from the East. Some of these had been grafted onto the state religion long before the crisis period under review, but their numbers and popularity greatly increased as the level of anxiety continued to rise. It is probably significant that one of the earliest and most durable of these, the worship of the Phrygian Mother of the Gods, had been adopted during the Hannibalic crisis of 204 B.C. As that war had dragged on, various Roman rituals and gods had been secretly and then openly abandoned and strange ceremonies instituted in their stead, culminating in human sacrifice and the triumphant entry of the Great Mother into the capitol where she had been greeted by the city's matrons and suitably enshrined.

The spiritual expedient set a pattern that persisted even after this particular crisis had passed, for the real crisis was far greater than Hannibal, and the traditional means were proving inadequate to it. So we find the importation and establishment of one mystical cult after another—Isis, Ma, Serapis, Mithras—and the gradual exclusion of traditional Roman religion, despite periodic imperial attempts to revive it. Dodds defines the heart of mysticism as the "belief in the possibility of an intimate and direct union of the human spirit with the fundamental principle of being," and then goes on to observe that mysticism and misery are related facts. Such a judgment seems applicable only to certain kinds of culture, but it clearly applies here where the mystery cults served the necessary function of shifting the attention of their converts away from this world, with its incremental disasters and decay, and onto another, eternal plane of existence. With their clamorous messages of contempt for this world and for the body, and in their transference of divinity to the infinite, the cults offered relief from the stresses of a disintegrating empire.

Frazer (and those who have followed him) speculated that the origins of these cults lie far back in the mythic rites of agricultural peoples among whom the unfathomed mysteries of seasonal changes, growth, and decay inspired a corresponding worship of dying and reviving goddesses and (later) their sons/consorts. More recent scholarship indicates that the significations were more complex than the famous dying/reviving symbolism Frazer worked out. Whatever the precise agricultural bases of these cults, in their Roman phases they were reinterpreted for the needs of the time and of a predominantly urbanized populace. Now the cults centered on direct communication with a cultic lord whose service was to redeem his worshipers with his own sacrificial death, a death that was merely temporary and a type of what the true believer could expect for himself. The convert was initiated with rites of purification and rebirth and achieved mystical union with the cultic lord through sacraments. Belief thus promised escape from the illusory cares of this life, but much more importantly it promised redemption from death into everlasting life. With the smell of decay so strongly in the air such a promise must have been powerfully attractive, so much so that some Romans apparently belonged to two or even three competing cults at once so as to be certain of salvation.

It is obvious that Christianity belonged to that welter of mystery cults that competed so fervently for the devotions of an increasingly despairing people. A comparison of its myths and rituals with, say, the cult of Attis or that of Mithra shows that it was responding to the same condition in generally the same way. To observe this, however, is in no way to attempt to disparage this cult's authority or authenticity, nor is it an adequate way of accounting for the character of its charismatic leader, whose influence was to prove far greater than any rival's (an argument used by the early controversialists to demonstrate the truth of their faith). Indeed, though it was younger by far than most of its competitors, it was perhaps the first to assess correctly the true nature and extent of the crisis. The observation merely places Christianity in its appropriate historical context. It is only when Christianity is so placed that we can appreciate its historical fate as a crisis cult that evolved into a state religion. For it was a civilization ostensibly authorized by this religion that undertook to harmonize unknown portions of the globe with itself.

Two outstanding students of the phenomenon of crisis cults have been Anthony F. C. Wallace and Weston La Barre. Both employ what Wallace has called the "organismic analogy" in an attempt to describe what happens to cultures in extreme crisis. In this analogy the culture is conceptualized as an organism, complete with cells, organs of proprioception, and so forth. The organism's major imperative is to resist externally imposed changes and fluctuations so as to maintain a relatively stable environment for itself. Just so, a culture resists change in order to remain stable and historically coherent. Actually, as cultural anthropologists have noted, all cultures constantly experience change and constantly adapt to it, but usually these changes are of such a gradual nature and the adaptive measures taken are in such general harmony with precedent that members of a culture do not often perceive the changes as threatening. Even in times of considerable stress a culture usually has traditional means to meet the challenges—emergency structures or plans more or less adequate to the task. It is only when the culture is faced with massive and rapid changes that render even the most severe emergency measures patently inadequate that a strategy of a radically different nature may be tried. And even in such a deep crisis a culture may struggle to maintain its integrity without having to make major concessions to the stresses that beset it. Wallace calls this the "principle of the conservation of cognitive structures" and suggests that neither a culture nor an individual will give up a deeply held conviction of the way the world is until convinced that an adequate and congenial replacement is at hand.

It is clear that the Roman empire had reached such a level of crisis by the beginning of the third century—and quite possibly the period of great individual and collective stress ought to be pushed even further back. For if, as Julian Huxley has said, culture is the cumulative transmission of experiences, then by A.D. 200 there was an increasingly distressing gap in Roman life between the experiences transmitted from the honored past and the realities of the barbarized, plague-stalked present with its endemic political unrest, its enfeebled social bonds and baseless codes, its rampant inflation, its decaying and neglected shrines and public buildings. The time was ripe for the rise and popularization of new religions promising deliverance from this state of affairs.

Crisis cults, charismatic movements, "revitalization movements" (Wallace's all-embracing term)—whichever term is employed, the signification is the same: an organized, conscious effort by some members of a culture to construct a more satisfying life for themselves out of what they perceive as the ruins of the present. In addition to Christianity, there are dozens of well-known, well-documented analogs scattered across history and geography. Ikhnaton's monotheistic cult in ancient Egypt is such; La Barre considers Greek Platonism one; Islam, Methodism, Mormonism, and perhaps the American Black Muslim movement are others. So too the various movements among the native peoples of the globe to whom, in one of history's greatest ironies, Christianity itself was brought. One of these latter, the Ghost Dance among desperate North American tribes in the nineteenth century, will occupy our attention in pages ahead.

The studies of Wallace and La Barre lead to the conclusion that this is a recurrent phenomenon in the history of human cultures and that perhaps thousands of examples of varying magnitude could be found. La Barre goes so far as to state that there can be no cult without a crisis, that with the birth of each crisis cult we witness anew the origin of religion, and that each new religion is the "Ghost Dance of a traumatized society."

Typically such a movement grows out of the inspired meditations, the mantic transports and dreams, of a single individual who has some-how ingested the entire crisis afflicting the culture. What such an indi-vidual then enunciates in apodictic syllables is the resolution to con-flicts, the solution to dilemmas, the passage out of crisis: what early Christians referred to with impressive simplicity as The Way. Such are the sayings attributed in the Gospels to Jesus in the first century—statements of a numbing, seductive simplicity that are tinged with the sense of present and deepening crisis. By the end of the third century, when Constantine elevated the cult to the state church, these had the gathered weight of all the intervening disasters:

From that time Jesus began to preach, and to say, Repent: for the kingdom of heaven is at hand.

(Matthew 4:17)

.    .    .

He answered and said unto them, When it is evening, ye say, It will be fair weather: for the sky is red.

And in the morning, It will be foul weather to day: for the sky is red and lowring. O ye hypocrites, ye can discern the face of the sky; but can ye not discern the signs of the times?

A wicked and adulterous generation seeketh after a sign; and there shall no sign be given unto it, but the sign of the prophet Jonas.

(Matthew 16:2–4)

.    .    .

Now learn a parable of the fig tree; When his branch is yet tender, and putteth forth leaves, ye know that summer is nigh:

So likewise ye, when ye shall see all these things, know that it is near, even at the doors.

Verily I say unto you, This generation shall not pass, till all these things be fulfilled.

(Matthew 24:32–34)

.    .    .

And what I say unto you I say unto all, Watch.

(Mark 13:37)

There are abundant warnings here too against that multitude of false messiahs (such as Simon Magnus and Apollonius of Tyana) already in evidence and certain to swell as things grew worse. For it was not only the Jews who were suffering persecution and looking for their promised redeemer; all were suffering the fate of empire (Matthew 7:15–20; Mark 13:5–6, 21–23; Luke 21:8; Acts 5:35–39, 8:9ff, 13:4ff).

In order for a prophet's revelation to achieve its intention and provide relief through a dramatically new way of living, some primitive form of organization must be developed. The prophet must have assistance in the dissemination of the new way, and typically this involves the culling of a small group of special converts who become disciples and constitute the basic foundation out of which any further organization must grow. These are individuals who, touched by the charisma of the prophet, experience the kind of revitalization in their own lives that is promised to the culture as a whole. In Christianity, of course, these were the Twelve Apostles to whom was added a crucial thirteenth in the person of Saul/Paul.

At this point the original revelation becomes a cult, but it is a long way from becoming a church, for this latter implies a level of organization not yet achieved because not yet needed. What holds the cult together in these early days is the charismatic leader. As Weber has shown in his study of such movements, it is the power that radiates from such an individual that provides the minimal degree of cohesiveness required by the cult at this stage. There is no hierarchy, no officials, no sense of a profession. The leader and his disciples live together in what Weber has termed an "emotional form of communal relationship," divorced from normal routines and social ties, dependent on voluntary gifts.

This is the period of maximum spiritual fervor when revelation seems to spread like an electric current from the figure of the leader. We might say that the converts at this stage are *possessed, captives* of the new spirit. Ecstatic conversions are a common feature, often accompanied by speaking in tongues. There are also prophetic utterances in consonance with the new spirit but not rigidly circumscribed by any letter of law, for there is none and no one to enforce it. In sum, we might say that such a period is colored by the excitement of the revelation of a new and living mythology that seems to bring all who are touched by it into an intimate relationship with the very springs of creation.

Everything we have of both a canonical and noncanonical nature indicates that the Christian cult enjoyed such a period and that it persisted even after the death of its charismatic leader. The Synoptic Gospels, the earliest of which cannot have been transcribed much before A.D. 70 and the other two of which surely belong to the last third of the first century, are infused with an unmistakable inner light of conversion, conviction, and joy. So too the whole of Acts, which recounts the history of the cult after the crucifixion of Jesus. Here we find the sure signs of inspired teaching, of the free proliferation of prophets, and of a remarkably simple, communistic organization held together by a mythology of love. The Agape feasts seem to epitomize all this with their charity, their inspirational addresses, their sense of mission that has not yet degenerated into a crusade. As we read of this time in Acts 4:31–32:

> And when they had prayed, the place was shaken where they were assembled together; and they were all filled with the Holy Ghost, and they spake the word of God with boldness.
>
> And the multitude of them that believed were of one heart and of one soul: neither said any of them that ought of the things which he possessed was his own; but they had all things common.

And in the noncanonical tradition we find evidence of that celebration of all of life that is the essence of a living mythology and that sends spiritual energy shooting through the veins of the group; but that, when it is severely qualified or curtailed, becomes instead a pathogen to its organism. Thus in the *Acts of John* (one of several very popular pseudepigraphical works produced between c. 150–250) we find Jesus depicted more after the fashion of that ancient dancing sorcerer of the cave wall of Trois Frères than either the martyred death figure or the wholly ethereal one of mature Christianity. He is described herein as instituting a ring dance for his disciples with himself in the midst of the ring. And then as they hold hands and move in humankind's oldest pattern, he sings to them:

> The Heavenly Spheres make music for us;
> The Holy Twelve dance with us;
> All things join in the dance;
> Ye who dance not, know not what we are knowing.

Of similar significance is the delightful imagery of life in the "Odes of Solomon," a collection of songs that may be the compositions of first-century converts:

> They who make songs shall sing the grace
> of the Lord Most High;

And they shall bring their songs, and their
heart shall be like the day: and like the
excellent beauty of the Lord their pleasant song. . . .

Ode 7

And I became like the land which blossoms
and rejoices in its fruits:

And the Lord was like the sun shining on the
face of the land;

He lightened my eyes, and my face received
the dew; and my nostrils enjoyed the pleasant
odour of the Lord. . . .

Ode 11

Other visions of the period depict Jesus as a figure of myth—as a sun god, for instance, in which guise Constantine seems to have worshiped him; or as Orpheus who charmed nature and perished.

It may be, however, that such a state of affairs cannot permanently exist. There are, as Weber has pointed out, serious social and economic problems in maintaining the original, primitive character of such a cult. For one thing, in its early, ecstatic period the cult tends to sever its converts from most of the normal routines of life—family, friends, home, occupation—and in this sense it makes "impossible" demands on those who would follow it, as the Gospels abundantly illustrate (e.g., Matthew 6:25–34; 7:13–14; Luke 9:57–62). It is in the interests of the cult that the association be placed on a more stable basis, one that permits access to routine pursuits, for this alone can ensure that the cult will not go the way of so many and become merely a transitory phenomenon closed off to all but a handful of fanatics. So the cult must in essence violate its own character if it is to survive: it must submit its charismatic nature to the process of "routinization" or "traditionalization." Weber points out that this involves institutionalizing charisma into offices and officials and dividing the cult into clergy and laity. If this perilous transitional period is successfully negotiated, if the cult's spiritual force can survive translation into more defined, less personal terms, then the cult will have achieved the status of a church.

This, of course, is what happened with Christianity, and if the taxonomic analysis applied here elucidates persistent patterns of human behavior, then Paul and the rest of the Church fathers who guided the faith through its formative years to its enthronement as the state church under Constantine did only what they thought they had to. But from our vantage point in what has been designated the "post-Christian" era, we must ask how much routinization is too much? At what point in the process of institutionalization is profession of faith coerced more by the paraphernalia of the Church and less by its authorizing mythology? And then, from our vantage point here in the Christianized New World, we must ponder the ongoing effects of a civilization that expanded and invaded other mythological zones in the name of what had become in reality a dead letter, rather than a living mythology.

In addition to the socioeconomic considerations suggested above, the early Church faced various problems of practical theology. Chief among these was how it might distinguish itself from the many other cults (especially, at the beginning of the third century, the very popular Mithraism) and so broaden and deepen its hold in the Roman world. One way was to begin making its appeal to others than those who had up to now been its principal converts—the have-nots. It now would seek out the educated and influential, and the only way it could do this was to rid itself of the stigma of being, as Dods says, a religion of ignorant slaves. The pagan controversialist Celsus, in his first-century attack on the faith, castigated Christian proselytizers as infamous beggars working their tricks on "striplings or a crowd of slaves and fools." To make a different appeal, however, involved some reinterpretation of Jesus' message as reported by the Apostles (see, for example, Matthew 19:24), which seems more directed toward those without than those with authority. Now, though the routes to conversion through ecstatic seizure and blind faith were not entirely closed off, they appear to have been relegated to a lesser status. Henceforth, the Church would seek to expand by argument, and argument required the creation, firming up, and internal recognition of dogma. It also required the training and deployment of skilled controversialists who could advance this dogma against the positions of rivals. Increasingly, too, it was felt that *any* concessions to "paganism" (a dangerously loaded designation inherited from Judaism) could be fatal to Christian aspirations. At this period there were more similarities than differences between "error" and "truth," and if Christianity was not to remain just another mystery religion, if it was to take fullest advantage of the leverage that Constantine's visionary conversion had given it, then it would have to become increasingly self-conscious and authoritarian. If it was not to degenerate like Gnosticism into a dark confusion of esoterica, it would have to find a way of limiting speculation and revelation, and of regularizing the preaching of the faith.

The way nearest to hand was to insist on the primacy, the inviolable sacredness, of the first two generations of Christian experience. Thus the notion was gradually developed that all teaching and all worship must conform to that of the Apostles and Paul; that everything authoritative about the faith was to be found back there: no hidden news, no new revelations could now surprise the faithful and so expose them to the savage ridicule of pagan adversaries. For did it not stand to reason that if Paul and the Apostles had had significant information respecting the Messiah and his message, they would surely have imparted it in their time?

These are the practical reasons, then—the need to distinguish itself from its rivals, to expand its appeal, and to make that appeal firm and unchanging—that Christianity made a critical turn in its early years as a state church. It turned away from an implicit understanding of itself as a mythology and, with increasing consciousness and intensity, delivered itself to history. Jesus and his ministry had been historical events that could happen only *once* in historical time. The pagan cults, on the other hand, could not be true because their myths were only that, were not historically verifiable, and were thus only temporary, illusory reliefs. Augustine, Bishop of Hippo, brooding in North Africa on the ruins of Rome, saw the collapse of the city of man as the result of a foolish, vicious belief in these fables. As E. C. Rust describes the Church's developed position: "The mid-point of salvation history is the unique and unrepeatable event of Jesus of Nazareth, the God-Man, in whom the eternal God lifted up history into his own life and fully entered historical time to redeem it."

Christianity's turn from myth toward history may have an interior historical explanation, for its first converts were Jews, as were its first authors. And the Old Testament influences were very strong in these formative years; so strong in fact that for a considerable time outsiders in the empire made no distinction between Christians and Jews other than to conceive of the former as an extremist branch of the latter. As we have seen, the ancient Israelites had made a similar turn and had created an implacable barrier between themselves as a nation caught up in the drama of sacred history and the surrounding pagan cults still in one way or another bound in the toils of nature-based mythologies. So it might be that the Apostles and Paul as well as the members of the primitive cells already had some sense of The Way as existing in historical time. Certainly Paul was starkly aware of the chasm between his new faith and all nature-based religions. But whatever the degree of the

historical impulse in the earliest years, it was grandly enlarged by succeeding generations of theologians, and it is hard not to feel that the
Messiah was betrayed, not in his person as by Judas, but in his message
by the fathers of the Church. For that message seems to be of the
divinity that dwells within and that is present in all creation, and of
how to live in accordance with this. Yet in the historical interpretation
of Jesus in the vast industry of Christology, the emphasis shifted from
the message to the historical figure, who was now presented as having
once intervened in the history of this world only to show it as the contemptible thing it truly was; who had assumed flesh only to show that
it was possible to live beyond the body, to ruthlessly subordinate its
desires to the utterly different needs of the spirit. This development has
had incalculable consequences for the history of the West and, through
exploration and settlement, for the rest of the globe. But if the consequences defy calculation while yet we suffer them, at least now their
formidable outlines have emerged.

Ironically, whereas early Christian controversialists claimed that rival
cults with their myths offered only temporary, illusory relief from the
fate of empire, we are now in a position to see that it has been Christianity in its turn away from myth rooted in nature that has offered
only temporary relief—a relief to those very few who shared the gusty
joys of the earliest days and to those who managed, like certain gifted
mystics, to find their solitary ways back there. For such a turn as this
can only gather force as the gap widens between that sanctified moment
of the historical past and the moving present that advances relentlessly
into its linear future. Thus the crisis cult that had arisen to deliver its
believers from the fate of the Roman empire turned them over to the
greater, and more inescapable, "terror of history."

Primitive cultures, says Mircea Eliade, enact measures to escape this
terror of history, the existential loneliness of the linear march of events
toward annihilation. Such an escape must be partially a failure since
consciousness of the passage of time is inevitable in deaths, births, natural disasters, and the other phenomena that willy-nilly record duration.★ But the escape is also in some true way successful, and it is so to
the extent that any tribal mythological system operates with integrity,

---

★ Indeed, as we shall see "primitives" long supposed to be entirely outside time and history have been acutely aware of their own history—though it is history of a very special
sort.

for the myths and the rituals that communally enact them are the measures of escape.

The effect of such a fully functioning mythological system is two-fold. First, to the extent that the system connects the culture with the rest of creation and thus with the process of growth, decay, and metamorphosis, it relieves its worshipers of the tension and anxiety of feeling themselves isolated by a special destiny. Theirs is the common destiny, theirs the fate of all that is. Death is still the great anxiety, but myth-bound peoples who live in a sacrilized world with a myriad of mutually existing forms of life seem better prepared to accept the belief that there are really no endings, only changes. "Dead, did I say?" an aged Native American chief asked himself and his white listeners. "There is no death, only a change of worlds." Or, as the great primitive poet Walt Whitman put it, having struggled to reinvent a mythic America:

> The smallest sprout shows there is really no death,
> And if ever there was it led forward life. . . .

Second, myth and ritual, as Eliade has explained, have the function of arresting time. If the myths record the beginnings of things, then every recitation or enactment of them in ritual becomes coeval with the primordial past. Thus what has transpired since the last such performance is abrogated, the chronometer slips, knowledge of the passage of time toward the end of time—surely a great burden—is evaded, and the world is re-created anew on the original model.

This would explain that feeling of a "timeless world" so frequently commented on by those of Christian civilization who have encountered the primitives in their native habitats: a dreamlike circumambience in which, though events occur, they are perceived as recurring in accordance with the changeless patterns announced in myth and confirmed in nature. Thus Jung, so sensitive to these matters, remarking on his travels in the mythic zones of Africa:

> The deeper we penetrated into the Sahara, the more time slowed down for me; it even threatened to move backward. The shimmering heat waves rising up contributed a good deal to my dreamy state, and when we reached the first palms and dwellings of the oasis, it seemed to me that everything here was exactly the way it should be and the way it had always been.

In such a world (as William Faulkner has one Indian say to another) tomorrow is just another name for today. And the difference between such a world and the one we know seems largely attributable to Christianity's turn away from the timeless cycle of myth and nature and its insistence on rendering the Messiah into historical—and thus entirely unique—terms. Robbed in this way of old comforts, and unable to feel reattached to the great events sealed off by subsequent history, the Christian West had to live onward, set its face resolutely forward in the hope of recovering in an apocalyptic future what it had once had in the past. The historical interpretation of Christian mythology thus became the very engine of history.

It is therefore of ironic significance that the mechanical measurement of time arose in part out of the routine of the Christian monastery. St. Benedict, Lewis Mumford tells us, added a seventh period to the round of daily devotions, imposing a Roman regularity on them, and a bull of Pope Sabinianus decreed that the monastery bells be rung seven times every twenty-four hours.* So the regular striking of the bells marked the passage of time and served to regulate and synchronize the actions of humans. Mechanical time became "dissociated from organic sequences," and the Christian world came increasingly to live by a mechanical contrivance that had no relationship to organically derived needs. Centuries later, invaded, surrounded, but still unconquered by Christianity, the Pueblo Indians' sacred clowns would carry with them alarm clocks in their dances as a way of ridiculing the whites' enslavement to the marching minute hand.

All of this suggests a source of that burden, that anxiety, so frequently sensed by primitives who have been encountered by the whites on their long march outward. As an aged Pueblo Indian once pointed out to Jung:

> "See . . . how cruel the whites look. Their lips are thin, their noses sharp, their faces furrowed and distorted by folds. Their eyes have a staring expression; they are always seeking something. What are they seeking? The whites always want something; they are always uneasy and restless. We do not know what they want. We do not understand them. We think that they are mad."

* Michael Grant writes that it was Benedict's regularizing and straitening of monasticism that made "western monks into missionaries, explorers, cultivators and preservers of inherited culture."

Surely one of the brilliant expositions of this condition is Norman O. Brown's *Life Against Death*, and to a lesser extent his *Love's
Body*. Though never avowedly a critique of the civilization Christianity
has created, Brown's arguments are based on what he and others have
observed of the life around them in the West, so that when he writes
of "civilization" and "man" he is really speaking of the Christianized
world we know.

Following Freud, Brown argues that the basis of history is man's
repression of his instinctual nature—the body, its needs and drives, its
rhythms, its desire to achieve atonement (at-one-ment) with the world
about it. This turn away from nature, in others words, is what makes
man the only history-making animal. Seen this way, history is the
steadily lengthening chronicle of mass neurosis, maybe even a kind of
suicide note, with mankind seeking restlessly, unconsciously, for a way
to end it all and be finally at rest. For repression can never be wholly or
finally successful, and the drive of the animal to seek the proper conditions of its existence can never be overcome. Only a reconciliation of
life (which Brown identifies as sexuality and the desire to feel at one
with the world) with death (as the natural end of things) can bring
about a state of rest. And this state is not stasis, the cessation of all activity, but the harmonious functioning of activity in which the pleasure
of repetition replaces the restlessness of novelty. Then history would
have a stop, for the desire to become would be absorbed by the ability
simply to be.

A second consequence of the historical interpretation of Christianity
is what might be called the slow starvation of the soul. This has come
about through the Church's gathered, settled refusal to admit new revelations.

Whereas revelation is a continous possibility in a living mythology
still tied in important ways to the natural world, Christianity sealed off
revelation at the end of the Apostolic era as if the stone had been rolled
back into the entrance of the Messiah's tomb. So, for example, it appears now as a historical inevitability that white observers of North
American tribal life should be baffled by the frequency and freedom
with which revelation (visions) occurred to shamans, chiefs, warriors,
or ordinary folk, and by the seriousness with which they were generally
received. To the Christians this seemed but further evidence of the fal-
sity of the tribal myths and of the shallow and credulous nature of savage belief. Yet as it becomes clearer to us that such revelations were not

violations of the tribal mythologies out of which they sprang but continuous reaffirmations of them, we can see that such possibilities are a
necessity if true belief is to be nurtured and sustained. If a tribal
member announced to the group that he or she had had a vision requir-
ing a personal decision, even a radical change in life pattern, this was no
threat, no blasphemy—though, indeed, it might prove trouble-
some—but living proof that "The Great Spirit sees me." Nor was it evidence that the tribal mythologies were "unorganized" (an anthropological euphemism that succeeded cruder judgments but amounted to
little more) so much as that they were open to new disclosures, personal experiences with the spirit world that manifested itself through
creation. Through vision quests, hallucinations, dreams, or in the accidents of waking life any aborigine might be brought to confrontation
with the Powers, and it is the sense of this possibility that gives such
richness to the tribal mythologies—and that confers such posthumous
tragedy on the pathetically mutilated records of them that the Christianized world has cared to keep.

Not so with Christianity, which early hastened to circumscribe revelation, however come by. There is evidence that many in the primitive
Church were converted through the power of such visions, through ecstatic trances, in dreams; and that new disclosures, like new conversions, were welcomed as further unfoldings of the messianic message.
*Prophetai* who spoke new words of divine truth were likewise an accepted feature of Christian life, as both Acts and the Pauline epistles indicate. Yet by the latter part of the second century such activities were
already being perceived as injurious to the ambitious designs of the
Church, and thus one of the earliest and most crucial of what was to
become a long chain of kindred controversies was that aroused by
Montanus. This converted pagan dared to speak in his trances for the
Holy Spirit. Taking his authority from a somewhat obscure passage in
John (14:16), Montanus gathered to him a group of converts who were
evidently delighted to hear the good news again. He spoke as the Para-
clete or intercessor, as apparently promised by the Messiah, but it was
not this claim, nor even the "irrational" basis of his appeal, that was
most threatening to the hierarchy. At the heart of the controversy was
the Church's determined refusal to admit any further testaments. It was
successful here, but at a price, for it lost one of its most brilliant controversialists, Tertullian, who defected to the Montanists. In vain did
Tertullian (and later, others) warn that the Church was substituting
dogma and ecclesiastical hierarchy for true belief. If it excluded revela-

tion and prophecy, it would cut itself off not only from an endlessly renewable mythology but also from the humble faithful who needed to feel themselves involved in a recurrent and unfolding religion rather than one that had been given, once and for all time, long generations before theirs.

Perhaps it was the Montanus affair that suggested to the hierarchy that the Holy Ghost was now a potential hazard, for "he" was by definition irrational, primitive, and incapable of being circumscribed by dogma or historical boundaries. In any case, as Dodds observes, it was now felt that the Holy Ghost (like the *prophetai*) had outlived his utility:

> He was too deeply entrenched in the New Testament to be demoted, but he ceased in practice to play any audible part in the councils of the Church. The old tradition of the inspired *prophetes* who spoke what came to him was replaced by the more convenient idea of a continuous divine guidance which was granted, without their noticing it, to the principal Church dignitaries.

Against this hardening tendency not even the important tradition of Christian mysticism was able to influence general practice or belief. For the mystics were by definition and official treatment *special*, and the hierarchy had no intention of allowing them to become the standard for lay behavior. Indeed, Henry C. Lea has shown that in Spain, when mysticism showed signs of becoming widespread, the Church took the fiercest actions against the mystics as part of its Inquisition.

Evelyn Underhill's classic study of the tradition of Christian mysticism reveals it as a series of solitary attempts to find some way out of the prison of the soul that Christianity had become, back to the primitive joys of the sacred period. Thus even here the historical character of the matured religion is revealed. In retaining the authority to admit certain mystical experiences and activities while often taking punitive action against others, the Church reaffirmed its commitment to doctrinal uniformity at the expense of spontaneous conviction and continuous regeneration. No doubt, too, the examples of those whose mystical experiences had run foul of what could be officially permitted helped the Church to keep the tradition the rarity it was.

But there was another, interior, reason why Christian mysticism proved incapable of significantly altering the character of the faith. This was its acceptance of the body/soul, world/heaven, man/nature dualisms that are the heritage of this historical religion. With but a few bright exceptions Christian mystics are characterized more by their denial of large aspects of creation rather than by any joyful acceptance of the same; by negative desires instead of positive; by the imagery and love of death rather than commitment to life. Even if we accept the postulate that the death of physical distractions opens the way to the true life of the soul, it seems unfortunate that so terribly much of human existence should have to appear under the guises of encumbrances, pitfalls, snares, or other terms of opprobrium.

Especially the body. The mystical tradition is so replete with examples of twisted, cankered sexuality and physicality, of hatred and fear of the necessary conditions of animate existence, that this alone vitiated whatever spiritually regenerating influence it might have had. Thus some of the greatest mystics, cited by Underhill with knowing approval, have been guilty of the greatest sins against creation—racking themselves on wheels, hacking religious emblems in their breasts, causing themselves to be partially buried in graves or hanged on gallows, burning themselves on ovens, licking up the vomit or drinking the blood of diseased patients. "Nought must be too disgusting" as Underhill glosses the essential axiom of mystical mortification.

Here it is useful to draw on Dodds' distinction (and one fully supported by ethnographic data) that this hatred of the body is fundamentally different from those widely observed primitive practices of fasting, abstinence, ritual scarification, or torture testing. For such practices do not emanate from a denial of the necessities of animate existence. Rather these practices, however strenuous, even bloody, are *temporarily* adopted strategies for becoming better attuned to the spirit life, for strengthening the will, or for gaining or renewing spiritual power. Nothing, it is sure, would have seemed more alien to the Sun Dancers of the North American Plains, with bone skewers thrust through the muscles of their backs or chests, than the notion that by behaving so they were punishing the contemptible prisons of their souls.

So it is only here and there that we can find solitary positive examples among these solitaries of the faith. Jacob Boehme, the German mystic of the late sixteenth and early seventeenth centuries, was one. Brown quotes his lament for man's loss of fellow feeling for the rest of creation:

"No people understands any more the sensual language, and the birds of the air and the beasts in the forest do understand it according to their species.

Therefore man may reflect what he has been robbed of, and what he is to recover in his second birth. For in the sensual language all spirits speak with
each other, they need no other language, for it is the language of nature."

Another was the Poor Little Man of Assisi, Francis, in whom the all-
embracing primitive joy bubbled forth again as from an underground
spring. Accompanying his singing on an imaginary viol, exhorting
his men to serve as God's minstrels, who could convert others through
demonstrations of the deep pleasures of the faith instead of its trials, he
sensed anew the life instinct in all creation, even wood and stones. He
died predicting infamy for his order, which indeed it was to earn for its
speedy betrayal of his ideals and its zealous role in the Inquisition.

But Christian mysticism, a potential spring of regeneration, was too
often arid by reason of its sponsorship by and identification with anti-
mythological orthodoxy. Instead of tapping new sources of life, it
generally chose mortification and death. Underhill quotes the German
mystic John Tauler:

"This dying has many degrees, and so has this life. A man might die a
thousand deaths in one day and find at once a joyful life corresponding to
each of them. This is as it must be: God cannot deny or refuse this to death.
The stronger the death the more powerful and thorough is the corresponding life; the more intimate the death, the more inward is the life. Each
life brings strength, and strengthens to a harder death.

.    .    .

". . . whatever it may be, to which he has not yet died, it is harder at first
to one who is unaccustomed to it and unmortified than to him who is mortified.

". . . a great life makes reply to him who dies in earnest even in the least
things, a life which strengthens him immediately to die a greater death; a
death so long and strong, that it seems hereafter more joyful, good and
pleasant to die than to live, for he finds life in death and light shining in
darkness."

Maybe all of this helps explain the largest significance of what Jung
intuitively grasped in his childhood. In his autobiography he records a
secret and profoundly disturbing vision he then carried within him of a
giant feces that in being discharged shattered the dome of heaven.
Sometime later, at his first communion and with his country pastor father officiating, the meaning of this vision and of his father's life became clear to him: the ceremony before him seemed not a celebration of
life but the institutionalization of death:

I was seized with the most vehement pity for my father. All at once I understood the tragedy of his profession and his life. He was struggling with a death whose existence he could not admit. . . . [He was] hopelessly . . . entrapped by the Church and its theological thinking. They had blocked all the avenues by which he might have reached God directly, and then faithlessly abandoned him. Now I understood the deepest meaning of my earlier experience: God himself had disavowed theology and the Church founded upon it.

Jung's way out of the trap that had snared his father and in which an entire civilization had thrashed for centuries, unconscious of the nature of its predicament, was to find a course back to the holistic wisdom of primitive myth. He could only watch helpless as his father perished in that trap, a cancer growing in his stomach like some grotesque parody of gestation.

# Hecatombs

A third consequence of the turn of Christianity from myth to history is of such significance for this essay that it requires separate treatment. But the reader should not be misled by the exigencies of literary organization into thinking of what follows as in any true way separate from what has immediately preceded it. Indeed, what follows is but a selective arrangement of some of the outward manifestations of the inner development sketched above. That these manifestations are well known in other contexts and that they have seemed the very stuff of Western history might in itself constitute a kind of evidence of the decay of Christianity. For we have learned to take such phenomena as the Crusades, the Inquisition, and kindred forms of religious persecution as more or less normal stages in the growth of our civilization, however attended by violence they have been. Generally we have not thought these might be symptoms of a deep spiritual pathology that has prevented us from experiencing more authentic forms of renewal.

I take as my chapter title a word from the Greek that denotes a large public sacrifice to the gods, usually of a hundred oxen. More generally, a hecatomb is any large-scale sacrifice. All religious offerings seem imbued with the idea of sacrifice as a necessary condition of purgation/renewal, and the mythic motif of new life purchased with the death of the old—god, king, scapegoat, prior self—was familiar long before the advent of Christianity. So what I suggest here is not that Western Civilization invented the idea of sacrificial death as a way of renewing itself, but rather that the persistence of this idea and the scope of its operations throughout the centuries nominally presided over by the Church is a consequence of the decay of the Christian mythology into mere theology.

More specifically, it seems to me that aggression against the body, against the natural world, against primitives, heretics, all unbelievers; and the vain, tragic, pathetically maintained hope of thus winning a lost belief or paradise: this is the terrific burden Christian history has to bear. It is the classic reaction of those who have lost true belief (or have been robbed of it) that they must insist with mounting strenuousness that they *do* believe—and that all others must as well. For as social psychologists have shown, if the bereft can thus succeed in harmonizing the world with themselves, then the inward gnawing doubt might be stopped and the intolerable condition of spiritual inanition alleviated.

To begin with the violence was turned inward, and in this Paul showed the way; later he would be invoked for the outer-directed violence of the Inquisition. Fanatic persecutor of Christians, struck down on the road to Damascus, he turned then to persecuting himself. Acts ends something in the New Testament, and Romans begins something else. For even though the figure of Paul is increasingly important in the latter passages of Acts, yet when Paul speaks in his own voice for the first time we hear the anguish of a being who has learned to despise part of himself; who would teach others to equate conversion with a kind of ongoing self-slaughter. Here begins in New Testament scripture the fatal divorce between body and soul, between nature and religion, that has come to seem the very essence of the faith (Romans 6:6, 12–13, 19; 7:18–25; 8:12–13). In passages such as these the great advocate effectively reinterprets the message of the Messiah. All that persecutory violence now turned inward, Paul preaches to the fledgling churches and to those who would later read his words and be guided by them —Tertullian, Origen, Cyprian, Clement, Porphyry, Chrysostom, Jerome, Augustine—of the prison house of the body, of the fatal, earthy attractiveness of women, and of the high, solitary virtues of continence and mortification.

And since Paul is next in importance in the New Testament only to Jesus himself, it seems not happenstance that an early and crucial manifestation of violence-as-regeneration took an internalized form in the mortified lives and legends of the Desert Fathers. This eremetical movement, which began in the third century with Anthony, is a kind of prism through which we can see how Paul and his followers raised to higher powers those antinature tendencies already noted in ancient Judaism and reinforced by the mystery cults of the later years of the Roman empire.

The conscious effort of these Desert Fathers, as well as the legends, orders, and systems of devotion that the Church would subsequently fashion from their example, was to recapitulate the desert trials of the Israelites, only this time to be equal to them through the New Dispensation of the Redeemer (Matthew 4:1–11; Luke 4:1–13). The aim once again was to resist the temptations and terrors of unhallowed lands. " 'Here abide men,' " as Helen Waddell quotes one of their chroniclers, " 'perfect in holiness (for so terrible a place can be endured by none save those of absolute resolve and supreme constancy). . . .' " This wilderness, like that of old, was the fitting abode of all dark things unsponsored by God, perhaps ruled over by the Evil One. Thus Anthony was once confronted by a Hippocentaur, a bestial being that ground out some barbarous answer to a question the saint put to it:

> And indeed whether the devil had assumed this shape to terrify him, or whether (as might be) the desert that breeds monstrous beasts begat this creature also, we have no certain knowledge.

Macarius' daybreak meeting with the Evil One suggests that the first possibility is more likely:

> . . . and the Evil One feinted at him with his sickle, but could not reach him, and began to cry out on Macarius for the violence he did him. "Yet whatever thou dost, I do also and more. Thou dost fast now and then, but by no food am I ever refreshed: thou dost often keep vigil, but no slumber ever falls upon me. In one thing only dost thou overmaster me." And when the saint asked what that might be, "In thy humility." And the saint fell on his knees—it may be to repel this last and subtlest temptation—and the devil vanished into the air.

But there were even greater temptations than this because in the silence of these wastes they were so much nearer. Anthony at the desert's edge beyond Cairo was visited constantly by the Devil, who reminded him that his body was weak and time was long, and who visited him with lascivious thoughts. At night, as the eremite lay locked in the iron sweat of his resolution, the Devil would come to him in the form of a ravishing woman, conversing of fornication. Legend tells us that this saint became so ashamed of having a body at all that each time he ate or satisfied any other bodily need he would blush.

Or hear Jerome, this maker of the Vulgate Bible, violating himself, set so hard against so much of life, echoing here Job in the deserts of the soul:

> "Oh, how many times did I, set in the desert, in that vast solitude parched with the fires of the sun that offers a dread abiding to the monk, how often did I think myself back in the old Roman enchantments. There I sat solitary, full of bitterness; my disfigured limbs shuddered away from the sackcloth, my dirty skin was taking on the hue of the Ethiopian's flesh: every day tears, every day sighing: and if in spite of my struggles sleep would tower over and sink upon me, my battered body ached on the naked earth. Of food and drink I say nothing, since even a sick monk uses only cold water, and to take anything cooked is a wanton luxury. Yet that same I, who for fear of hell condemned myself to such a prison, I, the comrade of scorpions and wild beasts, was there, watching the maidens in their dances: my face haggard with fasting, my mind burnt with desire in my frigid body, and the fires of lust alone leaped before a man prematurely dead. So, destitute of all aid, I used to lie at the feet of Christ, watering them with my tears, wiping them with my hair, struggling to subdue my rebellious flesh with seven days' fasting.
>
> .    .    .
>
> "I grew to dread even my cell, with its knowledge of my imaginings; and grim and angry with myself, would set out solitary to explore the desert. . . ."

In various ways these men took terrible vengeance on themselves in the hope of winning holiness, and the Church conferred sainthood on them and so made example of their mortifications. Some sat atop pillars for years; others wore hair shirts or heavy chains that blazed in the desert sun. Dodds records the melancholy finding of a skeleton in the Egyptian desert, clad only in its girdle of iron links, its last prison house. Others, after the manner attributed to Origen, castrated themselves. Some immured themselves in packing cases, caves, tombs. Some starved while their fellows fed like animals on wild grasses and lived without shelter. Fools for Christ's sake, they said they were, or God's athletes in strenuous training. A fourth-century canon law was necessary to check the growing popularity of self-mutilation, and at the end of that century, so Waddell tells us, a traveler through Egypt and Palestine estimated that these desert dwellers about equaled the population of the towns. And so far had the faith traveled from its mythological beginnings, so fiercely did it now set itself against the natural and all religions grounded in nature, that one of these Desert Fathers, Rufinus of Aquileia, could describe his encounter with cave drawings of ancient animal deities as an engagement with the very essence of that evil he had been sent to the desert to combat.

By the end of the eleventh century this violence had turned outward. This development may have been, as J. M. Wallace-Hadrill has suggested, because Augustine was now the dominant influence in the Church, and his militant we/they attitude and uncompromising hostility toward all sorts of unbelievers formed the basis of expectations of a literal victory over the enemies of Christ. In the Crusades and in campaigns against heretical sects, Jews, and suspected and/or purely imaginary thaumaturges, Christian civilization embarked on its first concerted effort at regeneration through sacrificial violence, offering the mutilated bodies of its enemies as gifts pleasing to a god whose son had once appeared as the Prince of Peace.

Conversion campaigns launched by the ex-barbarians Clovis (c. 465–511) and Charlemagne (c. 742–814) might be cited as preliminary exercises in extroverted violence, and the *Song of Roland,* though composed later, shows us that by the time of the First Crusade the familiar, dread oppositions, perhaps as honed by Augustine, were already in use: the Christians are white, and the land of civilization is the land of light; while the enemy is "Blacker than pitch, nor showing any white/Unless of glistening teeth," and their land one of "grim defiles" where

> The sun shines not, nor rain nor gentle dew
> Fall from the heavens and not a grain of corn
> May ripen; every stone is black, so black
> That some say the Devil put them there.

It is the Crusades that truly commence the pattern of large-scale, international Christian violence against all unbelievers that at last bears its cindered fruit in the ruins of Tenochtitlán. Once again, clouds of crisis hung heavy over civilization, and this seems to have played a part in the outward movement. Barbarians, plague, and economic and social disruption worked upon the populace. Around 1000 apocalyptic views were common, and though civilization persisted through the dreadful visitations of plague and even absorbed the harassing barbarians, apocalyptic views also persisted because the real crisis was inner. Norman Cohn's brilliant study of the mood of this time, *The Pursuit of the Millennium,* places the emphasis where it should be: the root of this crisis feeling was not merely economic or social (though certainly the period's severe economic and social dislocations were important) but fundamentally spiritual. As Cohn shows, many of the leaders of the dozens of millennarian sects that sang, danced, and flagellated themselves in expectation of the imminent Second Coming were of the upper strata and not of the rootless poor who formed the ranks of these sects, as they did of the crusading armies. What we encounter here is a general, shared condition of the poor in spirit.

This much seems to have been admitted by Pope Urban when he preached the First Crusade at Clermont at the end of November 1095. William, Archbishop of Tyre, late in the twelfth century wrote that when the pope devised the idea of the Crusade he did so as much in response to the deplorable state of Christendom as to the defilement of the Holy Land. The pope, he wrote, "was extremely anxious as to how he might counteract the many monstrous vices and sins which were unfortunately springing up and involving the whole earth." He went on to record that "Admonitions from on high were everywhere being disregarded, the doctrines of the Gospels were despised, faith had perished, while charity and every virtue were in danger."

We might, of course, see such a dire assessment as the typical rhetorical resort of the cleric who sees any age as sinful in comparison with the faith of the fathers, but here William's spiritual gloss appears an accurate one. The spiritual climate of the time was poor indeed. In Urban's own France and throughout much of Europe warfare was a permanent condition, robbery a substitute exercise in arms, and fighting the chief expectation of the feudal male. Against this the Church-instituted Peace and Truce of God was an utter nullity, partly because the Church itself was deeply stained by its involvement in secular affairs, its widespread practice of simony, and the notorious sexual misconduct of the clergy and the orders. Leaders and troops alike were devout Christians, to be sure, but as Sidney Painter observes, they were "little troubled by Christian ethics." Feudal husbands, Painter adds, also made savage warfare on their own wives and ignored the obligations of the marriage bond whenever convenient. William reports Urban as addressing the general condition in these words:

> "Turn the weapons which you have stained unlawfully in the slaughter of one another against the enemies of the faith and the name of Christ. Those guilty of thefts, arson, rapine, homicide, and other crimes of a similar nature shall not possess the kingdom of God. Render this obedience, well-pleasing to God, that these works of piety and the intercession of the saints may

speedily obtain for you pardon for the sins by which you have provoked
God to anger."

Although the pope's preaching of the Crusade apparently came as a
surprise to contemporaries, the immediate, spontaneous acclamation it
received and the speed with which it was prosecuted show that this was
an idea whose time was ripe. Contemporary sources report that when
Urban had finished speaking, a thunderous shout went up to heaven,
*Deus lo volt!* (God wills it), and the fields outside Clermont shook with
the stamping of eager feet. Even before the first official forces could be
mobilized, a motley horde, thirsty for blood, moved off in the spring
of 1096 under the nominal leadership of Peter the Hermit.

Recruiting devices made the Crusade no less attractive than the
pope's plea to save the Holy Land, its historic shrines and relics, from
the desecration of Satan's children. It is hard not to be cynical when we
read of the lavish inducements the Church offered those fighting men
to do unto outsiders what they had been doing unto one another; one is
saved from cynicism only by reflecting on the tragedy of the spiritual
situation and on the fate of the Crusaders and their victims. The second
canon of the Council of Clermont stated that for all who were truly
motivated the Crusade would take the place of all penance, and it seems
likely that Urban added to this a full remission of sins if one died fighting for the faith. Further, the pope decreed that all the possessions of
the enemy would lawfully become those of the Christians. At the same
time, it was apparently implied that only those who actively bore arms
would be so indulged. Unarmed, pacific pilgrims were not of consequence in such times.

In sum, the pope and the Church seized upon the expedient of a war
to recover the sites of the sacred past as a means of regenerating present
faith. God, so it was said, had looked down upon the degenerate state
of faith and had instituted a new route to salvation. It is revelatory of
the civilization's spiritual condition that such an expedient was so
widely and immediately accepted. It is as if all had tacitly agreed that
such a drastic solution was necessary.

Only a generally felt spiritual poverty, though unsuspected in its
causes, seems adequate to explain the savagery subsequently unleashed
against the enemies of the faith. For these soldiers of Christ did not wait
to blood their swords on the Saracens but rather in the spring of 1096
attacked and slaughtered Jewish communities at Worms, Mainz, and
Trier. This death in the springtime, a ghastly perversion of ancient re-

generation myths, was, as Cohn has said, the beginning of a tradition that came in historical time to include in its insatiable need increasingly disparate groups: Jews, Albigensians, Saracens, witches, Africans, and at last the primitives of unsuspected azoic zones. Here it is enough to observe that in successive expeditions many Crusaders felt themselves unworthy of the high work of destruction in distant lands until they had hung that first Jewish scalp to their belts on the way out.

What else can explain the gang warfare of the Crusaders once they had gotten beyond their geographical limits, released into spaces unsanctified by Christian history? At Antioch an entire city razed and its inhabitants murdered to the last infant. At Nicaea the heads of the slain enemies hurled by catapults into the city as part of the general assault. An offering of sliced thumbs and noses sent to the Byzantine emperor. And at Jerusalem, their goal, in July 1099, after a solemn religious procession around the city's besieged walls that culminated in an ascent of the Mount of Olives, the host fell upon the holy city with a ferocity that beggars language. "Regardless of age and condition," wrote the Archbishop of Tyre,

> they laid low, without distinction, every enemy encountered. Everywhere was frightful carnage, everywhere lay heaps of severed heads, so that soon it was impossible to pass or to go from one place to another except over the bodies of the slain. Already the leaders had forced their way by various routes almost to the center of the city and wrought unspeakable slaughter as they advanced. A host of people followed in their train, athirst for the blood of the enemy and wholly intent upon destruction.

So frightening was this massacre that even the victors experienced sensations of horror and loathing:

> It was impossible to look upon the vast numbers of the slain without horror; everywhere lay fragments of human bodies, and the very ground was covered with the blood of the slain. It was not alone the spectacle of headless bodies and mutilated limbs strewn in all directions that roused the horror of all who looked on them. Still more dreadful was it to gaze upon the victors themselves, dripping with blood from head to foot, an ominous sight which brought terror to all who met them. It is reported that within the Temple enclosure alone about ten thousand infidels perished, in addition to those who lay slain everywhere throughout the city in the streets and squares, the number of whom was estimated as no less.

Still, it went on to its appointed ending:

The rest of the soldiers roved through the city in search of wretched survivors who might be hiding in the narrow portals and byways to escape
death. These were dragged out into public view and slain like sheep. Some
formed into bands and broke into houses where they laid violent hands on
the heads of families, on their wives, children, and their entire households.
These victims were either put to the sword or dashed headlong to the
ground from some elevated place so that they perished miserably. Each
marauder claimed as his own in perpetuity the particular house which he had
entered, together with all it contained. For before the capture of the city the
pilgrims had agreed that, after it had been taken by force, whatever each
man might win for himself should be his forever by right of possession,
without molestation. Consequently the pilgrims searched the city most carefully and boldly killed the citizens.

A second such expedition was mounted almost immediately and, if
anything, it was costlier in lives than the first. Indeed, as the Crusades
went sporadically onward, if a certain amount of official favor was
silently withdrawn from them, the violence of the campaigns seems to
have increased both in randomness and intensity until with the so-called
Shepherd's Crusade of 1320 the very existence of Christian civilization
itself seemed threatened.

Yet all this while there is clear evidence that these sacrifices had not
achieved their purpose of spiritual renewal. Throughout the period of
the Crusades and after, the millennial sects continued to crop out and
spread briefly like contagions, only to be exorcised by the Church. In
the forms these sects took, in the allegations made, and in the
ruthless measures taken against them, Christians, both renegade and
orthodox, made testaments to the state of their faith. For it is clear that
the sects arising in Germany, France, the Netherlands, Italy, Spain, and
England were ironic successors to the crisis cult Christianity had once
been. The crisis now, however, was not the fate of empire, nor the survival of Western Civilization. The crisis was in the religion itself, bereft
of its vitality but presiding with its symbols, sacred texts, liturgical calendar, and ecclesiastical hierarchy over the states of the West. The bells
in the towers over the towns might dictate the activities of the people,
but these signs of the times indicate that the bells did not tone in the
people's hearts.

After the example of orthodox Christianity, some of these sects took
their cues from the more militant passages of scripture, especially Revelation and The Book of Daniel. These sects too expected victory over
the ungodly (the Church) and looked for the regeneration of the world

through the violent, swift destruction of Christ's enemies. Thus we have the blood-drunk anonymous "Revolutionary of the Upper Rhine" who brought this aspect of medieval millennarianism to its nadir. Cohn quotes from his *Book of a Hundred Days,* where he urges his followers to rise up and slaughter the fornicators in fine clothes: "Go on hitting them from the Pope right down to the little students! Kill every one of them!" This messiah estimated it would be necessary to execute twenty-three hundred clerics a day for four and one-half years to rid the earth of these vermin. It was, of course, the Church that accomplished most of the executions, here as elsewhere showing itself ruthless in its opposition to popular religion and zealous to proclaim each repression a new victory for Christ.

The allegations involved in this phenomenon add their telling comment on the deep disease of this antinatural religion, for with obsessive regularity the Church charged the sects (as it did all unbelievers) with sexual perversions and identified the heretical ones with all things natural, earthy, animal. They were beasts, toads, fornicators, sodomites, lovers of bestiality. The term "bugger," used to refer to heretics and/or sodomites, may be a corruption of "Bulgar," from which area one of the more offensive sects, the Cathari, were thought to have entered and infected civilization. It was routinely charged that the sects were presided over in their frolics by that arch-sodomite, the Devil, depicted (again) as a beast of earth and perhaps even possessed of a forked penis enabling him to commit fornication and sodomy simultaneously.

In point of fact, however, with a few striking exceptions the only bodily excesses these renegade groups seem to have been guilty of were those of self-torment, for flagellation was often associated with these movements. The spectacle Cohn gives us of hundreds of worshipers beating the blood out on their bodies with spiked scourges and then drinking the spilled gore as the wine of a new communion is a pathetic illustration of the lengths humans will travel when their religion robs them of the earthy, the fleshly; when they feel compelled to strive, in whatever twisted ways, to *feel* again the earth, flesh, blood.

True, some sects like the famous Free Spirit, the Ranters, and the sixteenth-century revolutionaries of Münster did give the Church warrant for its characteristic accusations. Generally, however, the sexual conduct of the dissenters was more exemplary than that of the orthodox, and it is possible to see in at least some of these movements conscious efforts to reestablish that very primitive Christian notion of Agape love that the Church had so systematically extinguished.

Both the efforts of these various groups at regeneration and the Church's persecution of them toward that same end span the period from the beginning of the Crusades to the flood tide of exploration. Throughout this long period the Church became ever more hysterical in its accusations and its punishments of paganism, thaumaturgy, and witchcraft and in its attempts at massive sexual repression, until finally it sought to purge and renew Christendom in the fires of the Inquisition, a tragic perversion of a primeval busk.

At issue during these centuries was the shape and content of the psychic geography of the West. In the same way that civilized men had cleared the earth, pruned back the forests, planted villages, towns, and cities, so had Christianity stripped its world of magic and mystery, and of the possibility of spiritual renewal through itself. In cutting down the sacred trees in the mystic groves, in building its sanctuaries on the rubble of chthonic shrines, and in branding all vestiges of ancient mythic practices vain, impious superstition, the Church had effectively removed divinity from its world. But its victory here was Pyrrhic, for it had rendered its people alienated sojourners in a spiritually barren world where the only outlet for the urge to life was the restless drive onward, what Norman O. Brown has called the desire to become. Eventually this drive would leave the religion itself behind.

Meanwhile, the old pagan practices died hard, so rooted are they in our nature. This was especially so in the European outback where women and men were still daily entangled in the immemorial, ahistorical cycles of budding, molting, mating, and harvest. So one thinks that in some intuitive way Margaret Murray was right when she argued that an ancient fertility cult underlay the witchcraft persecutions. There must have been at least some rude, residual attempt to keep close to the ancient vision of the way things are, though it was probably never the organized effort complete with covens that Murray claimed it was. There is too much archeological evidence of animal fertility images to imagine that such deeply rooted beliefs died easily, and there are too many reports of medieval dances that appear to be associated with animals and fertility to dismiss all of them as fabrications or hallucinations. Such a suspicion gains further ballast from such stray bits of evidence as a seventh-century injunction that reads:

> If anyone at the Kalends of January goes about as a stag or a bull; that is, making himself into a wild animal, and putting on the heads of beasts; those

who in such wise transform themselves into the appearance of a wild animal, penance for three years because it is devilish.

Two recent studies of witchcraft in England show that at least where that country is concerned, Murray's thesis is baseless, and it may prove to be so for the Continent as well. But even these studies, by Keith Thomas (*Religion and the Decline of Magic*) and A. D. J. Macfarlane (*Witchcraft in Tudor and Stuart England*), show that as late as the seventeenth century a great many people still retained a clandestine emotional attachment to the old mythic world that Christianity had come to uproot. It seems true also, as Frazer and Freud claimed, that what has been superseded by subsequent developments in a culture may persist as a negatively charged, hidden survival, especially if it was once sacred. As Henry C. Lea puts it in his study of the Inquisition, "The sacred rites of the superseded faith become the forbidden magic of its successors."

The clearest illustration of the spiritual plight of the Western world before it reached out beyond itself into the wildernesses is the Inquisition. For just as the fervor of the Crusades was dying (in the last years of the thirteenth century, Pope Nicholas IV found it necessary to offer full remission of sins for all who either sponsored Crusaders or went at another's expense), this new plague of violence was visited upon the first of what may have been well over a million people before it was at last concluded. It was as if the culture now abhorred a vacuum of violence. Or, as critics of the Church have it, the hierarchy recognized that yet another mass movement was required to bolster a faith flagging in all but formalisms. Lea argues in his huge *A History of the Inquisition of the Middle Ages* that in its drive to supreme power the Church had created so enormous a chasm between itself and the daily lives and practices of the laity that it had almost forced itself into such drastic measures. Like Spengler, Lea uses the towering medieval cathedrals (and abbeys) as a symbol of this situation, for these were constructed with terrific sacrifices of labor and money of the laity to support the opulent, often decadent tastes of the clergy. While some (like Henry Adams) have seen these edifices with their glories of stained glass refracting heaven's rays, aspiring Godward from the tilled land, as symbols of the strong faith of a unified civilization, others (like Lea) regard them as symptoms of a dangerous disease that in the Inquisition became virulent. Like the faith itself, they were formal and imposing and costly, but inwardly they were empty.

The doctrine of "justification by works" had by now degenerated into an empty formalism in which so much could be atoned for through mere repetition of set words. Preaching was neglected and the consequence was spiritually starved congregations. Indulgences were granted the well-to-do in return for gifts to the Church, there was a trade in blank excommunications, and clerical concubinage was flagrantly practiced. All these things had bred a mixture of cynicism and desperation for which violence appeared as the cure to hand. But it would be a mistake to believe that the Church by itself could foment and prosecute so massive a phenomenon as the Inquisition. For this it had to have the willing support of the general populace, as it did for the Crusades. And if that support came for different reasons, the results were fatally the same.

Beginning with Innocent III in the last years of the twelfth century and continuing in a mounting-and-receding pattern throughout the following four centuries, the clerical and secular arms of the West worked in coordination to purge civilization of the vast hidden army of Christ's enemies. Only by such actions, it was asserted, could Christianity be saved. Gregory the Great unwittingly exposed the truth behind this when he stated that the bliss of the saved in Heaven would be incomplete unless they could gaze across the abyss and behold the sinners tormented in hellfire. Perhaps this explains the great popularity of the Dives and Lazarus theme in medieval funeral statuary: as if only by exposing, torturing, or killing others could one win salvation—or belief.

By the sixteenth century the full force of thwarted desire had been reached, and the enemies burned by the thousands. A bishop of Geneva, that quiet, saintly city, burned five hundred in three months. A Bamberg bishop burned six hundred. A bishop of Wurzburg, nine hundred. In 1586 the spring was late and its warmth weak in the Rhinelands; nature's power to be renewed had obviously been tampered with. Thus the Archbishop of Treves burned 118 women and 2 men in these fires of spring. Regeneration was thought to rise from such sacrificial ashes.

Is it any wonder that at this same time the mines of New Spain were littered with corpses?

# Loomings

The Inquisition was a spiritual nadir in the West. The primitive mythology of love had in its maturity become a punitive theology of fear and death that could only regenerate itself in such life-denying ways. Yet to abandon it was unthinkable because to do so meant to abandon civilization for paganism or even savagery. For Christianity *was* civilization, and its theology, symbols, clergy, and churches underpinned, towered over, and authorized the daily order in Portugal, Spain, Italy, France, and England. Still, there was this death at the center of the civilization's life, and though all was retained—texts, symbols, church bells, even certain deeply held values reinforced by the very shape and tenor of the European world—human energies began to turn elsewhere. Like those Desert Fathers repressing their bodily desires in mortification, the West turned to exploration as both a "palliative remedy" (Freud) and a way of harmonizing the rest of the world with itself. If it could succeed in making the map of those spaces beyond itself match that of the Christian West, then certain torments of the spirit, certain cognitive disconfirmations (as psychologists would have it), might be better borne.

The half-legendary figure of Raymond Lully can serve here as an odd sort of microcosm of this entire development. Lully was born in Palma, the capitol of Majorca, in 1235. His family was aristocratic, and Lea tells us that Lully learned the ways of his class at the royal court where he eventually rose to the post of seneschal. He married and fathered children, but after the fashion of the time was entirely cavalier about the marriage bond; in the legend subsequently developed around him much is made of the reckless, debauched life he led. One of his amours of the moment, a Leonor del Castello, he pursued with special abandon. Indeed, on a certain Sunday he followed her on horseback right into the midst of services at the church of Santa Eulalia. But here the legend makes a turn, for the episode transmogrified the rakehell knight into a fanatic soldier of Christ. Leonor, to discourage Lully's further attentions, silently exposed her breast to him, and he beheld all that lovely mortality "ravaged by a foul and mortal cancer." Here in a moment was revealed the shocking home truth of the Christian message: of the betrayal of the body, of the worthlessness of human life, even of that seductive snare that is woman. Like Paul, Lully was instantly converted, and with the same passion with which he had pursued pleasure, he now punished himself. He undertook a ten years' penance during which he mastered first his body, then his mind.

When he had finished with himself, he turned his efforts to the unconverted of the world. For forty years he harangued popes and princes with proposals for new Crusades and schemes for the conquest of the Moslem world through armed invasion or naval blockades. He has even been credited with suggesting that the way to the East was around the tip of Africa, which stretched, God knew how far, into gloom.

At last, starved and stoned by the Moors, his wasted frame was brought ashore for the final time at his native Palma. And as Lea tells us, immediately "it shone in miracles and the cult of the martyr began." In 1487, as Christopher Columbus was shuttling about Europe importuning kings and queens with his scheme of outreach, the bones of Lully were enshrined in a Franciscan chapel.

So it appears neither accident nor native genius that impelled Western Civilization to embark on the conquest of the globe. It was need, and that casts a different light on this spectacular achievement. Though it is not thus robbed of its heroic aspects, since desperation and thwarted desires may fuel heroic behavior of a kind, we can no longer attribute European exploration to "the vigor of the Renaissance," or "the optimistic mood of expansion," or "the natural and healthy competition of the nation-states." We can retain the common explanation of it as a consequence of missionary zeal, but because of what we know of that zeal and its history, we must regard it differently from the traditionalists. And of the ultimate results of that zeal, accomplished over centuries on many fronts by many hands, who shall yet presume to say?

In the way of such things, the civilization's inner needs fed its ability to reach out beyond itself. Thus, for example, the Crusades and the second excursion of the Polos, both of which are manifestations of those needs, conferred essential gains in geographical knowledge, which in time made possible farther probings, vaster designs. By comparison with the Crusades, the Polos' trip was incidental: they represented to the newly elected pope, Gregory X, that the Great Khan was eager for instruction in the faith. And so the pontiff attached two priests to their trading expedition in 1271. In a seemingly more major way the Crusades opened the eyes of all Europe to the treasure houses of the East and greatly extended the civilization's knowledge of the globe. This came about not only through the actual invasions but also through written accounts based on them such as that by Marino Sanuto, a Venetian geographer and military strategist, who wrote a manual for Crusaders. But it is possible that that small, incidental trip of the Polos, and especially the book made out of it, had more lasting effects.

Twenty-seven years they were gone from the known world, moving, either by accident or the designs of the pope, ever deeper into the heart of Asia until at last they stood before the Great Khan himself at Shandu. At home their influence began the moment when (so legend tells) they returned to Venice and convoked a gathering of kinfolk and friends. Here the weathered ambassadors from the unknown opened wide their cloaks, disgorging a shower of gold and gems before the dazzled eyes of onlookers. This was tangible evidence of the fabulous East with its rumored riches, its muslins, silks, spices, and fruits, its precious stones and metals.

Marco Polo wrote of all this, and the "Milione" of his name, according to one strand of the legend, referred to his thousands of traveler's tales. On his deathbed he was beseeched by a kinsman to recant and so save the family from further ridicule. "I have not told the half of it," the old man breathed, and died faithful to his tales.

He writes of Noah's ark; of the land of Gog and Magog; of Indo/Tartar bandits so skilled in enchantments they could bring a covering darkness across the sun's broad face; of the Old Man of the Mountain and his gang of hashish-eating assassins (Joinville in his account of the Seventh Crusade mentions this figure also); of rubies found in veins of silver and of the Valley of Diamonds. There were horses descended from Bucephalus; tailed men and others with heads like mastiffs who fed on human flesh. And still others who prefigure those aboriginals found by the whites when they penetrated the tall grasses west of the Mississippi: hardy horsebacked nomads who traveled enormous distances on only mare's milk and dried curds, who dressed in flame-hardened animal skins and fought with an abandoned disregard for life. When they are much reduced, writes Marco, they open the veins of their mounts and drink the blood.

In Marco Polo we find a mingled terror of and resistance to the wilderness that had all the gathered force of Judeo-Christian tradition behind it. Poised on the edge of the Gobi Desert, Marco retails what awaited the unwary Christian in those devilish spaces where none lived but *jinns*. Accompanied by drums, music, and movement as from a vast, unseen caravan, these desert demons called out to a man in his own name luring him away from his friends into trackless silences and unhallowed death.

Whatever the precise mixture of fact and embroidery in its pages, it is certain that Marco's book excited the interest of the Christian world and remained the most influential single source of geographical information about the East until the Portuguese went to see for themselves. In his wake the first missionaries were dispatched against the citadels of idolatry. They wrote back of mass conversions, of bishoprics carved out of heathen spaces, of Tartars, troglodytes, and naked, long-haired outlaws who lived like goats, and, again, of the Satanic terrors of the wilderness. Thus C. Raymond Beazley quotes Odoric of Pordenone, a Franciscan who early in the fourteenth century traveled much the same route as the Polos:

"As I went through a certain valley lying upon the River of Delights, I saw countless bodies of the dead, and heard divers kinds of music, and especially that of drums, marvellously beaten. And so great was the noise that extreme terror fell upon me. Now this valley is seven or eight miles long, and if any unbeliever enter it, he dies forthwith. And going in (that I might see for good and all what this matter was) I saw in a rock upon one side a man's face, so terrible that my spirit seemed to die within me utterly. Wherefore I continually repeated with my lips, *The Word was made Flesh*. Close up to that face I never dared to go, but kept always some seven or eight paces from it. And at the other end of the valley, having climbed up a sandy mountain, I looked round everywhere, but could see nothing but those drums, which I heard played upon so wondrously."

The missionaries were but the first of those who followed the blazes of Marco's book. Beazley says that the famous Catalan map of 1375, recording the excursions of Catalan sailors along Africa's west coast, contains a faithful reproduction of Marco's conception of the East.

Prince Pedro of Portugal presented his brother, Henry the Navigator, with a copy of the book, together with a map based on it, and this served as a highly valued weapon in the navigational arsenal Henry built at Sagres. A Latin version of the book was owned by an even greater navigator, Columbus, and heavily annotated in his hand. Despite subsequent disclaimers, it is certain the Admiral was trying to get nowhere other than to Marco's Cathay and Zipangu where the king's palace was roofed with gold. Nor does the book's remarkable influence end here, for while reading Marco's remarks on the lands of the far north beyond Russia, Prince Rupert is said to have been inspired with his scheme for the Hudson's Bay Company.

So, there was a chain of entailments to this trip and its book. But to insist too much on the specificity of Marco Polo is to obscure the important fact that he was only an advance man for a civilization bound outward on a search for the resolution of that history it had to create continuously. It is clear that both Henry the Navigator and Columbus took Polo only representatively.

If we were to adopt a kind of rude economic explanation for the influence of the Polos (or, for that matter, of the Crusades) and say that Marco's reports of the opulence of the East inflamed the cupidity of a whole civilization, still we would have to ask ourselves whether such a desire for stones and metals is not nurture rather than nature: whether this lust for *things* that will automatically confer power on their possessors is not another symptom of the petrifaction of the faith. For, as the questing Christians were soon to discover in the spaces into which they thrust themselves, the native peoples who lived amidst vast, unexploited lodes of these very things often regarded them as mere sparkling parts of an infinitely larger and more beautiful design. Maybe no single aspect of the cultural difference between Christians and natives is more revealing of the differences between a civilization ruled by a dead mythology and people animated by vibrant ones than this contrast of attitudes toward stones and metals. Maybe this is how to measure the truth of that distance we had to travel to get to those palmy shores, vine-tangled river banks, and mountain gorges of other worlds beyond.

Not many years ago the mysterious writer B. Traven invented a speech on this subject for an Indian chief addressing a gold-mad white man. Traven had lived for some time among the Indians of Chiapas and, moreover, knew his New World history, so the fictional speech is really a composite of the many we have on historical record. I do not need gold, nor want silver, the chief says:

"The soil bears rich fruit every year. The cattle bring forth year in, year out. I have a golden sun above me, at night a silver moon, and there is peace in the land. So what could gold mean to me? Gold and silver do not carry any blessing. Does it bring you any blessing? You whites, you kill and rob and cheat and betray for gold. You hate each other for gold, while you can never buy love with gold. Nothing but hatred and envy. You whites spoil the beauty of life for the possession of gold. Gold is pretty and it stays pretty, and therefore we use it to adorn our gods and our women. It is a feast for our eyes to look at rings and necklaces and bracelets made out of it. But we always were the masters of our gold, never its slaves. We look at it and enjoy it. Since we cannot eat it, gold is of no real value to us. Our people have fought wars, but never for the possession of gold. We fought for land, for rivers, for salt deposits, for lakes, and mostly to defend ourselves against savage tribes. . . . If I am hungry or my wife is hungry, what can gold do, if there is no corn or no water? I cannot swallow gold to satisfy my hunger, can I? Gold is beautiful, like a flower, or it is poetic like the singing of a bird in the woods. But if I eat the flower, it is no longer beautiful, and if I put the singing bird into a frying-pan, I can no longer enjoy his sweet song."

Gold, silver, and stones, like technology, are pathetic substitutes for a lost world, a lost spirit life, and to the extent that they rule a culture we may infer its inmost health.

So it is not surprising that well before the Polos' fabulous tales developments were in process in the West that would make possible the acquisition of the riches of far lands. As early as the eleventh century there are signs of that remarkable gathering and resynthesis of the technical leavings of others that Mumford has detailed. Significantly, he connects the first great wave of technical interest to the weakening of other institutions in the West. "This early triumph of the machine," he writes, "was an effort to achieve order and power by purely external means, and its success was partly due to the fact that it evaded many of the real issues of life and turned away from the momentous moral and social difficulties it had neither confronted or solved." Mechanical invention, he observes more pointedly, became the answer to a dwindling faith.

By the turn of the thirteenth century there are clear traces of interest in movement outward through unharmonized spaces. The English scholar Alexander Neckham was toying with an ancient Chinese device—the magnetic needle, Adelard of Bath had already given the Christian world Al Kharizmi's tables of latitude and longitude, and Roger Bacon would take the measure of the globe and come strikingly close to its actual circumference. By the Polos' time Westerners had absorbed most of what their Arab teachers had to give them, both of geographical knowledge and maritime technics such as the lateen sail.

Here truly was a civilization that would take such teachings and technics farther than any other. The Arabs themselves feared the Atlantic as the "Green Sea of Gloom" whereon to sail was sure proof of madness. But they and other non-Christians were willing to put their mathematical and astronomical skills at the service of these driven, wind-screwed men. So Pedro X of Castille funded the redrawing of the contours of the Mediterranean by Saracen and Jewish astronomers, who corrected Ptolemaic errors. This led to a startling and revealing development of the pre-exploration period: the new science of cartography, which allowed proximate movement through distant spaces. Here again the Christian world took the clues and hints of others to scarcely guessed limits.

These others, of course, had had maps long before the first modern ones appeared around the middle of the thirteenth century in connection with the Crusades. A jagged bit of baked clay incised with ridges, circles, and lines reminds us of our debt to the ancient Near East, for it is a map of northern Mesopotamia c. 3800 B.C. and includes a representation of that very Haran that may have been Abram's original home, from which his god set his face toward a distant land. Greeks and Romans too had their maps, and Leo Bagrow tells us that the roots of "cartography" are Greek and Latin words meaning carved in stone or metal—as indeed the early maps were.

Even "primitive" peoples of the world have had maps, and the generic difference between these and the maps developed in the West is of interest. Bagrow includes in his *History of Cartography* illustrations of the navigational devices of Eskimos and Marshall Islanders. As you look at the beautifully carved wooden charts and relief maps of the Greenland natives, or at the delicate, poised, bamboo-and-shell constructions of the Micronesians, you are impressed anew with that *tactile* sense of territory such peoples possess, their intimate engagements with stretches of coastland, with wave and wind patterns. Their devices are, like our maps, abstractions from the physical territories. They are, again like our maps, attempts to "control" those territories by representing them, since representation as we saw earlier is an attempt at mastery. Yet both the level of abstraction and the impulse to control are minimal, and these maps, like the sailing natives, keep close to the realities of their worlds.

In this one sees not so much the technological inferiority of these peoples as the possibility that their mythologies were still their basic technologies, providing them with spiritual as well as spatial orientation, answering the questions all humans pose of the uncharted spaces beyond experience. Spatially as well as technologically, their worlds remained closed, limited ones, and often these limits imposed severe hardships upon those within. But, as Spengler observes, space terrifies because of its cognate sense of annihilation, and one thinks of certain comforts to be had in circumscribed worlds, comforts either forgotten or else remembered only in dreams, accidents, or passions by the people of the West.

When this wider world of ours burst into the fragile perimeters, the little worlds of the primitives shattered as a Micronesian bamboo map might in a hurricane gust. After years of contact with the West, the Marshall Islanders ceased making their navigational constructions, and in time (and history) forgot they ever knew how.

If, as Joseph Campbell has said, myth is a picture language as well as a means of orientation and control, then the maps that emerged in the fourteenth century might be conceived of as pictures of a dead myth. For maps are mental and spiritual projections as well as physical ones, and psychologists and anthropologists have for some time been using the map as a concept to suggest the ways in which a culture provides psychic orientation for its members. So the map in the West reveals to us a psycho/spiritual geography in which lines are projected across spaces devoid of life and essence, except for the demonic spirit that might lurk there, as the Israelites had discovered of old in the Sinai. In either case, whether the strange lands were dead or demonic, the Christian traveler could only move through them as an alien sojourner, every land a strange one and he a stranger to them all. As E. R. Dodds quotes a patristic description of Christians:

> "They live in their own countries, but as aliens; they share all duties like citizens and suffer all disabilities like foreigners; every foreign land is their country, and every country is foreign to them."

PART TWO

# Rites of Passage

# Mythic Zones

At the end of the fifteenth century the lines of cartographic projection could reach only so far. The farther out from the Anglo-European homeland, the more distorted and fictional the shapes of the world became, for beyond geography as mapped there was nothing recognizable. Mapping, as we have seen, is psychic as well as spatial, and the Western tendency was to consider *terras incognitas* as either empty or demonic.

Sailing for Oriental civilizations and unconscious of either true destination or the motives that drove the sails, Columbus and his successors broke in upon mythic zones wholly unsuspected. It is impossible to overemphasize their error. What soon became known as the "New World" was in fact the old world, the oldest world we know, the world the West had once been. Now the onward press of Christian history brought a civilization into contact with its psychic and spiritual past, and this was a contact for which it was utterly unprepared. The ensuing conflict was so deep that it has yet to be resolved or even understood.

No, this New World was not empty. It was off our charts but it was surely *there,* lying not in darkness, nor in the white color of terror it would assume as the lines of the charts reached out to account for it. It existed in its own light and colors, its own tides, seasons, flocks, and flowers. Circumambient, beautiful, violent, and pacific, even as nature is all of these, the New World teemed with its native life. It teemed also with the nature-inspired speculations of its humankind, the spectacular petals of myth. Here were those still captivated by the phenomena of nature, still the celebrants of it. Here those strange and strangely familiar fictions yet lived and beat in blood-pulse drums, chants, and rituals, and wholly informed the lives of the dark millions. Fragments of their myths have survived destruction to tell us what they can of a world our invading ancestors could not accept.

In the slim arc of islands now known to us as the Antilles a gentle and defenseless group of migrants from a vast mainland celebrated their origins in a narrative of emergence from an underground cave into the light of this world. They told too of the origin of the sun and moon in another cave, which they decorated with paintings in commemoration and adoration. They worshiped a moon goddess named Atabeyra who presided over the tides of woman and of the sea and over the generative process. A statue shows her crouched in the act of birth, her elbows clenched to her sides, the child emerging between her thighs.

They told how they had been guided from their river-and-plains homeland to these islands by the grand god Yocahú, who had provided them with the gift of the manioc plant and had showed them how to extract a nurturing substance from its lethal root. As they had pushed off from where the mouth of their huge river yawned into the limitless sea, they had been given manioc plantlings by the god, and the promise of lands somewhere on the waving horizon. And when the first volcanic peaks arose out of the dipping distance, the people knew that Yocahú had indeed provided. Ashore, on one island after another, they planted their manioc, reaped their harvests of fish, and reverently acknowledged Yocahú's continuing protection in conch-shell carvings that represented the god's volcanic dwelling place.

Farther in, along that huge river system (the Orinoco) that reaches into the heart of the mainland from whence these migrants had come, a neighboring tribe told how disease and death had come into their world. There was a fisherman of long ago, they said, who wished continually for a wife from the Water Spirits who lived in the dark river bottoms and along the floors of the sea. He wished so much for her that one day while fishing alone she appeared to him, rising water-smooth and beautiful, waist-high above the surface. "Would you like me for your wife?" she asked, and of course he answered yes. So they went ashore to his home where the bride told him that he must never offend her with the offer of fish to eat but give her animals and birds instead.

The next day the fisherman and his bride went fishing again in his corial, * and as he worked, the woman dropped over the side and disappeared beneath. After a time she arose again with a message from her

_____

* A dugout canoe.

father below who said he would be very glad to receive his son-in-law. But the fisherman was afraid until his wife took his hand in hers, and together they filtered down and down, away from light, until at last they arrived at the father's house. The ancient being greeted the fisherman warmly and to put him further at ease gave him a bench to sit on. It was really a large live alligator, which is why the people still make their benches in this form. When the fisherman had sat on it for some time the ancient spoke to him, saying, "I sent you my daughter for your wife, and now for your part you must promise to live a good life and to send down your sister for my son." The fisherman readily agreed, and after a meal the couple arose once more to the waiting corial, held in its place by a servant boa snake, and then they turned toward home.

On the way the wife spoke to the fisherman thus: "Remember this," she said, "when I am moon-sick you must not follow the custom of your people and banish me to the *naibo-manoko**∗ but instead allow me to remain with you. If you should insist on my going, I shall die, and my father shall take vengeance on you and your people." But when her time came, the custom of the people was stronger than the wife's warning. The women of the village spoke so threateningly that at last the unhappy man gave in to them and sent his wife to the *naibo-manoko*. And it was as she had said, for on the next morning they found her dead.

After several days the solitary fisherman returned to his work, but there on the broad stillness of the great river his loneliness took him so that he plunged into the water and kicked down and down, away from light, seeking the home of his lost wife. He found it and saw therein her dead form lying along the far wall. Then the grieving father spoke to him, saying, "Why did you not listen to my daughter? You can see now what you have done. From this day sickness, accident, and disease shall visit among your people, and what is more, if any of your women should travel along our surface when they are moon-sick, my people will draw their shadows from them."

The fisherman was stricken by this judgment, for not only had he lost a wife, but he had become the cause of anguish to the people, since before this the only illness they had ever known had been the women's moon-sickness. However, like most he was of short memory, and after

---

∗ A specially designed hut to sequester women during menstruation. At this time women were felt to be charged with a fearsome power; thus their separation from the community.

a time he proposed to his friends and relations a voyage on the sea, forgetting the ancient being's warning.

They started in two large corials. When they were well out in deep water, so deep that to the Water Spirits below them the two craft looked like parrots aloft in the sky, the Water Spirits released their round-knobbed arrows, hit the corials, and sank them to the bottom. There the Water Spirits unleashed the sharks they keep as dogs, and the unfortunate people were torn to bits. These were the first deaths.

Along another river system (the Amazon), this one even grander, the people live amidst flat tall-treed plains broken here and there by bush country and scrublands and pluvial lagoons. The region bursts with wildlife—jaguars, tapirs, giant otters, wild dogs, jabiru storks—and an efflorescence of fabulous narratives. One is of a hero who followed his late friend to a festival of the dead and then came back to tell the people how it was in that other world.

The hero's name was Aravutará. As for his friend's, it does not matter except to say that while he was in this world the two were inseparable. They hunted together, fished together, walked together, and as they did so, they would talk of this closeness that must not be broken—even by death. Thus they promised each other that if one died, the other would go in search of him, bringing a bow and arrows so that the dead one might hunt in the sky world.

One day Aravutará's friend fell ill and, foreknowing his end, sent for Aravutará. "Friend," said he, "I am ill and shall not recover." Aravutará gazed in shock upon that well-loved face, now drawn and already strange, and he tried to cheer his friend with hopes of recovery. But from these the man turned away as from a light too bright, and shortly it was as he had predicted.

Fulfilling his promise, Aravutará searched everywhere for his friend's spirit. He visited their old haunts daily, hoping for an encounter, but in vain: the spirit was nowhere to be found. Then one day the sun went out, and Aravutará recalled in darkness that his friend had told him to make a special effort on just such an occasion. He bid his mother farewell and took a trail out of the village to a lonely spot where he sat waiting. Night fell and dawn rose on the waiting man but nothing appeared, and so as the sun moved across the sky he went sadly home.

Very early on the following day he took to the trail again, following its bends ever farther from the village until he was dropping with exhaustion. He began to whimper now, just a little and softly, for he was so far away and so lost and lonely. Just at this moment he heard the tramp of feet and then ghostly laughter: it was them, the mamaés.* Soon he saw them filing by on some unimaginable errand, consumed by some unearthly joke. Aravutará sat silent by the trail's edge, witness to something unseen before by the living. At last, failing to find his friend among these spirits, he spoke to one of them, inquiring for him. "He's back there," said the mamaé, "bringing up the rear." And there at the end was the friend, and the two were instantly united in eager conversation.

"Where is this group going?" Aravutará asked.

"They go to a festival to do battle with the birds," he was told, and he begged to be allowed to go with them, so greatly did he feel his loss. The friend consented to wait there on the trail while Aravutará returned to his village to get his bow and arrows and some bast mats to keep the captured bird feathers in. When he returned the friend was waiting, and soon they had caught up with the spirit file.

That night they slept in a camp and the next day continued on toward the festival engagement. During the long march the living man three times proved his worth to his spirit companions, saving them from the assaults of plants and creatures. Now because of his deeds the people know that the mamaés may not be wounded in any way, otherwise their spirits die and vanish completely.

At last they saw in the distance the village of the birds. At that same moment the birds spied the mamaés and in terrific, bristling splendor they flew to the attack: parrots, macaws, toucans, fierce eagles. Instantly the mamaés began to drop all about the two friends. Bitten, scratched, and hairless, these spirits fell beneath the beaks, claws, and bludgeoning wings of the birds. But once again the living man, Aravutará, came to their rescue, playing a song on his flute until the birds, routed by this weapon from another world, flew off, leaving behind the remnant attackers. Aravutará gathered between the bast mats the feathers of the birds he had killed until the mats bulged. As he did so the birds were clearing the festival ground of the bodies of the slain mamaés. Each corpse they carried to the Giant Eagle and in an instant it disappeared within the snapping beak.

"Look, my friend," said the spirit companion, "now you may see and understand how it is with us. This is the way all of us end." And with this the spirit file turned back.

---

* Spirits of the dead.

When they had reached that place where Aravutará had once waited so long, the spirit friend took his final leave. "Do not look back at me," he warned, and then he was gone. Aravutará listened to the receding footsteps, and then there was only silence and the sounds of the great forest. Alone again, he could but retrace his way to his village.

At nightfall he entered it. At the door of his home he could hear his mother weeping within, but when she saw him returned, she dried her tears and listened as her son began to tell of his long and strange journey. Soon Aravutará too began to weep as he talked of his friend and of their final leave-taking. Then the weeping turned to vomiting, and then the sojourner fell into a deep, deathlike faint.

The shamans brought him back, and it was then that Aravutará knew clearly why he had gone with the spirits to the festival battle with the birds. It had been so that he could come back to this village and tell the people how it was with the dead, how when the sun disappears in the day the *mamaés* are keeping their fatal appointment with the birds, and how it is the Giant Eagle who finishes the spirits' lives forever. Handing out his captured feathers to the villagers, Aravutará told them all this, and so it was that the people came to know the Fate of the Dead.


Of life and its sacred obligations there is equally much to tell, and throughout the continents the tribes passed down along the generations such narratives and enacted them in rituals. Far to the north of Aravutará's village there was another village—a town, really—sitting on its hilltop above the flat "plain of warm mists," ringed about by long low ridges to the west and south and by the steep cliffs of Thunder Mountain to the east (Zuñi, New Mexico). The plain beneath the town was lush in its season with melon vines and the waving tasseled tops of corn. Here is a story the people told of a time before history when they had forgotten gratitude for the plenty of life that nurtured them as children at the breast.

For so long had spring rains and rivulets washed the plain floors and the warm mists kept out the breath of ice that steals life that the people of Há-wi-k'uh had come to rest easy on the heaped bounties of grains and fruits. They had forgotten want and had grown a new crop all their own—arrogance. So it was that one day the chief Priest of the Bow excitedly described to the elders a plan of his that would show all the neighboring peoples how proud and luxurious life had become at Há-wi-k'uh. "We shall stage a sham battle for our guests," he said, "but it shall be fought with food: sweet mush, bread, cakes, tortillas, and all seed foods. Then they shall see the wealth of Há-wi-k'uh and marvel at it!" Instantly the elders approved, and that evening from the roofs of the highest houses the priests announced the festival, explained how it should be conducted, and dispatched swift runners to the neighboring towns. Then the people fell to grinding, cooking, and baking so that the whole town was filled with the sounds and smells of preparation as for a great feast.

Now far down the valley to the south there lived among the White Cliffs two beautiful goddesses, the Maidens of White Corn and Yellow. When they saw the wasteful festival in preparation they were saddened yet determined to give the people a chance to show gratitude for their blessings. So, disguised as two ragged and hungry travelers, the Maidens visited Há-wi-k'uh on the day before the festival, and in front of them came the warm, misty rain that always announces the arrival of these Mother-Maidens from Summer-Land. But even of this omen the foolish people took no heed, feeling instead that the rain was the gods' humble recognition of their town's exalted position. Only a boy, his infant sister, and their uncle bothered to notice the bedraggled and dusty strangers as they passed slowly through the streets and by the doorways now heaped with baked things of steamy goodness, and they offered corn cakes to the disguised Maidens but were sharply rebuked by voices from within their doorway. "Let such strangers labor for their food as we have," said the elders, "instead of following the scent of cooking pots like coyotes!" And so the Maidens passed unbidden, unacknowledged, through the whole length of Há-wi-k'uh until they had reached the battered house of a lone old woman at the foot of the hill. Round about this doorsill there were no heaped baked goods—she was much too poor for such waste—but instead heaps of rubbish that the people above had flung down the hill in their carelessness.

Strange to tell, it was just here that the disguised Maidens found a welcome, for the poor old one of this house bade them come in and gave them the little she had. As the Maidens sat before their meager bowls of coarse mush, the old woman busied herself about the room, pretending that she herself had no need of food and that there was plenty. Then the Corn Maidens revealed themselves, and the old one covered herself in fear before their radiant beauty. But they soothed her, saying that only two small children and their old uncle, silent and sad before the spectacle of the people's profligacy, had taken note of their presence. But here, they said, we have found welcome, and you have given us all you had. They gave her honey bread, tiny and juicy melons, and beautiful mantles to hang in her storeroom. "Hang these," the Maidens said, "upon your blanket poles, and on the following day you shall have plenty. No longer shall you be poor." And saying this, they passed out of the town, again unattended except by the two children and their uncle, who thought they saw two beautiful strangers pass through the lower streets just at sunset.

That night as moonlight flooded a valley in the White Cliffs, the Squirrel and the Mouse convoked a meeting of the animals, for these little creatures had been forewarned by the Corn Maidens of coming calamity. "We must organize an expedition to Há-wi-k'uh," they said, "for tomorrow those wasteful beings are to throw away most of their food, and after this there shall be famine." So they spoke together and arranged for the rescue and storage of all that would be wasted in the festival.

Then came the festival itself. Before the astonished eyes of the guests the people fought with food. They pelted one another with hard bread, stopped one another's mouths with dough, and besmeared their clothing and hair with batter. All day long they fought, and as dusk came and the wondering guests trailed away, a mood of sour disgust settled on the people, and they went to their beds with hollowness.

With the moon came the expedition of the little seed-eaters, and all that night they toiled in silence, carrying off and storing away all the refuse of the great sham battle. They even entered the corn rooms through tunnels made by the gophers and stole large quantities of grain.

When the sun rose the next morning, turning the adobe walls to blush, and the people had climbed out on their roofs to look about their town, not a trace of their mad and foolish wastefulness remained. The streets, doorways, and plaza all had a clean and hard-swept look. Some few were troubled at this, for they had expected to recover much of the spilled food after the guests had left. But there were many more who said, "Who cares? We have more stored than we could eat in a year!" Only the old woman beneath the hill found surplus food that morning, for as they had promised, the Corn Maidens had provided, and her storeroom was filled with piled cords of white and yellow corn.

Winter waned and with it the remaining stores of corn, and the people began to discover that the seed-eaters had somehow carried off a great deal more than usual. They longed for the rain-bearing winds from the south that would commence again the old cycle of birth and growth and plenty. But the winds did not come from the south, and if clouds arose over the blue surrounding ridges, they were soon enough swallowed into the burnished sky, and the ground grew hard and closed in upon the struggling little corn plants. The people danced their old dances, the priests sacrificed their prayer plumes, but there was no harvest. Then winter fell in great unfamiliar drifts of snow on the plain, on the roofs of the houses, and collected in the angles of the streets. The people wondered in their new poverty how the old woman below seemed to prosper through it all, but they were afraid to accept her offers of food, thinking she must be a sorceress trying to revenge herself upon them for their previous ill treatment.

At last, the people began to die, both old and young, and in desperation a delegation was sent to the neighboring Moquis* to ask for relief. After many days two strong Moqui runners arrived at Há-wi-k'uh with the news that the people would be welcome. There was a feeble sort of joy now in the houses and streets as the survivors hastily threw together their few possessions and made ready to depart. Then, without order, they began to straggle off. Just as the parents of the boy and his little sister were about to quit their home, the old uncle shouted down to them, "Do not forget the little ones!"

But the parents only looked back at the sleeping forms of their children and said, "Let them sleep on. They could not keep up, and we cannot wait now for children or anything else." The father covered them with a buffalo robe and left them there by the silent and cold hearth.

Imagine the fright of these two small ones when they finally awakened to their condition and gazed from their rooftop over the deserted town: utterly alone, abandoned, except for the old woman below the hill, and as she had no reason to go out, they could not know of her presence.

For some time they lived on the chickadees the boy caught with snares he fashioned from the hairs of the buffalo robe, but the little sister each day grew more waxen and silent, and at night she whimpered for some parched corn. One day—it was just to distract her—the boy fashioned from corn stalks and piths a tiny creature he hoped would resemble a butterfly; he made also a cage of corn straws to hang the creature in so that it might give his sister some small, brief pleasure as she lay in her bed. And indeed, this did seem to brighten her, and she fell to talking to it thus: "Dear little treasure, fetch me a few corn grains

---

* Zuñi designation for the Hopi tribe.

that my brother might toast. You have long wings and can fly to some distant place where the corn still grows."

One night as the sister slept and her brother lay wakeful in the moonlight that came through the sky-hole, he heard a strange whispering above him. Yes, it was the would-be butterfly, whirling in its straw cage. "Let me go," it whispered, and again, "Let me go." With his heart pounding, the boy unfastened the cage, and the creature hummed out into the room. Then it spoke to him again through the semi-darkness: "My father, your heart is better than many men's together, for see, you have given me a body where before I had none, and you have loved your sister faithfully and well. Open the sky-hole and release me so that I may help you both. Surely I shall never desert you." The boy obeyed, and quickly the creature mounted up through moonlight and was gone, winging westward in the night until it had reached the dark shores of a great lake beside a river. * Here it suddenly dived, plunging past the startled stares of those watching sentinel gods who there attend the arrival of men's souls. Down and yet deeper it went until it had reached at last the Dance Hall of the Dead where it flitted into the dazzling company of the gods and the happy souls of men. Here, darting from place to place, it told the story of the lone children of Há-wi-k'uh, begging the gods' indulgence and mercy.

"Yea, we will happily help these beloved little ones," they chorused. "And you shall teach them their new duties that we may do so." And promising this, they sent their *Hé-he-a-kwe* † to gather pouches of corn from the seed stores of the creatures of the White Cliff Valley so that these might nourish the children during their period of instruction, for it was now clear that these orphaned ones had all along been singled out for a special destiny.

It was no less than this: to teach again to the people the lessons of humility and gratitude for what the gods have chosen to grant, and to guard forever after against arrogance and its only product, waste. So the creature now taught them daily, showing the boy how to construct the forgotten shapes of the prayer plumes with which the vanished people had once supplicated their great gift-givers. He watched as the boy cut the willow wands to the proper length, painted them in the six sacred colors—yellow, blue or green, red, white, speckled, and black—and attached the bird feathers to them. When the boy had made his own prayer over these and had bound up some prayer dust and paint in

---

* The Colorado Chiquito of present-day Arizona.
† Runners of the Sacred Dance.

dried husks, then the little creature took the offerings westward through the night to the Dance Hall of the Dead, where he deposited them at the feet of the gods.

Now the God of Fire spoke: "Grandfather, return and cherish these little ones, and when the time for spring comes we shall waft the warm clouds over the vale of Ha'-wi-k'uh as of old, and our swift and unfailing runners shall plant in that plain from the seed stores of the gods themselves. And these little ones: they shall now become the mother and father of their people for generations, and so they shall welcome back their foolish people from exile."

And it was so. When the time for spring arrived, the coming of the Maidens of White and Yellow Corn was again announced by the warm and pregnant clouds drifting northward out of Summer-Land from the White Cliff abode of the Maidens, shadowing the plain below the all-but-deserted town. But this time the Maidens were welcomed, and in the barren rooms of the children's house they nursed these orphans, brother and sister both, who now drank of the flesh of the Mother-Maidens. Nor did these goddesses forget the old woman who had continued in their plenty. Visiting her, they announced the mission of the children and instructed the old woman to comfort the sister, as would a mother, until the people should return.

The rains came. The barren and close-locked soil of the plain opened, yielded in the night to the swift runners who once again sowed it with the seeds of corn and melon from the storehouses of the very gods. The next morning green shoots shone in the sun where only yesterday there had been but dust the color of dried blood. And the morning following this the corn waved its fluted leaves, and on the next, tassels. On the fourth, ears of corn were poking upward through the green, and fruit-heavy vines trailed the floor. Then the little creature spoke a last time to the boy-priest (for this is what he had now become): "Make from the stalks that are growing below another of my form, and send her forth to me that men may call us and our offspring 'Dragon-Fly.' Do this for me, for it has been through my work and through the milk of the Mother-Maidens that you and your sister have become father and mother to your people."

And the boy-priest did as requested and promised further in remembrance of all this to paint on sacred things the images of the black, red, and white dragon-fly and of the green female one that follows him. So it is that these bright, winged creatures announce with their coming the ripening of the corn as they dart from tassel to tassel.

It remained then only for the people to return to Há-wi-k'uh. In eight days from the appearance of the corn ears they appeared, trooping in order over the hills from the northwest, led by the old uncle, who carried a token ear of corn in his hand like a signal torch. They passed through the waving, glistening fields, passed beneath the house on which stood the boy-priest and girl-priestess and beside them the old woman. Here they paid homage as children grown slowly wise must to parents who have grown wisely old. And here too the boy-priest announced to these humbled ones the new order for the eternal care and conservation of the gods' bounty so that never again might the people forget their true dependency. Hence to this day the people must obey the orders of these, their Corn Priests.

A thousand miles north of the plain below Há-wi-k'uh lie other and vaster plains, stretching for a thousand miles from mountains on the west that rarely lose their mantle of snow eastward toward sunrise and lake country. In between, the land rolls slowly down from the westward foothills, and once most of it was covered with a tough grass that moved in running billows before the constant wind. These plains are broken by stony buttes fantastically scored and grooved by wind, and there are ravines and wooded river valleys that wind like green ropes through the region.

When this stretch was grass-covered, it was the range of enormous herds of buffalo that fed on that wind-waved grass. Until the introduction of the horse and the gun about two hundred years ago, these beasts enjoyed a favored position here and collected in vast herds that browsed slowly along seasonal trails. They were dangerous game to hunt on foot, but to the horseless tribes that ventured out after them they were an ideal source of food, shelter, and implements, so the risk was worth it. This is a story from that time of a heroic human journey into the buffaloes' world and of a return to the people bearing lessons of life.

The people had built a great *pis'kun,** so high and stoutly made that no buffalo could escape it. Yet, for some reason the herds always

---

* Literally "deep blood kettle" in Blackfoot language. In the days before the horse the Plains tribes' favored way of killing buffalo was to stampede a herd over a steep cliff into a corral at its foot. There the buffalo not killed by the fall could be dispatched within the enclosure. Above the cliff and leading to its edge the people built lines of rock and brush, converging to form a large V, and behind these they concealed themselves. A medicine man would entice these fatally curious animals to follow him into the V-shaped chute, and then the people would rise up, shouting and flapping their robes, precipitating a stampede over the cliff. This method was also used to trap and slaughter antelope.

turned aside instead of running over the cliff: at the last moment the leaders would swerve to the right or left, and the others would follow, racing down the slopes and across the valley to safety. So the people went hungry, and then they began to starve.

Early one morning a young woman went to get water, and as she stood by the stream she looked up and saw a herd of buffalo grazing right at the edge of the cliff above the *pis'kun.* "Oh," she cried out. "if you would jump over, I would marry one of you!" They were that desperate. Yet, of course, she never expected her words to be taken seriously, and her surprise was great when indeed the herd began to rumble over the cliff, falling one after another into the brushy enclosure.

Then she was scared, for a huge bull cleared the walls with a bound and came snorting for her. "Come," he said, taking her by the arm. The young woman cried out in terror and tried to hang back. Then the bull reminded her of her promise. "See," he said, "the *pis'kun* is filled." There was nothing she could say, and he led her up over the bluff and across the prairie.

Now when the people had finished slaughtering, they missed this young woman, and her relations sought her everywhere and were very sad. Her father, taking his bow and arrows, vowed to find her, and he went away from the camp and out across the plains.

After traveling a weary way he saw a buffalo wallow and a little way off a grazing herd. He was tired now and sat down at the wallow to rest and think. A magpie lit near him and began to peck about, and the father spoke to it thus: "Ha! beautiful bird: help me. As you travel about, look for my missing daughter, and if you should see her, tell her, 'Your father waits by the wallow.' "

So the helpful bird flew off, skimming over the grazing herd until he spied the young woman amidst their shaggy backs. Then he lit near her, pecking this way and that, always with his beady eye on her and edging ever closer. When he was right next to her he said, "Your father waits by the wallow."

"Sh-h-h! Sh-h-h!" replied the young woman in a tense whisper, for the great bull, her husband, was asleep but a few steps away. "Don't speak so loudly. Now go back to my father and tell him to wait." And the bird did so.

After a time the bull-husband awoke and commanded the young woman to fetch him some water from the wallow. At this the young woman was glad, and taking a horn from her husband's head, she went on her errand. When she got to the wallow she said to her father, "Oh, why have you come? They will surely kill you!" But the father was resolute, replying that he had come in search of her and now meant to take her back.

"No! No!" she argued. "Not now. They would chase us and run us into the earth. Wait at least until he sleeps again, and then I will try to sneak away." Saying this, she filled the horn from the wallow and returned.

But when the great bull had tasted the water from his horn he snorted. "Ha!" said he. "There is a person close by here."

"None but me," said the young woman, but her heart turned within her.

The bull drank more deeply, and when he had done so, he lumbered to his feet, tossed his huge head and bellowed, "Bu-u-u! m-m—ahoooo!" A fearsome sound! Up rose the other bulls with their short tails raised and their heads twisting this way and that to pick up an enemy scent. They tore up the earth with their hooves rushing about, and when they had reached the wallow they found there that poor man. They trampled him flat, they tossed him and hooked him with their sharp curved horns, and then they trampled him again so that at last not even a small piece of his body could be seen.

Then the daughter cried over the killing ground, torn and scraped and empty, *"Oh! Ah! Ni'-nah-ah! Oh! Ah! Ni'-nah-ah!* (My father! My father!) But her bull-husband silenced her cries, saying, "Ah, you mourn your father. You see now how it is with us. We have seen our fathers, our mothers, so many of our relations, tumbled over your rocky walls and there slaughtered by your people. However, I will take pity on you. I will give you this one chance: if you can bring your father back to life, both of you may go back to your people."

Then the young woman called out to the magpie, "Pity me! Help me now. Search in this trampled mud for just a piece of my father's body, and bring it to me."

The magpie searched the ground about the wallow, every hole and indentation, until, digging about with his sharp bill, he found something small and white. He pulled hard on it, and it came free of the mud. It was a joint of the father's backbone, and he flew back with it to the young woman.

She placed the piece upon the ground, covered it with her robe, and sang a certain song. When she uncovered the piece, there lay the entire form of the father, looking as if he had just died. Once again she covered the form with her robe, and once more she sang. This time when she removed the robe the father was breathing, and at length he stood up and opened his eyes. The herd was astonished, and the magpie flew about making a great racket.

"This day we have seen strange things," said the bull-husband, "for he whom we had trampled to death is alive again. The people's medicine must be very strong.

"Now before you return you must learn something more of us so that we all may continue to live together in this land. Therefore I shall teach you our dance and our song." And this he did, showing the humans the slow and stately dance of the buffalo and the chant that must accompany it. When he had finished, he turned once again to the father and said, "Go now to your home, and do not forget what you have seen here. Teach this to your people that their lives may be properly ordered. The medicine for this ceremony shall be a bull's head and robe, and they who wear this in the dance shall be called 'Bulls.' "

The people's joy was great when from out of the prairie the father and his daughter returned to the camp. And the man called a council of the chiefs and told them all that had happened. Then the chiefs selected certain young men to whom the father should teach the dance and the song of the buffalo people and their medicine also. Such was the beginning of the *I-kun-uh'-kah-tsi.*★

In that eastern lake country where the long spine of the Appalachians yawns seaward there was a domain of dense forests interspersed with flat meadowlands of good soil in which the people could supplement the harvest of the hunt with yields of maize and melon. It was a strong land, deeply etched by seasonal changes, and it demanded strong people to live in it. Such were the Five Nations of the Iroquois. One of the nations, the Seneca, told a story of a time when their land had spoken to them of itself, of its past, of a world that had existed long before the people had come to live here. This is what they said.

It was very long ago. In a Seneca village there lived a boy whose parents had both died when he was but a few weeks old. A woman who had been a relation of the parents took the boy to raise as her own and gave him the name Poyeshao$^n$, which means "Orphan."

The little boy grew quickly and well. He was active and keen-

★ The All Comrades Society of the Blackfoot, a benevolent and protective order consisting of more than a dozen secret subgroups, the oldest of which appears to have been the Bulls. One of their most important functions was the regulation of hunting practices.

minded, and it was not long before his foster mother saw that he could begin helping her with the business of living. One day she called him to her and gave him a bow and arrows, saying as she did so, "It is time for you to learn to hunt. Tomorrow morning take these into the woods and kill all the birds you can."

That evening the woman shelled corn and parched the kernels in hot ashes. She rolled these up in a piece of buckskin and the next morning gave them to Poyeshao[n] on his way out. Her last words reminded him that a good hunter always has a happy home.

The boy traveled into the woods. Before noon he had shot a good number of birds, and so he sat down to rest and eat his parched corn. Then in the afternoon he worked back toward the village while the woods shadows deepened and stretched. When he arrived, his foster mother praised him warmly.

The next day the woman sent the boy out again. Again he rested at noon, eating his corn in a small clearing with his string of birds beside him. And again he returned at evening with a goodly number. His foster mother thanked him and told him that now he had begun to help with the business of living. He felt happy at this and determined to become so good a hunter that soon he would be allowed to go after big game.

Each day he went out, farther and farther into the great forest. And each evening he returned with a longer string of birds until finally he was forced to pack them in bundles on his back. His happy foster mother now had enough to give away to her relations.

On the tenth day Poyeshao[n] hunted a part of the woods he had never seen. As he took aim at a bird, the sinew that bound the feathers of his arrow loosened and the boy lowered his bow to make repairs. There was a clearing just ahead, and in its midst a high, smooth stone, round and with a flat top. The boy scrambled up to this natural seat, lay his string of birds beside him, and began to retie the sinew. Just as he was rearranging the feathers along the shafts, he heard a voice right next to him. "Shall I tell you stories?" it asked. Poyeshao[n] looked up expecting to see a man standing there in the clearing, but there was no one. He clambered around the rock and looked behind it, but he seemed alone. So he sat down again and resumed his work. Again the voice spoke, right under him, it seemed: "Shall I tell you stories?" The boy looked up quickly but could see no one. But this time he determined to watch carefully to see who was fooling him, and so when the voice spoke again, asking its question, he found that it came from the stone. It was the stone that had been speaking to him.

"What is that?" he asked. "What does it mean to tell 'stories'?"

"It is telling what happened long ago," the stone replied. "If you'll give me your birds, I'll tell you stories."

"You may have them all," Poyeshao$^n$ said quickly, and then he put his head down in a listening attitude and waited. Now the stone began, telling one story of the long ago and then another one as soon as that was finished. The boy sat atop the stone in the woods clearing until nightfall, when the stone said, "We'll rest now. Come again tomorrow, and if any should ask where all your birds are, tell them that you have killed so many that now you have to go a long way to find even a few." As he went homeward through the gloom, the boy was able to shoot a few birds and give these to his foster mother along with the excuse.

The next morning he took his pouch of corn, his bow and arrows, and went into the woods. But his mind was on the stories, not the birds, and he shot only those that lighted on his way to the stone in the clearing. When he got there, he put the birds on the stone and called out, "I've come! Here are the birds. Now begin again." And the stone did this, telling story after story until it stopped at dusk. On the way home the boy looked for birds, but it was late and he could find only a few.

That night his foster mother told her neighbors that there was something strange about Poyeshao$^n$. When he had first hunted he had been very successful, but now, though he was in the woods from morning to night, he brought back but five or six birds. Perhaps he threw the others away, she thought, or else gave them to an animal. Or maybe he just idled his time away and did not hunt. She hired an older boy to follow Poyeshao$^n$ to find out what he was up to.

So the next morning the hired boy took his bow and arrows as if to hunt and followed Poyeshao$^n$ into the woods, keeping out of sight and observing that Poyeshao$^n$ killed a good many birds. Then about the middle of the morning, Poyeshao$^n$ suddenly stopped hunting and ran toward the east as fast as he could. The hired boy followed until he came to a clearing and saw Poyeshao$^n$ climb atop a large round stone and sit down. As he crept nearer, the hired boy heard talking but could see no one to whom Poyeshao$^n$ might be speaking. So he went boldly up to him and asked, "What are you doing here?"

"Hearing stories."

"What are 'stories'?" the other asked.

"They are tellings of things that happened long ago. Put your birds here on the stone and say, 'I've come to hear stories.' " The hired boy did so, and the stone began again and continued until sundown, the two boys sitting silent in the clearing with their heads down, listening hard.

When they got back to the village with but a few birds, the hired boy explained to the foster mother that he had followed Poyeshao[n] for a while and then had spoken to him. "After that we hunted together," he said, "but we couldn't find many birds."

The next morning the older boy told his mother that he was going hunting with Poyeshao[n]. "It's good fun to hunt with him," he said, and off they went. By the middle of the morning each had a good string of birds, and then they hurried with these to the clearing.

"We have come!" they called to the stone. "Here are our birds. Now tell us stories." And the stone commenced, one after another, the boys listening with their heads down, until at dusk the stone said, "We'll rest until tomorrow." The boys returned to the village, but as before, they had little to show for their day-long hunt.

Several days went by in this way until at last the foster mother said to herself, "These boys kill more birds than they bring home." And she hired two men to follow the boys when they went into the woods the morning after. The men watched the boys hunting eastward until each had a good string of birds. Then they stopped hunting and hurried to a clearing where the men saw them climb a large stone and sit there listening to a man's voice.

"Let's go there and find out who is talking to them," one of the watchers said, and they stepped quickly into the clearing.

"What are you doing here, boys?" they asked. The boys were startled, but Poyeshao[n] spoke up quickly, saying that they must tell no one. The men agreed, and then Poyeshao[n] directed them to climb the stone and sit with them.

Then he said to the stone, "Go on with the story. We are all listening." So there were four of them now, sitting with their heads down as the stone told a story and then another and another. When it was almost night the stone finished.

"Tomorrow," it said, "all the people in the village must come and listen to my stories. Tell the chief to send everyone, and let each bring something for me to eat. And you must clear away the brush so that the people may sit on the ground near me."

That night Poyeshao[n] related the stone's instructions to the chief, who sent a runner to give the message to each family in the village. Early on the morrow all the people followed Poyeshao[n] into the woods, carrying with them bread and meat as gifts for the stone.

When the gifts had been deposited and the people quietly seated in the clearing, the stone spoke to them: "Now I will tell you stories of long ago. There was a world before this one, and the stories I am going to tell happened in that world. Some of you will remember every word I say; some will remember part of my words; and some will forget them all. I think this will be the way, but each must do the best he can. Hereafter, you must tell these stories to one another. Now listen!" The people bent their heads, and the stone went on until nightfall. Then it told the people to come again in the morning bearing gifts.

The next morning as the people gathered again in the clearing and deposited their gifts, they found that those they had left the day before were gone. When all was quiet the stone began to speak and continued until sundown. And this time it said, "Come again tomorrow. Tomorrow I will finish the stories of long ago."

Early in the morning the people gathered around the stone, and when it was quiet the stone began again. It went on into the afternoon, and then it said, "I have finished! You must keep these stories as long as the world lasts. Tell them to your children and your grandchildren, generation after generation. One person will remember them better than another. And when you go to a man or woman who knows these stories well, take something along to pay for them, bread or meat, or whatever you have. You must keep these things up. I have spoken!"

So it has been from that time to this. From the stone the Seneca learned all they know of the world before this one. That must be why they are called The People of the Stone.

No, it was not empty or demonic, this New World, and these narratives, chosen not quite at random, suggest but the mere outlines of the rich tribal lives that tragically were to remain *terras incognitas* to Western whites even after the New World had succumbed to Christian civilizing, yielding to ax, gun, and plow until all seemed known and mapped and the last savages in the deepest portions of the remaining wildernesses were wards of the state and maintained in a kind of human zoo.

Somewhere the great radical primitivist D. H. Lawrence, in surveying such remnants of primitive myth, inveighed against scholarly attempts to analyze these fictions to live by. We may analyze them, he wrote, but if we think we have thus denoted their meaning, we only reveal ourselves as silly and partial. There is much to be said for this view since in some sense to believe our analyses can pin down the meaning of a myth is to recommit the older error of believing that once a territory is reduced to the one-dimensional surface of a chart, it is known and under control. Geography, as a Saul Bellow character remarks, is a bossy notion according to which once you have been there, you know all about it. Rational analysis, then, may amount to a disbelief in the power of myth and in its deep authority for us.

Still, in another way, it seems no violation of the truth and life of myth to remark on certain genotypical features of it in connection with American tribal lives. Thus I have chosen to retell here some narratives that illustrate a very large and important class of myths found in the Americas (and indeed throughout the world), a class that may have much to tell us of the realities and possibilities of New World life. These are the myths of heroes★ who journey to strange territories and return to this world to tell of their experiences. The Xingu narrative from the Brazilian Amazon and the Blackfoot myth from the North American Plains are the most obvious examples, though to lesser (or more recondite) extents the Warrau myth from the Venezuelan Orinoco, the Zuñi myth from New Mexico, and the Seneca narrative also dramatize such transformative experiences. So extensive in fact is this class of myths that one of the earliest attempts to analyze Native American literatures was based solely on hero stories. This was Daniel G. Brinton's *American Hero Myths* (1882), a pioneering volume rendered partially invalid through this scholar's tendency to reduce all such narratives to "poetic" descriptions of the conquest of night by day. Now the classic work on hero myths is Joseph Campbell's *The Hero with a Thousand Faces,* a profound and beautifully conceived cross-cultural study of these narratives of journey and transformation.

The pattern Campbell discerns in these narratives is clear and dramatic: separation, initiation, and return. It begins often in blunder or through mere accident, as in the Seneca myth, the effect of which is to separate the individual from the community and to confront him with a

---

★Less for convenience than in the interests of accuracy the central character of these narratives will be referred to as a "hero" even if, as is often the case, the character's gender is feminine. In living tradition no distinction would have been made between hero and heroine of myth: the actions, by whomever performed, were heroic, and this is what counted.

challenge, a call to herohood. Stumbling somehow out of the wonted paths of life, the individual senses dimly the presence of a wholly different order of things, even a different world. The choice then is either to go on and explore this otherness or to return to the light of normal life. It is safer, of course, to choose the latter course, for the Other is dark, unfamiliar, and seems to speak in vague syllables of a kind of annihilation. Acceptance of the call amounts to acceptance of the possibility of so radical a personal alteration as to constitute a death. So Campbell finds that the hero is essentially that person who *submits* to this challenge, who agrees to the sacrifice of the self-as-it-is in the conviction that there is some higher, unknown state to be attained.

Between this stage of separation and the desired one of initiation lies the problematic journey to the threshold and the tests that await the sojourner on the other side. Typically the threshold is symbolized by some sort of barrier—mountain, moat, sea, cave, gate, or buffalo wallow—and it is guarded by creatures of appropriate strangeness or fearsomeness. Their function is to turn away all those unworthy of entry into what Campbell terms the "higher silences" beyond, either through shallowness of motive or inconstancy of resolve. Often these guardians are imaged as disgusting, loathsome, even bestial.

The hero's task is to see these guardians for what they are and not mistake them for adversaries. He must be equal to the challenge they present him with, firm in resolve and vision, ready to follow trustingly wherever they might lead or send him. If the hero faces this challenge successfully, then the guardians turn out to be guides, and on the other side of the threshold they speed the hero on his way toward rebirth.

Then there are the tests or trials within. As often as these are symbolized as tasks to perform, battles to be won, they are also shown as temptations to be resisted. Their cumulative effect is to break down the remaining resistances within the hero, to strip from him the last lingering attachments to the old way, to destroy the last vestigial remains of the ego, so that at last the individual arrives at the long-sought goal purged of pride, fear, and hope, utterly naked and thus fit for transformation.

Initiation in these narratives equals transformation, and the agent of this process can be either god or goddess, assume either anthropomorphic form or animal, or come in the guise of some natural phenomenon. In whatever shape, the Great Teacher's message is essentially the same wherever in mythology it is encountered. It is the inculcation of a profoundly altered vision of the self and cosmos. Now the hero sees that the divisions and dualities by which we commonly live are but the merest of illusions. At the same time, the hero discovers—as we do with proper guidance—that his long outward journey has all along been an inward one into the mysterious and awful regions of the human heart and that the godly wisdom or power lay nowhere else. Campbell writes that from this new perspective it can be seen that

> the perilous journey was a labor not of attainment but of reattainment, not of discovery but of rediscovery. The godly powers sought and dangerously won are revealed to have been in the heart of the hero all the time.

And:

> The two—the hero and his ultimate god—the seeker and the found—are thus understood as the outside and inside of a single, self-mirrored mystery, which is identical with the mystery of the manifest world. The great deed of the supreme hero is to come to the knowledge of this unity in multiplicity *and then to make it known.*

I italicize these last words since it seems to me that the return to the community bearing these lessons of life is the most crucial part of the entire process. It is what makes a circle of the pattern. Lacking this, the story is but another one of adventure, however thrilling, courageous, and compelling. For the hero's return to the community—to us—bearing a new experience with immemorial wisdom, is what justifies the great effort and its sacrifices. This return of the vanished one is what allows for what Campbell calls the "continuous circulation of spiritual energy into the world." Through the hero's words and whatever ceremonies and songs he may institute, the community at large becomes the beneficiary of those sacrifices and that ultimate submission. It too is renewed. This is why these narratives so often conclude with the hero standing in the midst of a conclave of chiefs or elders, telling of that Other from which he has lately returned. And this is why the deepest meaning of the word "hero" is "one who serves."

As Campbell makes clear, these narratives are really instructions for individual and group behavior in the grand rites of passage that mark the thresholds of life: birth, naming, puberty, marriage, and death. In the deportment of the mythic hero in confrontation with threatening newness and strangeness, in his successful return to and reintegration into the community, and in that community's eager acceptance of the heroic message, myth-bound peoples of the world dramatized a crucial ideal: the maintenance of stable and cooperative communities in the midst of the endless vagaries of existence through equally endless re-generations of spirit. The form of these narratives supplied the form of the communal rites in which the individual was temporarily cut off from the community, tested in isolation, and then returned to normal life transformed—a "sun door," as Campbell so beautifully puts it, through whom the community could hope to gain new strength of spirit.

In the huge and untamed New World the native communities were constantly in the presence of newness and strangeness, but where European sojourners felt terror and reacted with a hostility sanctioned of old, these natives often found spiritual sustenance. After his initiation ordeal in the windy solitude of the Great Plains, during which he had striven to emulate the heroic figures of myth, a Cheyenne brave, for example, was returned to the welcoming arms and eager ears of the village elders. "My heart," he remembered of that time, "was like the sun coming up on a summer morning."

# Defloration

The more we come to know of history, the more it reveals itself to be symbolic, as the discrete events, artifacts, and personages tend to lose something of their individualities and to become increasingly representative.

Consider the career of Columbus. It was symbolic from the moment of his birth. His Genoese parents gave him the name Cristoforo in honor of that giant pagan-turned-saint who carried the burden of the Western world on his unwitting shoulders. The legend tells how this pagan, hearing of Christ, went in search of him but failed to find him. Then on the advice of an anchorite he took up his vigil on the shore of a broad and bridgeless river across which he carried poor travelers, all the while waiting for Christ to take notice of his good labors. One night, asleep in his riverside hut, the pagan heard a child's voice calling to him through the darkness. "Christopher! Come and take me across the waters!" With staff in hand, the pagan took the child upon his massive shoulders and waded into the water. As he pushed along, he felt the weight of his little passenger inexplicably increase so that before he had gained the far bank it was all but insupportable.

On the far side he set the child down, saying as he did so that if he had carried the weight of the whole world on his shoulders, it could not have weighed more than this little child. It was then that the child revealed himself as Christ, saying to the pagan, "Marvel not, Christopher, for thou hast indeed borne upon thy back the whole world and Him who created it. I am the Christ whom thou servest in doing good." Henceforth the saint was known as "the Christ-Bearer."

Who knows what designs the mother and the weaver father had in mind for their child? But it is certain that in so naming him they participated in the vast design their civilization was weaving for itself—and others. It is certain too that in manhood Columbus would think of himself within the context of this saint's legend as fated to carry the burden of the Christian message across a watery waste to implant it on foreign shores where it would blossom into the gorgeous flowers of conversion.

As with the lives of so many heroes, there is no childhood here. Without fumbling efforts at speech and movement Columbus enters history fully formed, bent upon his destiny, armed with a private and intense devotion to the Holy Trinity. Somewhere on that Mediterranean where his civilization took its first tentative maritime lessons, he learned his ropes, rudders, and sails. By the age of twenty-five he had mastered all the navigational lore available to one of his time and place.

In that year of 1476 the second event in the unfolding symbolic career occurred: as a crew member of a Flemish merchant ship bound for Portugal, England, and Flanders, Columbus was forced overboard when his vessel was sunk by French raiders. This was off the coast of Portugal, and the young man drifted on an oar to the port of Lagos.

For more than half a century Lagos had been the launching place for Europe's first lengthy thrusts outward into the gloomy, unknown Atlantic. In 1418 Prince Henry (later styled "The Navigator") had established himself just up the coast at Sagres. There on the barren, rocky point, the continent's farthest probe seaward, the prince had gathered the men and materials for long-range, large-scale exploration. Fittingly, the spot had formerly been the site of a Druidic temple, and the classic geographer Strabo tells us that the ancient Iberians believed it to be the nighttime gathering place of the gods. Under Henry its archaic spirit yielded to the imperative of outward-bound Christian history.

Henry must appear to us now as Columbus's spiritual and historical ancestor. In this solemn-faced, black-robed figure who for the love of God girded his loins with an abrasive sash of silicon, the West found its anchorite of exploration. He appears to have lived for nothing but to push Christian civilization beyond its geographical confines, outflanking Islam and reaching at last those fabled treasure houses of the East (he too had consumed his Polo). After achieving knighthood against the Moors, a celibate Master of the Order of Christ, he retired to Sagres to take up his own kind of heroic mission. Here he built a chapel and an observatory—a civilization's elemental structures—and with his recruits conned old and recent travelers' narratives, drew new maps and corrected older ones, and fitted out ships down at the port of Lagos. Like a lodestone the operation at Sagres attracted mathematicians, cartographers, and professional maritime adventurers from all of Europe, including Columbus's Genoa.

Outward and ever outward the prince urged and cozened his captains. Africa—the way to the East, to its infinite stores of metals, spices, and heathen idolators—must have an end somewhere to the south. But the sailors were afraid to pass Cape Non on the African west coast. "Who passes Cape Non," they told one another, "must turn again or else begone." Or if, by some chance, you could pass this cape, you would be turned black as a sign of God's vengeance for meddling in zones not meant for Christians. Nonsense, said the Master of the Order of Christ, and sent ship after ship coasting southward until at last in 1434 the southernmost tip of the continent was doubled and the wood of the caravels floated on easy seas. The way had been opened.

From Lagos the first penetration of the African landmass was made. Nine years before the Genoese birth of the Christ-Bearer who was miraculously cast up on its wharves, ships had returned here with the first tiny trickle of African treasure. It was gold. But in 1444 another sort of treasure returned here to port: six caravels sailing in under the banner of the Order of Christ, carrying in their holds the huddled tangle of captive flesh. Prince Henry was there to greet them and to supervise the distribution of this first fatal fruit out of the dark wildernesses beyond. His contemporaneous biographer, Gomes Eannes de Azurara, described the disgorgement of the cargo:

On the 8th of August, 1444, early in the morning on account of the heat, the sailors landed the captives. When they were all mustered in the field outside the town they presented a remarkable spectacle. Some among them were tolerably light in colour, handsome, and well-proportioned; some slightly darker; others a degree lighter than mulattoes, while several were as black as moles, and so hideous both in face and form as to suggest the idea that they were come from the lower regions. But what heart so hard as not to be touched with compassion at the sight of them! Some with downcast heads and faces bathed in tears as they looked at each other; others moaning sorrowfully, and fixing their eyes on heaven, uttered plaintive cries as if appealing for help to the Father of Nature. Others struck their faces with their hands, and threw themselves flat upon the ground. Others uttered a wailing chant, after the fashion of their country, and although their words were unintelligible, they spoke plainly enough the excess of their sorrow. But their anguish was at its height when the moment of distribution came, when of necessity children were separated from their parents, wives from their husbands, brothers from brothers. Each was compelled to go wherever fate might send him. It was impossible to effect this separation without extreme pain. Fathers and sons, who had been ranged in opposite sides, would rush forward again towards each other with all their might. Mothers would clasp their infants in their arms, and throw themselves on the ground to cover them with their bodies, disregarding any injury to their own persons, so that they could prevent their children from being separated from them. Besides the trouble thus caused by the captives, the crowds that had assembled to witness the distribution added to the confusion and distress of those who were charged with the separation of that weeping and wailing multitude. The Prince was there on a powerful horse, surrounded by his suite, and distributing his favours with the bearing of one who cared but little for amassing booty for himself. In fact he gave away on the spot the forty-six souls which fell to him as his fifth. It was evident that his principal booty lay in the accomplishment of his wish. To him in reality it was an unspeakable satisfaction to contemplate the salvation of those souls, which but for him would have been for ever lost.

In such a fashion did the virgin prince advance his civilization and its faith. As C. Raymond Beazley notes with perhaps unconscious accuracy, not since the high tide of the Crusades had the West been so expansive. At the prince's death in 1460 a line of Portuguese trading factories was in the works, stretching around the African coast toward India and the Far East.

So Lagos was *the* place in all the West for the young mariner Columbus to drift ashore. He spent eight years in Portugal, learning the lessons only the tiny Lusitanian kingdom could teach. He absorbed "many things from his Portuguese shipmates," writes his biographer, Admiral Samuel Eliot Morison: "how to handle a caravel in head wind and sea, how to claw off a lee shore, what kind of sea stores to take on a long voyage and how to stow them properly, and what sort of trading truck goes with primitive people." Indeed. And along with this last, Columbus learned the lessons of his civilization about how such primitive people were to be dealt with. The lessons learned long ago in the Near East by those within the cities who faced with terror and hostility the shifting, barbarous hordes without; the lessons engraved by the iron finger of the Old Testament mountain god and imparted to the chosen ones; the lessons lately recited in the deeds of the Crusaders and even now in those of the Inquisitors—all went into the inexorable construction of those outposts of progress that began to appear like leprous sores on the white spaces of the expanding charts.

In 1481 the new king of Portugal, João, authorized the construction of a trade factory on the Gold Coast. The pope, following papal precedent of the Crusades, granted full indulgence to all Christians who might die in this affair deemed so essential to the health of Christendom. Columbus was one of those who set out from Lisbon late that year under the command of Diogo d'Azambuja. On the Gold Coast the Christians worked on into 1482, constructing a high castle of thick masoned walls with a moat surrounding it, facing on one side the endless sea and on the other the equally endless jungle. Inside were storerooms and dungeons. São Jorge da Mina, St. George of the Mine (Cape Coast Castle, as it is now known), was, of course, a slave castle where the perishable produce of the slavers was held in fetid storage until its transport out of this wilderness into the usable light of civilization. Here Columbus could not have avoided witnessing the spectacle of the dungeons and the first of the many generations who would leave their relations, tribal identifications, rituals, blood, and tears in the wide low rooms that gradually acquired a humus of leftover skin. Ten years later, writing from a wilderness in another hemisphere, the Admiral's letters to his sovereigns bear witness to the depth of his learning here.

Consider the matter of Columbus's marriage. He met the woman at a convent belonging to the military order of St. James, the patron saint of the nation that would ultimately fund his scheme. Here the order provided shelter for the wives and daughters of those knights away fighting the heathen in other lands. The woman herself had symbolic entailments. Felipa Moniz Perestrello was the daughter of Bartholomew Perestrello, first governor of Porto Santo in the Madeiras, islands that had once served as outposts in the Atlantic. She was also the granddaughter of Gil Moniz, one of Henry's most illustrious captains. Old Perestrello had introduced rabbits to the unsuspecting island, and these creatures quickly reduced the vegetation to rags. So Perestrello, first governor, was also popularly known as original despoiler, and the island to which his daughter and son-in-law now moved was a barren accusation of his maladministration.

The old man bequeathed to Columbus through his widow all the documentary flotsam of a life spent in the midst of exploration. The widow, knowing of her son-in-law's consuming interest, unfolded to his eager eye the old packets, charts, and written rumors of lands elsewhere. On the denuded island the Christ-Bearer read voraciously.

The Madeiras were a curiously appropriate spot for him. On these islands off the mainlands of Europe and Africa rumors and speculations of lands westward washed ashore like the utterly foreign visitations reported of the Azores. There, so the colonists reported to Columbus, when the wind was from the west, weeds and pines of a strange sort were found on the beaches. Once on the island of Flores two corpses were discovered with faces unlike those of civilized men. Here death came from the west, but in the rush of speculation the omen went unnoticed.

Like many others of his time Columbus had moved beyond the terrestrial conceptions of Prince Henry. The goal was still the East and the outflanking of the rival monotheism of Islam, as indeed it had been since the days of Raymond Lully. But increasingly the cosmographic speculators had turned their thinking to the west as a way of getting east. (Just why this civilization had to get anywhere else had for centuries ceased to be a serious question.) Among such questing intellects the spherical shape of the earth was generally accepted—in covert defiance of the lingering Churchly position that it could not be so shaped. The only remaining major point of dispute was just how far west ships would have to go to get east. Columbus thought it could not be very far.

In this error he was probably somewhat influenced by a fellow Italian, Paolo Toscanelli, though perhaps in a more important sense by the ubiquitous Marco Polo, whose book Columbus read and heavily annotated in his Porto Santo stay. Toscanelli was a Florentine who did service as physician, mathematician, astronomer, and cosmographical theorist in the days when the sum of Western knowledge was about as small as he thought the globe. He himself had been heavily influenced by Polo, and when Columbus heard of him he was in correspondence with the canon of Lisbon Cathedral as part of the prelate's campaign to suggest to the Portuguese crown that there was a shorter way to the Indies than around the tip of Africa. Columbus wrote to Toscanelli and received a gracious reply, which included a map based in part on Polo. Perhaps this was but the final piece to be put in place; perhaps Columbus did not need Toscanelli except as additional confirmation.

When he returned from Porto Santo with his wife and son in 1484, his scheme of east-by-west was formulated, and in that same year he gained an audience with João. Columbus was what Jung would call the fully modern man and what others would call the hero of history: an individual who has somehow divined the urgently felt needs and aspirations of his time and intuited a means by which these might be satisfied.

Such an individual, says Jung, has not only mastered the mechanics, facts, and theories of his age, but has gone beyond them so that he must seem to his contemporaries eccentric, misguided, even dangerously deluded.

As we know, Columbus in his time seemed all of these and, as we shall see, he later became as deranged as his detractors always thought him. Now, in 1484, the Genoese mariner and cartographer possessed the essential mind of the West with all its twisted religiosity, its background of classical geography and medieval folklore, and its recent acquisitions of technical skills that were already crowding the other contents. Beneath all this, out of touch like those Iberian caves that lay as yet undiscovered, was that deepest substratum of man—the organism's need to establish and celebrate its spiritual identity with the phenomenal world and the cosmos.

The audience with João was not a success, in part just because this foreign petitioner was so much the man of his time that he seemed a dreamer. Also, he had the offensive manner of many a self-taught genius. And so when João's court theorists advised the king that Columbus's computation of the distance between Europe and the East was grossly underestimated, he believed them. He was right to do so, as it turns out, since Columbus had calculated it at 2760 miles whereas more than 12,000 miles lie between Portugal and Cipangu (Japan). The Portuguese theorists told their king that ships could never make it to Cipangu by sailing westward, and they were right. No one could imagine what lay between.

A year later, his wife dead and his plans rejected, Columbus packed his charts and books and moved on to Spain, considerably in debt. He was to wait another seven years in that country while a list of higher national priorities was attended to, but in retrospect it seems obvious that sooner or later someone would fund him, for the plan was so ineluctably right for a civilization in need of reaching beyond itself. If not the Spanish crown, then maybe a noble of great means like the Duke of Medinaceli, or even England's Henry VII, with whom Columbus's brother had apparently treated when the Spanish sovereigns turned at last in Columbus's direction. This was early in 1492, and though Columbus must have been buoyed as he hastened back along the roads from the town of Pinos to his meeting with the sovereigns, the conditions of his belated success were ominous for those peoples and lands he would now be empowered to discover and annex.

One of those conditions was the final defeat and expulsion of the

Moors after a protracted and savage racial/religious war of eight centuries and three thousand battlefields that left enduring scars on the Hispanic character. The Spanish had become inured to the inevitable cruelties of warfare and intolerant of religious and skin differences; they even deemed the very existence of these last as threats to civilization. Another condition was the sizable group of men-at-arms suddenly bereft of adversaries. Many of them were unfit for steady labor and sought refuge in the Church or else hung on in the army awaiting further adventures. A third was the fate of unbelievers in the Spanish midst. Some attention has already been paid the zealous and benighted work of the Inquisition, which at this time was working with a terrible efficiency. And there were now new targets of intolerance, the Jews, who in this very Spanish spring of 1492 had been issued an official ultimatum: convert instantly or begone. As many as a quarter of a million of them were forced from their homes, often under the cruelest of circumstances, their houses and possessions forfeit to the Christians.

Such conditions as these did not bode well for those Columbus and his Christian expeditionary force were now to seek out. Nor did the royal expectations of the undertaking. For the lands the newly designated Admiral of the Ocean Sea was being sent to secure for the Spanish royal couple were the mainland of China and adjacent territories, including Cipangu (Japan). With the desperate confidence of those who do not truly believe in their cause yet fear more than anything to question it, the crown assumed that the Great Khan and other Oriental potentates would immediately recognize the superiority of the Europeans and turn into vassals. We might term such an assumption insane were it not plain that so much of the subsequent history of the West reveals that we are the products and practitioners of just this assumption. No people has ever easily assented to the "natural" superiority of another, and they are less likely to do so when the invaders' actions betray internal doubts of their vaunted superiority.

That there was such doubt in the Christian West is attested to by the heavy air of apocalyptic expectation that hung over Europe in these years. Here is Admiral Morison sketching in the mood of the West during Columbus' first voyage:

> With the practical dissolution of the [Holy Roman] Empire and the Church's loss of moral leadership, Christians had nothing to which they might cling. The great principle of unity represented by emperor and pope was a dream of the past that had not come true. Belief in the institutions of their ancestors was wavering. It seemed as if the devil had adopted

as his own the principle "divide and rule." Throughout Western Europe the general feeling was one of profound disillusion, cynical pessismism and black despair.

And he adds that the *Nuremberg Chronicle* of 1493 placed that year in the Sixth or penultimate age of the world and left but "six blank pages on which to record events from the date of printing to the Day of Judgment." But, says Morison, at the very moment that the *Nuremberg Chronicle* was being seen through the press, "a Spanish caravel named the *Niña* scudded before a winter gale into Lisbon, with news of a discovery that was to give old Europe another chance."

It was as "another chance" that Europe was to think of this discovery of the "New World," once the news of it had gotten around. And so it might have been, since regeneration often may spring from an unexpected confrontation with radical newness. But Europe was wrong, and the confrontation was wrong from the start. The hand that signed the sailing orders for the crown, Juan de Coloma's, was the same that had just signed the ultimatum to the Jews, and the winds wafting the flotilla westward bore an old and foreign pestilence to those far, flowered shores. Doom, not regeneration, shook from those sails.

The flotilla's threshold to the great beyond was the Canary Islands. Precisely ninety years earlier these storied bits of green in the expanse of gray and blue had been the targets of the first cannonading of the wildernesses by Westerners. But before the French adventurer Jean de Bethencourt had led his motley crew in an invasion of the Canaries in 1402, they had been sung by the classic poets as the Fortunate Isles, or the Isles of the Blest, where favored souls repaired for the afterlife. As such they existed unmolested in a zone of myth until the restless spirit of the peoples to their north and east blew to their shores casual coasters who stopped for goat's flesh and then for the human flesh of captives. Then as Prince Edward of England and King Louis of France were leading the last major Crusade overland against the unbelievers, a Genoese fleet under one Lancelot Malocello touched at Lanzarote Island and established a fort there. Little more is known of this landfall until the Bethencourt expedition, which, its clerical chroniclers tell us, was organized on the familiar and fatal assumptions:

> Inasmuch as, through hearing the great adventures, bold deeds, and fair exploits of those who in former times undertook voyages to conquer the

heathen in the hopes of converting them to the Christian faith, many knights
have taken heart and sought to imitate them in their good deeds, to the end
that by eschewing all vice, and following virtue, they might gain everlasting
life; in like manner did Jean de Bethencourt, knight, . . . undertake this
voyage, for the honour of God and the maintenance and advancement of our
faith, to certain islands in the south called the Canary Islands, which are
inhabited by unbelievers of various habits and languages.

In the course of three years Bethencourt and his men accomplished
great slaughter among the natives, established colonies, and converted
the islanders to their satisfaction, if not to any great profit.

From these instructive isles, now Portuguese possessions, Columbus
set his prows into the unknown on September 8, 1492. A sixteenth-
century engraving by the Flemish iconographer of the New World,
Theodor de Bry, renders for us the spiritual situation of that fall day as
the ships made slow headway west in heavy seas. The armored Admiral, his feet planted resolutely on the planks of the deck, stares outward.
Cannons flank him, and in the cross-hatched waves he leaves behind in
his passage are the outmoded, outlived figures of myth—mermaids and
gods bowing and bobbing aside to the keel of history. The flag the Admiral holds depicts on its stiff surface the Redeemer hanging from his
cross, sacrificial, regenerative.

They sailed onward, out of sight of all land—always a terror to these
sailors since the charts were coastal and to lose sight of the coast was to
lose all track and be lost utterly. They worked their primitive mecha-
nisms through those unvarying days—the hourglass, the crosspiece,
and the wood chips they tossed off the bows to gauge speed by their
passage sternward—while they marked their onward passage through
time and history with the formal observances of the faith: a religious
ditty at dawn, *Pater Noster, Ave Maria, Salve Regina.* The Admiral kept
two logbooks, one to calm the crew, and the other to compute how far
they had really traveled.★

On the 23rd of September, a Sunday, the ships swam in mild seas
and milder winds and the men made a low undertone of complaint
against this man who had brought them so far into the wilderness. For,
they said, if no winds out here blow us onward, neither will any blow
us homeward. But then great swells rose beneath them and sent the
ships onward, and the Admiral noted in his logbook, drawing hard
comfort from the Old Testament analog: "I was in great need of these

★ Admiral Morison, retracing the first voyage, found the public log the more accurate.

high seas because nothing like this had occurred since the time of the Jews when the Egyptians came out against Moses who was leading them out of captivity."

Perhaps some deep memory of myths in which birds are omens and messengers made them begin to note the flights of boobies, terns, and petrels. All during the night of October 9 they heard birds passing, but dawn brought no relief of land and again the men grumbled, while the Admiral explained to them the terrible logic of the logbook: it was farther back than onward.

On the eleventh in rough seas they saw green reed and more petrels, and the men on the *Pinta* saw a cane and a small stick apparently shaped by human hands. All breathed easier as sunset came on. In the darkness two hours before midnight the Admiral, standing on the sterncastle, saw a light, but it was so indistinct he could not tell whether it was from land. He called out the steward, Pedro Gutiérrez, and then the royal accountant, Rodrigo Sánchez de Segovia, and all three strained through the dark for that little light. Then they saw it again, rising and then blotted out like a wax candle. They could not have known that on an unseen island the natives were lighting smudge fires of palmetto leaves to ward off the stinging sand flies and mosquitoes. As the wax of the leaves dripped into their fires, the light sprang up to alien eyes and then subsided in protective smoke. Four hours later all hands saw the island two leagues away, and they lay close-hauled waiting for morning.

Friday, October 12. The Spanish cleric-historian Bartolomé de Las Casas, who saw and made a transcription of the now lost logbooks, described the scene:

> This was Friday, on which they reached a small island of the Lucayos, called in the Indian language Guanahani. Immediately some naked people appeared and the Admiral went ashore in the armed boat, as did Alonso Pinzón and Vicente Yanez his brother, captain of the Nina. The Admiral raised the royal standard and the captains carried two banners with the green cross which were flown by the Admiral on all his ships. On each side of the cross was a crown surmounting the letters F and Y [for Ferdinand and Ysabela]. On landing they saw very green trees and much water and fruits of various kinds. The Admiral called the two captains and the others who had landed . . . and demanded that they should bear faithful witness that he had taken possession of the island—which he did—for his sovereigns and masters, the King and Queen.

Thus did Christian history enter once again the zones of myth.

The armored ones on that beach we know; the naked ones were those gentle Arawaks who had migrated to these islands more than a thousand years earlier, carrying with them their manioc plantlings, their great deities Yocahú and Atabeyra, and their rituals of renewal. Well built and graceful, they accepted with evident pleasure the strangers' gifts of red caps, glass beads, hawk's bells—that "trading truck" the Admiral had learned went with primitive people—and then swam in the dazzling clear water out to the waiting ships, bringing their own gifts of parrots, balls of cotton thread, and spears. Neither group could effectively converse with the other, for the fleet's official interpreter, Luis de Torres, had been selected for his knowledge of Arabic, a language the Christians had encountered everywhere else they had gone. Columbus described the people as the color of Canary Island natives.

They were so gentle, so ignorant of the arts of war, he noted, that when he showed them swords, they took them by the edges and so spilled their first drops of bright blood against Western steel. Addressing the sovereigns, he wrote:

> They are so affectionate and have so little greed and are in all ways so amenable that I assure your Highnesses that there is in my opinion no better people and no better land in the world. They love their neighbours as themselves and their way of speaking is the sweetest in the world, always gentle and smiling. Both men and women go naked as their mothers bore them; but your Highnesses must believe me when I say that their behaviour to one another is very good and their king keeps marvellous state, yet with a certain kind of modesty that is a pleasure to behold, as is everything else here. They have good memories and ask to see everything, then inquire what it is and what it is for.

And the lands! As yet no gilt-gabled cities as described by Polo, but as they coasted amidst these islands, the water was so clear the bottom could be seen even at great depths, and when the breezes blew offshore the scent of the flowers and trees delighted them all and maybe even lulled them into ancient dreams of rest here. Of one island the Admiral wrote:

> Though all the others we had seen were beautiful, green and fertile, this was even more so. It has large and very green trees, and great lagoons, around

which these trees stand in marvellous groves. Here and throughout the
island the trees and plants are as green as Andalusia in April. The singing of
small birds is so sweet that no one could ever wish to leave this place. Flocks
of parrots darken the sun and there is a marvellous variety of large and small
birds very different from our own: the trees are of many kinds, each with its
own fruit, and all have a marvellous scent.

What secrets, what unforethought-of initiations waited here amongst
these islands, their flowers and trees and generous, trusting inhabitants?
What entrances into mystic groves, cycles of simple contentment? Pre-
lapsarian visions hovered like tropical clouds—and like tropical clouds
were dispersed by the burning, relentless sun: gold. Where was it?

> I have watched carefully to discover whether they had gold and saw that
> some of them carried a small piece hanging from a hole pierced in the nose. I
> was able to understand from their signs that to the south . . . there was a
> king who had large vessels made of it and possessed a great deal. I tried hard
> to make them go there but saw in the end that they had no intention of
> doing so.

Two days later the Admiral broke off a description of the natural delights of these islands to remark that he had not really explored them
because he was "anxious to find gold." He was certain that if these
trusting creatures did not have gold themselves, they must be the vassals of a great golden lord of the yet-to-be discovered mainland. And if
they were his vassals, then surely by right they were the slaves of the
Christian majesties. In his *very first island entry* (October 12), Columbus
advanced this suggestion:

> They should be good servants and very intelligent, for I have observed that
> they soon repeat anything that is said to them, and I believe that they would
> easily be made Christians, for they appeared to me to have no religion. God
> willing, when I make my departure I will bring half a dozen of them back to
> their Majesties, so that they can learn to speak.

And two days later in a characteristic chiaroscuro passage he went even
further:

> . . . these people are very unskilled in arms, as your Majesties will discover
> from the seven whom I have caused to be taken and brought aboard so that
> they may learn our language and return. However, should your Highnesses

command it all the inhabitants could be taken away to Castile or held as slaves on the island, for with fifty men we could subjugate them all and make them do whatever we wish. Moreover, near the small island I have described there are groves of the loveliest trees I have seen. . . .

On his second voyage he would act upon his own suggestion, bringing back with him a cargo of slaves along with spices, woods, and precious little gold. He would eventually suggest to the crown that a steady traffic in slaves be instituted to help defray the costs of colonization, and he would bring back from his third voyage fifteen hundred such captives. Alas, it was learned too late for these that such inoffensive islanders did not keep well in the holds. Better to work them where they were.

To each bit of land he saw he brought the mental map of Europe with which he had sailed. Anciently, as George Stewart has shown, place names arose like rocks or trees out of the contours and colors of the lands themselves. Many of the aboriginal place names of the New World were doubtless references to what could be seen by the naked eye. And as a group took up residence in an area, that area would be dotted with names commemorating events that took place in it. Or a place might be known by a vision had there, or by the spirits of the indwelling gods. Where one tribal group supplanted another, it too would respond to the land, its shapes, moods, and to tribal experiences had there. Now came these newest arrivals, but the first names by which they designated the islands they coasted were in no way appropriate to the islands themselves. Instead, the Admiral scattered the nomenclature of Christianity over these lands, firing his familiar names like cannonballs against the unresisting New World. He named his first landfall San Salvador after that savior whose light he now brought to this darkness, whose burden he bore in his own name. He named other islands after the shrines of the Virgin he knew in Spain: Santa Maria de Guadalupe after the Virgin at Estremadura, Santa Maria de Monserrate after Loyola's monastery near Barcelona, Santa Maria La Antigua after the Virgin in the Cathedral of Seville. One group was called Los Santos because the Christ-Bearer sailed past them on All Saints' Day. Later he would give darker names to match the color of his perceptions and experiences. An armored Adam in this naked garden, he established dominion by naming.

Nature and negligence combined to provide the symbolic ending to Columbus's first voyage. In the first hour of Christmas Day, the current ran the *Santa Maria* on the rocks off the island of Hispaniola; the rudder had been left to the ship's boy while all the others slept below. Before dawn it was clear to Columbus that the ship could not be saved, and that day, with the generous assistance of the island's cacique, Guacanagarí, the salvageable materials were removed to the *Niña*. So "Holy Mary" went down on her son's birthday. The day following, the Admiral went ashore as the guest of Guacanagarí and rested for eight days. He was feted with sweet potatoes and manioc and—most blessed— some small sheets of gold the natives brought from another island.

When Guacanagarí saw how much these shiny trinkets pleased his guest, he promised more. And when, in his turn, the Admiral displayed to him the destructiveness of his magical weapons, a bargain occurred to both. The stranger would leave behind a garrison to protect the natives from the raids of the cruel Caribs, who had already driven the Arawaks here from islands toward the south; in exchange for this the stranger would be given more of the great gold. Now the stranger would return in his two remaining ships to his king for instructions and a new supply of gifts, and, as the Admiral's bastard son, Ferdinand, later put it, shipfuls of "reinforcements" and people "to advise him in all matters respecting the occupation and conquest of the country."

This is how this first red and white treaty of the New World was sealed: On those verdant and pearly shores, the Admiral ordered a lombard shot fired through the remnant timbers of the surf-washed *Santa Maria*. There was a sound like dull thunder and then a stone ball hissed for the first time over those waters to smash through the flimsy wood. A small white cloud of smoke rose in sudden tribute and drifted with its shadow over the water toward the island, but the natives never saw it, for with the sound and the smash they had fallen on their faces in the sand. It was the kind of triumph Columbus had in mind, because it had inspired fear. He could now in confidence weigh anchor from Puerto de la Navidad, Christmas Harbor, leaving behind a first beachhead: Pedro Gutiérrez and Rodrigo de Escobedo and thirty-six Christians in a fort made of the timbers of the fabled *Santa Maria*.

Such were the beginnings of the "new chance" vouchsafed a civilization enervated by its commitment to a history that precluded genuine renewal. Yet to understand the tragedy of the missed opportunity it is necessary to believe in the sincerity of the Christ-Bearer and his associates and successors. Poised off the Azores for his triumphant return to civilization in the early spring of 1493, Columbus wrote this to his sovereigns:

So all Christendom will be delighted that our Redeemer has given victory to our illustrious King and Queen and their renowned kingdoms, in this great matter. They should hold great celebrations and render solemn thanks to the Holy Trinity with many solemn prayers, for the great triumph which they will have, by the conversion of so many peoples to our holy faith and the temporal benefits which will follow, for not only Spain, but all Christendom will receive encouragement and profit.

This is a brief account of the facts.

If, in view of our knowledge of Columbus's private notations and subsequent actions, we regard such a document as evident sophistry, we miss the truth of the tragedy. For not only were the Arawaks and other cognate island tribes about to be victimized by the restless onwardness of these Christian soldiers, but the soldiers themselves, in failing to sense the opportunity for what it might have been, condemned their civilization to go on with the march. Cortés, Coronado, Soto, Raleigh, Hudson, La Salle—the list goes on, nor does it end with the dead planet of the moon. No, Columbus was wholly sincere, believing to the end of his days that the enslavement, exploitation, and extirpation of these naked ritualists conferred strength and new vitality upon Christian civilization. And so were the others, though, as we shall see, it was *possible* then to conceive otherwise.

Meanwhile, amidst majesterial receptions and parades through wondering street throngs who regarded the specimen "Indians" the Admiral brought back as people from outer space, preparations went forward for a second voyage. No flimsily funded three-ship operation this time. Seventeen ships assembled at Cadiz, three of them transports, with twelve thousand to fifteen thousand men and horses and other domestic animals for settlement. A constellation of famous men shipped out to add their somber glows to this emerging history: Diego Columbus, the Admiral's younger brother; Juan Ponce de León; Juan de la Cosa, the famous cartographer; Alonso de Hojeda, an athlete of arms fresh from the Moors; the father and uncle of the historian Las Casas; Dr. Diego Chanca, whose long letter is one of the expedition's principal documents. As they weighed anchor at Cadiz, September 25, 1493, some Venetian galleys happened into the harbor and joined their cheers to the others so that it seemed in this moment to a contemporary observer that all Europe was united in its hopes.

But if Europe was still hopeful, the New World was not. The white men's reentry was a grim one. In the first days of November they cruised off the shores of Guadalupe, marveling for a moment, maybe even against their wills, at a magnificent cataract that poured itself almost into the surf. They saw houses there, and the Admiral ordered a light caravel to land and investigate. No welcoming beach throngs here, however, nor smiling swimmers shiny with brine. The natives fled before the approach of the boat, and in their houses the foraging Europeans found dismembered arms and legs. These were the dreaded Caribs from whose cannibal raids the Admiral had sworn to protect the Arawaks.

They sailed on toward the colony of La Navidad on Hispaniola, the Admiral naming the islands as they went—St. Ursula, the Eleven Thousand Virgins, San Juan Baptista. Then they came to Hispaniola, but before reaching the colony, they anchored two leagues distant at a harbor the Admiral called Monte Christi. The crew went ashore to inspect the land, river, and shoreline as a potential village site. Near the place where the river debouched into the sea they found something: two corpses partially interred in the sand, mud, and water. They had been there long enough for their racial identity to be blurred, and they were naked. However, one had a Castilian rope around his neck, and the other—Santa Maria!—had his arms tied to a pole "as if on a cross."

This was on the first day. On the next, they went farther upstream in a landscape that suddenly seemed devoid of life and movement, but for all that, full of menace. Here they made another find: two more corpses, "one of which was so well preserved that it was possible to see that he had been heavily bearded." So wrote Dr. Chanca, adding that they all began to suspect the worst.

They went on up the coast, anchoring off La Navidad as night came on. The Admiral ordered two lombard shots fired as a signal to bring the Christians down to the beach. Nothing, no response. No fires could be seen through the breaks in the foliage. Some three hours later a canoe came out to the Admiral's ship, and in the lantern light the Admiral recognized a kinsman of the cacique Guacanagarí and allowed the Arawaks aboard. What they were able to relate through signs and a few catch words was that some of the Christians had died of disease and because of quarrels that had arisen among them. Also, the village had been attacked by two hostile caciques and Guacanagarí had been wounded. Then the Indians left in their canoe, returning through the darkness to the island. That night was like a shroud.

On the next day the Admiral ordered a small reconnaissance, and the party found the ragged, scattered remains of that first New World outpost, the stockade made of ship timbers burnt, the village empty, bits of Spanish clothing scattered amidst the grasses and charred wood. Unlike those friendly natives of the first voyage, the few they now saw flitted behind cover, nor would they come to Spanish summonses.

The next day the Admiral himself conducted a fuller investigation, and some natives, lured closer by presents of hawk's bells, beads, and other trinkets, showed where eleven Christian bodies lay beneath a thin blanket of sod. Within a few days the story was reasonably clear. The colonists left behind as an outpost of progress had become utterly abandoned. They had gone into neighboring territories in search of gold and native women and had taken what they could find—three and four women apiece. This first exploitation had its inevitable ending. Caonabo, first of the Americas' patriot chiefs, took summary vengeance, slew the whites to the last man, and burned the settlement. It was unclear to the Admiral whether Guacanagarí had had a part in all this. So when those cannon shots had been fired across the evening waters toward the colony named in honor of Christ's redemptive birth, there could have been no answer. La Navidad had borne no such hopeful issue here, only the sterile emptiness of rape.

There was nothing to do but begin again, though the new beginning would obviously have to be on different terms. Such native "treachery" had removed the enterprise's flimsy cover of amity. Close to Monte Christi, where the first telling corpses had been found, the Admiral supervised the construction of a new town, Isabela. Then he organized the first systematic searches for gold. Reports came to him of great finds in the adjacent province of Cibao, and he now began to think in terms of a third voyage in which he would bring back mining equipment and personnel. If the gates of perception had ever been opened, they were now bleakly shut. Exploitation showed itself nakedly as the whites roved through the native territories and villages as if these were uninhabited. When they did take note of the native presence, it was with the curled lip of disgust.

These people, so like animals, say they wish to become Christians, noted Dr. Chanca, "although they are actually idolators. There are idols of all kinds in their houses. When I ask them what these are they answer that they belong to *Turey,* that is to say the sky. I once made a show of wanting to throw these in the fire, which so upset them that they were on the point of tears." A Ramon Pané, who described himself as a "poor anchorite of the Order of St. Jerome," provided a primitive ethnographic report on the native culture; it is chiefly of use to suggest how hopelessly unequipped these Westerners were for this strange encounter. Pané gave garbled renditions of the Arawakan gods and mythology, including brief mention of Yocahú and Atabeyra, and in the fashion that was soon to become standard, his own inability to understand these ancient signposts came to be interpreted as *prima facie* evidence of the foolish, devilish nature of the native beliefs.

The Christ-Bearer too made his ignorant remarks on the native religion, attempting interpretation of those conical carvings (the cemies or zemies) that represented Yocahú and other members of the pantheon. He noted the brutish resistance of the natives to the Christians' attempts to desecrate these and the houses of worship. "It once happened," he wrote,

> when they were most suspicious of us that some Christians entered an image house with them. A *cemi* [carving] emitted a shout in words of their language, and this revealed the fact that it was hollow. A hole had been cut in the lower part, into which a trumpet or speaking tube had been fitted, which communicated with a dark corner of the house obscured by branches. Behind these branches stood a servant who spoke whatever words the *cacique* wished, as we do through a speaking tube. Our men suspected this, kicked the *cemi* over and discovered the contrivance we have described. When the *cacique* saw that we had discovered the trick, he most insistently begged our men not to reveal it to his subjects or anyone else, since by this device he kept everyone obedient to him. We can therefore say there is some tinge of idolatry, at least in those who do not suspect the tricks and devices of their *caciques,* since they believe that it is the *cemi* that speaks and the deception is fairly general. The *cacique* alone knows and preserves their superstitious belief, by which he exacts all the tribute he desires from his people.

This is not cynicism—not yet. It is instead pathetic evidence of the Western divorce from the lost world of living myth. As if blinded, the Christ-Bearer and his men could not see in these cemies, and the shamans who stood behind them interpreting their messages, vital adaptive responses to the realities of these island worlds, responses that had allowed these people to live in simple contentment for a thousand years, disturbed only by the predacious Caribs. Nor were they prepared to understand that the mythological mind is fully capable of conflating the human behind the mask with the presence of divinity. The Christians could only attribute to the shamans and caciques the same base motives of those churchmen who were the custodians of their own dead mythology.

And then the Admiral really was blinded. Returning from Jamaica and Cuba (which he had *forced* his crew to swear was the mainland of China), he was, his son wrote, "afflicted by a serious illness, something between an infectious fever and a lethargy, which suddenly blinded him, dulled his other senses and took away his memory." It was the onset of a madness that intermittently possessed him to the very end and from which he was never wholly free. It could hardly have come at a worse moment: the little community of Isabela was coming apart like a rotting shoe. The town, placed in fatal ignorance of the land, was a pesthole, and malaria was at large in its muddy streets. So were the mutinous designs of men who had come out from civilization in expectation that the wilderness would instantly yield its riches so that they could return home and enjoy them in comfort. Individual parties of settlers were fanning out through the island, looting and raping, while native resistance was organizing behind the caciques Caonabo and Guatiguaná. Through it all the leader lay powerless, until finally in the spring of 1495, weak though he was, he momentarily unified the community by directing all its hostility outward. In the first major engagement between whites and natives in the New World, Columbus led his forces down into those very gardens that had once seemed so paradisal. With blaring trumpets, drums, savage dogs, and harquebuses, the mounted men swept in and crushed the Arawaks, here as against the Caribs simply too gentle. Caonabo was captured and resistance crumbled to nothing. Still the forces went relentlessly inland, establishing a line of garrisons, demanding tribute, and at last enslaving.

In the wake of this complete victory Columbus established a system of labor and tribute that is the precursor of the *encomienda* ★ system that was to take the lives of millions of natives here on the islands and on the mainland: in the province of Cibao, where the richest gold fields were presumed to lie, each person above the age of fourteen was required to pay "a large belly-full of gold dust" every three months; elsewhere they were to pay tribute in cotton. Of this Ferdinand Columbus wrote:

> . . . in order that the Spaniards should know what person owed tribute, orders were given for the manufacture of discs of brass or copper, to be given to each every time he made payment, and to be worn around the neck. Consequently if any man was found without a disc, it would be known that he had not paid and he would be punished.

★ Royal land grants to be worked with resident slave labor.

The results were immediate and spectacular. Following his father's notes, Ferdinand writes that the fortunes of the Christians improved markedly, and for the first time since the return to La Navidad it was possible for a Christian to go about this country without fear of ambush. In fact, wrote Ferdinand, the Indians would carry him on their backs wherever he wished to go. The Admiral, he concluded, attributed this peace to God's providence, since without such divine help the tiny band of Christians could hardly have subdued so numerous a people. Plainly, God had wanted these natives "beneath His hand."

As for what remained of the symbolic career, call it madness, mutiny, and despair. There was a third voyage in 1498, which ended when the sovereigns, alarmed by persistent rumors of chaos in their new kingdoms, dispatched Columbus's dispossessor, Francisco de Bobadilla, to investigate. But when Bobadilla entered the New World, he saw no signs of opulence; instead he saw the huddled huts of white men and amidst them a high gallows from which swung several Spanish corpses. The Admiral had had to turn cruel dictator in order to enforce his enfeebled will. Shortly thereafter Bobadilla completed his investigation by sending the Admiral and his brothers home in irons.

However severe a humiliation this was for the man who seven years before had returned to Spain in triumph, riding the streets at the right hand of the king himself, it was nonetheless clear to almost everyone else connected with the enterprise of the Indies that Columbus was used up. His mind, which in its ordered arrangement had seemed so uncannily that of a whole civilization, now in its derangement revealed something more of that civilization. On this third voyage he had brushed against the edge of the mainland and there had been pushed seaward by the might of the Orinoco as it emptied its vast system into the ocean. Yet he had refused to recognize so sure a geophysical sign for what it was, writing instead to his sovereigns that this was probably the mouth of one of the four rivers of the earthly paradise in which God had placed the Tree of Life. The shape of this new hemisphere is not as others claim, he continued, but instead is rather like a woman's breast. I am sailing gently up it but do not expect that any man can get to the nipple, for that is Paradise itself, from which we are barred, as is said in the Old Testament. Yet the highnesses would attain a sort of temporal paradise through this exploring, for this is a land of huge population, "all of whom should pay tribute."

Incredibly, two years after his return in disgrace he went back again, and the nightmare deepened. Disregarded, broken in mind and fortune, barred from even entering the harbor of the colony he had founded, he was sent where he could do damage only to himself and his crews. For a whole year he lay marooned at Jamaica, the hulls of his ships besieged by a world massed against its first violator. While the teredo worms ate at the wood of the Old World, Columbus raved on to Ferdinand and Isabella about the nearness of the Ganges River and Cathay, classical and medieval geographical theory, seas of blood boiling like a cauldron on a mighty fire, conversations with the Almighty about his holy mission to these Indies, a mountain with a sculptured tomb atop it, inside which a corpse lay embalmed, apes with human faces and hogs with tails like tentacles. He revealed that the bequest of gold from King David to Solomon had been gold from these Indies. Here Jerusalem and Zion should be rebuilt by Christian hands. But as for the actual settlements, "they are in a state of exhaustion. Although they are not dead their sickness is incurable or at least very extensive. Let him who brought them to this state produce the remedy if he knows it. Everyone is a master of disorder." He begged his sovereigns' pardon for such agitations. They were, he said, the inevitable results of the strains of discovery, of being surrounded by a million hostile savages full of cruelty, and all that gold. . . .

The historian of the first phase of the exploration of the New World, Bartolomé de Las Casas, wrote of Columbus that "since it is obvious that at that time God gave this man the keys to the awesome seas, he and no other unlocked the darkness, to him and no other is owed forever and ever all that exists beyond those doors." It is not clear from this passage whether Las Casas meant this as an indictment of the Admiral or as praise for his singular efforts. He certainly knew enough of the realities of the post-Columbian New World to have intended the former, for he was an eyewitness to developments in the islands from about 1500 on. In either case—indictment or praise—the attribution is too much. Even as Las Casas witnessed the establishment of Western rule among the native kingdoms, it was obvious that Columbus had been merely the instrument of great historical forces that long preceded his career.

Yet it is also true that under his fitful leadership and with his suggestions this first effort at colonization took on enduring shape. Such was the sense of urgency to fill up and exploit these wilderness spaces that neither the crown nor the Church nor the Admiral and his administrative successors took the slightest care for the character of the colonists. Columbus had approved the use of, and had sailed with, criminals and hoodlums who took more than the usual liberties with the new lands and peoples. To be sure, the Admiral had commented a number of times on the necessity of sending out colonists of good character and strong religious conviction. But always this desideratum yielded to the stronger imperative of personnel who could man and enforce the sprawling operation. Within the lifetime of the Admiral the Indies became the New World analog of the Crusades where, under the cover of righteous Christian outreach, criminal rapaciousness had sanction.

With greedy haste and abandon the unprepared colonists came out to the islands and hied themselves off to the hills with picks, pots, and pans in expectation of instant golden rewards. The trails to the goldfields swarmed with eager figures, and the graveyards of the settlements almost as quickly received them back again—victims of poor diet, worse hygiene, and the climate. Busy as the miners were, the priests were busier still, trying to keep liturgical pace with the death rate.

And yet, great as was the mortality rate of the whites, still more appalling was that of the natives, who were now being transformed into the machines of Western commerce. As laborers in the mines, as beasts of burden, and as field hands, they lived and died under conditions that might be called "inhuman" were they not so lamentably familiar a feature of human history. Let us then call them "ghastly" and observe that families were torn asunder, workers separated by long distances and longer intervals of time from their tribes and relations and worked ceaselessly on a meager diet of cassava and water. The native women fared little better under this regime than they had when that first group of rapists had thought to establish itself at La Navidad. Under such debilitating and demoralizing conditions the native birth rate, first on Hispaniola, then on Jamaica and Cuba, dropped steeply. When the slaves grew too weak to work efficiently, their masters summarily dismissed them to wander off and die on the roads or in the woods. One Spanish word the natives quickly learned was "hungry," though it did them no good. Starving, worn-out slaves were often encountered on the roads, pitifully and uselessly pronouncing it, *"Hambriento. Hambriento."*

Gradually, like a spreading stain, the slaving expeditions had to go farther and farther out from Hispaniola, Jamaica, and Cuba to find and catch natives to take the places of those so speedily used up. Now the practices developed by the Portuguese in Africa were put to New World use. The holds of the slave ships were stuffed with human cargoes and on the journey to the colonies corpses were thrown out onto the waters in such profusion that it was said a course could be charted by this floating refuse alone, a kind of dead reckoning. In a few years the Spanish had to go as far as the mainland, searching for replacements, but here the native resistance was too strong, the territory too large and unfamiliar, and they had to give it up for a time.

Around 1516 sugar mills were established on Hispaniola, Jamaica, and Cuba, and these began to compete with the mining operations for the drastically reduced labor supply. One solution was the importation of Africans. They seemed hardier than these Indians, more adaptable, less prone to lapse into suicidal despair, perhaps because they were so far removed from the heart-wrenching sights and sounds of their former freedoms, whereas the natives were still surrounded by vestiges of their pre-Columbian realities.

No one at the time seems to have found this new development a terrific indictment of what the New World had been turned into, not even Las Casas, who was one of those who suggested it—though he did so for complex reasons. When he had come out to the Indies in the still roiling wake of Columbus, he had settled on Hispaniola and had accepted his allotment of natives without question—the standard practice a scant eight years after initial contact with the New World. Ten years later, Las Casas played a part in the "pacification" of Cuba and was rewarded with a sizable *encomienda*. He was by this time a priest, but it was hardly unusual for a churchman to take an active part in the destruction of the heathen. At this same time, however, Las Casas's conscience was pricked by some Dominican sermons he heard preached against the universal practice of indentured, interminable servitude.

For four years Las Casas meditated on these unrecorded words and then in 1514 dramatically renounced his *encomienda* and preached his own sermon against the destruction of the Indians. Yet such was the nature of the faith that he could still advise the importation of blacks as a means of freeing the reds. Only when he had the opportunity to witness the same dehumanizing results with this new labor did he inveigh against it too, wondering as he did so whether his earlier ignorance would be sufficient excuse in the eyes of God.

For half a century then this solitary man buttonholed and petitioned princes, prelates, and colonial functionaries on this single subject, for he was one of the few of his time to understand the meaning of the New World. Las Casas had come to see the New World for what it should have been—itself—and to recognize it as an enormous spiritual opportunity. Just as he had come to understand that enslavement is destructive of *any* race of human beings, so he gradually moved away from feeling that the New World was an enormous evangelical opportunity toward the conviction that it was a mysteriously granted gift through which Christianity could recover its primitive vigor. Whereas he had once pitied the Indians in their gentleness and simplicity, increasingly he was drawn to an admiration for the *rightness* of their cultures. At last he would venture that it would have been better for their souls never to have been touched by the Christians—surely a blasphemous judgment for one of his position and time. God, he wrote, did not want the extension of Christianity at such a cost. And he would go even further in his late years and state that God's judgments against Spain could be seen in the inflation, political corruption, and anarchy of the colonies.

Though he could never bring himself to admit that there was something radically wrong with Christianity itself, Las Casas became adamant in his conviction that it was surely diseased in its New World practice. Though he could not admit that something in the history of this religion made men callously commit great cruelties in its name, he could name those cruelties and he came very close to discerning one of their causes: the divorce of the faith from the human body. The Christians despise these natives, he wrote, because they are in doubt as to whether they are animals or beings with souls. Such a distinction is foreign to the world of living myth, and if Las Casas could never make his way back to that, he was nevertheless quite certain that for all their attachment to nature, to the earth, these peoples had immortal souls. In 1551 he argued this very proposition against Spain's most distinguished theologian, Juan Ginés Sepulveda, who held the Aristotelian notion that the natives as brutes were fit objects for enslavement.

Thus, through Las Casas, we can see something of the specific beginnings for us in the Americas of that estrangement from place that is so odd and pervasive a feature of our common life. Attachment to place, or better, a sense of the *spirits* of places, was spoiled by such beginnings as those Las Casas had participated in and later bore witness against. His people cared nothing for the lands or the peoples they moved through; even their own settlements were chimerical and ephemeral to them. "A Spanish town," Las Casas observed, lasted only so long as the Indian population, and the Spaniards abandoned it the minute they saw an open door, such as going to San Juan, Jamaica, Cuba or the continent to wage cruel wars against the Indians and cast them in mine pits, only to start again elsewhere after exhausting both gold and captives.

From such an outset the New World became a haunted place of violated spirits, of the ghosts of destruction. Isabela, where Columbus had wished to begin anew after the abortion at La Navidad, was abandoned in its turn and became the first of our many "ghost towns," those rotted monuments to greed and haste and estrangement. Residents of Hispaniola, said Las Casas, were terrified to go near Isabela after it had been abandoned:

> Reports circulated about the horrible voices and frightening cries that could be heard day and night by anyone who happened to pass near the town. There was a story about a man walking through the deserted town and coming upon people lining the streets on both sides who were dressed like the best of Spanish courtiers. The man, awed by this unexpected vision, greeted them and asked where they came from. But they kept silent, answering only by lifting a hand to their hats as a sign of greeting, and as they took off their hats, the whole head came off so that two files of beheaded gentlemen were left lining the street before they vanished altogether.

Despised and ridiculed in his time as a fanatic and troublemaker, and castigated in our own as a pathological liar, ideologue, and irresponsible exaggerator, the creator of the "Leyenda Negra" (Black Legend of Spanish cruelty), Las Casas remains for us a vivid testimonial that it was possible for an individual to sense his way through soldered armor of his faith to the green springs of renewal that await true discovery. The existence of Las Casas, Justin Winsor wrote, is the indictment of Columbus. Let us ease the Christ-Bearer of some of that burden if we can, and say that the existence of Las Casas is the indictment of the Christian West.

# Penetration

At the outset of his iconoclastic study of the historiography of the Americas, Edmundo O'Gorman quotes the phenomenologist Martin Heidegger to the effect that only what has been conceived can be seen, and that which has been conceived has been invented. Thus, argues O'Gorman, the key concept for the historian is "invention" rather than "creation," since the latter implies the production of a something out of a nothing. In the history of humankind, he writes, nothing has been produced *ex nihilo*. Each new phenomenon, whether mechanical contrivance or mental concept, is the product of its past. Parturition may be by accident or design, but in either case the new thing will have been prepared for and will have to be understood within the context of that preparation. And if some new thing, happened upon by accident, should indeed be wholly new, unexpected, and unlooked for, it will be forced to fit a previous conceptual mold, yielding its own nature to the human imperative to account for the unknown always in terms of the known. The Americas, he writes, were not so much discovered as invented, the Admiral to the end of his days professing that he had discovered the "Indies."

O'Gorman is surely right about the Admiral. Following those jagged lines of uncharted coasts where the waters changed from enchanting aqua to impenetrable white to formidable black, he had to invent it all. And he invented out of what his civilization had bequeathed him— the classic geographers, his Portuguese predecessors, and behind all, Christian history. There was little in this to prepare him for what he found, and still less could it have been adequate viaticum for those lesser men who now rushed in where he had led.

Though it became progressively obvious during the Admiral's last years that these islands were not the Indies, it was by no means clear what else they might be. Though the life blood of Christianity had been leeched away, the Church remained an imposing authority, and the Church had no way to admit these new lands and their natives for what they were. The Genesis narrative of Adam, Noah, and the dispersal of their kind over the habitable places of the earth made no allowance for the New World. What O'Gorman shows is that gradually the mind of the West opened wider to accommodate the New World by tacitly ignoring the theological obstructions, just as it had earlier ignored the official attitude toward the shape of the earth. Sailing, as it were, around the landmass of Augustine, the West had reencountered the speculations of the ancients, who had divined the existence of lands westward. Eventually the Church itself had to get around the obstruction of Genesis, and the route it took was conversion. The New World was conceived of as a battlefield whereon the True Faith would once again have the opportunity to vanquish evil and wrest from its grasp souls otherwise lost.

So, when the Indians were admitted to be what they seemed—non-Orientals—and their lands for what they were—tropical islands—and the landmass was dimly perceived for what it was: when all this *had* to be admitted, then the New World was invented. As O'Gorman says, it came to be a feature of the European mental landscape, emerging out of the seas of ignorance just as Columbus's flotilla had once raised it from their dripping bows. But though O'Gorman wants us to see this invention as an inspired, supreme effort of the Western mind, the brute facts of the invention tell us something else. Let us think here on those deflowered islands and on our contemporary archeologists picking down through the silences of the old aboriginal sites to exhume if they can the pathetic, broken remains of Arawakan cemies and construct from them tentative theories of the preinvention past. On all those islands whose green groves once whispered unintelligible things to the Admiral and his men, not a single aborigine remains to guide those who would understand a lost world.

In the earliest stages of the defloration of the islands Hernán Cortés grew to young manhood as an unemployed man-at-arms. Apparently, he had been destined by his parents for the law, but after a brief period of study at Salamanca, he quit and returned to his home in Medellín. According to his secretary-chaplain-biographer, Francisco Lopez de Gómara, the young man at this time was "a source of trouble to his parents as well as to himself, for he was restless, haughty, mischievous, and given to quarreling, for which reason he decided to seek his fortune." There were, so Gómara says, two ways open at the time for this: one was to go to Naples under *el Gran Capitán,* Gonzalo Fernandez de Córdoba; the other was to go out to the Indies under Nicolás de Ovando. But owing to an injury incurred in an amorous adventure and to a quartan fever, Cortés missed his tides to both places, and it was not until 1504 that the nineteen-year-old made the trip to the Indies. When he arrived in Santo Domingo some still believed Cuba to be an arm of the mainland, a mainland that was then but a strong rumor. Like most settlers, Cortés had come out believing that gold nuggets lay scattered like stones through the land, and he was determined to pick up his share and go back. Like most, he endured a period of want while trying to become wealthy.

In this same year of 1504 Cortés had the opportunity to glimpse the Admiral himself as he passed through the port on his last return, maddened, mocked, issuing suggestions and orders that no one would heed. It is not known whether the young man was chastened by the sight of this figure, broken by the weight of his find, but it is probable that he was, since he patiently bided his time for some years thereafter. He accepted with grace his allotment of lands and Indians, and rose through the ranks of colonial administration. He was notary public of a small town, then secretary to the island's governor, Diego Velásquez, and at last treasurer of the king's island income. At twenty-six Cortés was a man of consequence who had proved his right to respect on the battlefields of punitive expeditions against the remaining pockets of native resistance, as well as in the rooms of counsel and counting. Meanwhile he observed the ill-prepared thrusts at the mainland by Diego de Nicuesa and Alonso de Hojeda. These accomplished nothing except to strengthen the emerging impression of the immensity of the lands beyond the islands and the military competence of their natives. Whatever else this New World might be, it was a place of death and terror where Christians might be swallowed into the maw of the wilderness just as the Israelites of old had once been threatened in their passage to promise through the desert.

The tales of these trips brought back to the islands were of such a tone as to inspire dread and hostility. On the attempt to slave-raid and colonize near Cartagena, Hojeda had lost more than seventy men, including the famous cartographer Juan de la Cosa, who, Las Casas says, was found by a search party tied to a tree and so swollen with poisoned arrows that he looked like a porcupine. Of the Nicuesa part of this undertaking it was said that a remnant of the party, wandering lost along a blistering coast, had become so reduced that they had eaten the corpse of an Indian they had found. Fatal New World food! They all died of it.

Such things could not have been lost on Cortés, a man of singular shrewdness. But then neither could the persistent rumors of the land's great wealth, and so deeply did these filter into him that he dreamed of sharing in it. Salvador de Madariaga in his biography of Cortés quotes an old source for this visionary experience: "One afternoon, as he had fallen asleep, he dreamt that suddenly, having shed his old poverty, he saw himself covered with rich cloth and served by many strangers who addressed him with words of great honour and praise. . . ." Cortés is reported to have told his friends subsequently that he would either eat with trumpets or die on the gallows.

So, he waited while further expeditions organized by Governor Velásquez went out under inferior commands and came back leaky, battered, and empty-handed, yet carrying rumors of wealth like a low-grade infection. One of these expeditions, under Juan de Grijalva, appears to have made unwitting contact with ambassadors of a powerful inland potentate. These ambassadors carried "cotton wraps" on which were depicted various indecipherable exploits that seemed in some way to include the white, bearded strangers to whom they were shown. But almost all of this was lost in linguistic gropings, and the expedition's priest recorded that Grijalva had said that all he wanted was gold.

Like a child shooting one arrow after another, hoping to find the lost one, Velásquez dispatched still another expedition to prowl the coasts of the mainland and to find out what had happened to the two groups still away from port. For this one, he chose Hernán Cortés, now a hardened Indies veteran of thirty-three. Velásquez, fat and inept, could not have known that he had happened upon the man who would thrust the West's iron point into the heart of the New World.

Whatever the governor may have ordered, Cortés thought beyond the terms of any commission. He thought in terms of invasion and conquest, and we have the word of one of his soldiers, Bernal Díaz, that as soon as Cortés had been appointed captain general, "he began to search for all sorts of arms, guns, powder and crossbows, and every kind of warlike stores which he could get together. . . ." And he had two standards and banners made bearing the cross-bordered legend, "Comrades, let us follow the sign of the holy Cross with true faith, and through it we shall conquer." In the dawn hours of November 18, 1518, they sailed.

First an extraordinary piece of business occurred on the island of Cozumel off the Yucatán coast. One of the charges of the commission from Velásquez (who had been left impotently raging on Cuba at the stealthy leave-taking of his captain general) was to search for certain Spaniards believed left behind in the earlier expeditions of Nicuesa and Balboa. Strange that in all the mad haste to plunder the wealth of the New World there should have been this great concern for a handful of missing white men—strange especially in the somber light the old accounts shed on the cruel treatment the Spaniards visited upon one another. The way to understand this, I think, is in terms of the deep spiritual resistance the whites had formed to this wilderness they had blundered into looking for the cities of the Great Khan. To be *lost* in this! The idea thrilled them with horror.

One day a large native canoe approached Cozumel from the direction in which Cortés had made inquiries after the rumored whites. The canoe beached, and one of its occupants cried out in Spanish as rusty as a disused saber, "*Dios y Santa Maria de Sevilla!*" Andrés de Tapia brought the man into the presence of the captain general, but before Cortés would recognize this creature as a man of his own world, he ordered him clothed in a shirt, doublet, and drawers. Bernal Díaz records that the man's name was Jeronimo de Aguilar, a man of holy orders who had been captive in these parts for eight years until he had received word of the mission to find him and his Indian master had graciously allowed him leave to rejoin his own kind. It was a bit of Providence to find one of the missing whites in this unfeatured landscape, and just how large a bit quickly became clear to Cortés when he learned that Aguilar could do the expedition immense service as an interpreter.

But there was another Spaniard in this same area, and this other one, hearing of the search for him, refused to come out. His name was Gonzalo Guerrero, and he was now a cacique. When Aguilar had brought him news of the search party, Guerrero had replied that he was now married, with sons, and that his people considered him very brave. Besides, his face was tatooed and his lip pierced. He would stay where he was. And, Aguilar added, he thought that a year ago this Guerrero had advised the natives to attack some Spanish vessels at Cape Catoche. How could this be understood? How could a Christian refuse deliverance out of darkness? It was for shame, says Gómara, shame at having married into the tribe. But how could such a one turn renegade, advising war on his own kind?* Surely this was madness, the proved temptation of the devil in his wild haunts. "I wish," exclaimed Cortés, "that I had him in my hands for it will never do to leave him here." But there was no way to forcibly bring him out, and so on March 14, 1519, they weighed anchor and sailed northward, leaving Gonzalo Guerrero behind, a precursor of that curious class of willing captives.

Eight days later Cortés made his landfall on the mainland at a place known from the previous coastings of Grijalva. Close by was the town of Tabasco, and its inhabitants now crowded the shores and the banks of a river, armed and fully intent that the strangers should not land. But these strangers were even more determined to do just that. Before the inevitable hostilities the Christians felt compelled to spell out across the tense distance the precise terms of the encounter. The instrument of this was the so-called *Requerimiento,* an official weapon drafted by the Spanish Council of the Indies for the arsenal of exploration. Las Casas renders its essence as follows:

> In the name of King Ferdinand and Juana, his daughter, Queen of Castile and León, etc., conquerors of barbarian nations, we notify you as best we can that our Lord God Eternal created Heaven and earth and a man and woman from whom we all descend for all times and all over the world. In the 5000 years since creation the multitude of these generations caused men to divide and establish kingdoms in various parts of the world, among whom God chose St. Peter as leader of mankind, regardless of their law, sect or belief. He seated St. Peter in Rome as the best place from which to rule the world but he allowed him to establish his seat in all parts of the world and rule all people, whether Christians, Moors, Jews, Gentiles or any other sect. He was named Pope, which means admirable and greatest father, governor of all men. Those who lived at that time obeyed St. Peter as Lord and superior King of the universe, and so did their descendants obey his successors and so on to the end of time.
>
> The late Pope gave these islands and mainland of the ocean and the contents thereof to the above-mentioned King and Queen, as is certified in writing and you may see the documents if you should so desire. Therefore, Their Highnesses are lords and masters of this land; they were acknowledged as such when this notice was posted, and were and are being served willingly and without resistance; then, their religious envoys were acknowledged and

---

* The word "renegade" is a part of this history. Originally it meant any Christian who converted to another faith, especially to Islam. In Spain, many Christians had been forced into such a conversion during the centuries of Moorish domination. The word came to be applied to those who in the wake of reconquest failed to reconvert, as many did, especially in Granada.

obeyed without delay, and all subjects unconditionally and of their own free will became Christians and thus they remain. Their Highnesses received their allegiance with joy and benignity and decreed that they be treated in this spirit like good and loyal vassals and you are under the obligation to do the same.

Therefore, we request that you understand this text, deliberate on its contents within a reasonable time, and recognize the Church and its highest priest, the Pope, as rulers of the universe, and in their name the King and Queen of Spain as rulers of this land, allowing the religious fathers to preach our holy Faith to you. You owe compliance as a duty to the King and we in his name will receive you with love and charity, respecting your freedom and that of your wives and sons and your rights of possession, and we shall not compel you to baptism unless you, informed of the Truth, wish to convert to our holy Catholic Faith as almost all your neighbors have done in other islands, in exchange for which Their Highnesses bestow many privileges and exemptions upon you. Should you fail to comply, or delay maliciously in so doing, we assure you that with the help of God we shall use force against you, declaring war upon you from all sides and with all possible means, and we shall bind you to the yoke of the Church and of Their Highnesses; we shall enslave your persons, wives and sons, sell you or dispose of you as the King sees fit; we shall seize your possessions and harm you as much as we can as disobedient and resisting vassals. And we declare you guilty of resulting deaths and injuries, exempting Their Highnesses of such guilt as well as ourselves and the gentlemen who accompany us. We hereby request that legal signatures be affixed to this text and pray those present to bear witness for us, etc.

Whatever we may think of these assumptions and of their terrible consequences, however cynical they may seem, it is clear that at the levels of law and historical consciousness they were implicitly accepted by captains and spear carriers alike. These men marched to this Christian version of human history, fought under its banner, believed themselves agents of its destiny. The *Requerimiento* was not merely a grotesque artifact of the Hispanic legalistic temper, nor an old-fashioned scheme for real-estate development. Rather it was a milestone in the onward march of a religion fleeing its primitive beginnings and poised for a civilizing mission in these strange, vast lands.

Of course, there *was* cynicism in this document and its use in the Indies. These fierce, bearded men knew that the Latin language of the instrument could not be understood by those to whom it was read except in those very rare instances when an interpreter was available. They must have known too that the natives' incomprehension would inevitably make them "delay maliciously" in complying with the document's terms, and that this would furnish the pretext for attack. But such cynicisms were ancillary and must be understood as a consequence of the decay of Christian myth, for where myth is rationalized into history, cynicism may be expected to appear also.

Here on these shores the Tabascans had the benefit of the interpreter Aguilar, but the translation of this new version of their existence only seemed to anger them more. So the next morning Cortés drew plans for the attack of the town from two sides. Seeing the Spaniards' determination, the natives attacked in their canoes and out of the shoreline mangroves, sounding their drums and blowing their trumpets. The Spaniards got ashore with difficulty, and the future conqueror of the Americas' most powerful native empire arrived on the mainland minus a shoe he lost in the mud. Having achieved a beachhead, the Christians rallied under a hail of arrows to their own martial music, "Santiago, and at them!" They pushed the natives back from the beach and the river, and after a hard fight they drove them out of the town. That night they slept in the main square surrounded by the natives' temples and idols and their own sentries.

The battle continued the next day on a plain outside the town, the Tabascans reinforced by recruits from neighboring settlements. They fought desperately with their fire-hardened wooden spears and their paddle-shaped swords edged on both sides with obsidian and were beaten only when Cortés ordered the cavalry into action. The horsemen circled unseen behind the natives and then charged them across the flat plain. The warriors had never seen horses before, and they broke and ran, the Spanish crossbowmen and musketeers doing effective work in the rout. Because it was Lady Day in the Christian liturgical calendar and because they had gained this first victory through her, the conquerors christened the town Santa Maria de la Victoria. By the campfires they bound up their wounds and seared them and those of their horses with the fat of an Indian corpse.

Negotiations opened on the next day amidst the bearing away and burning of the hundreds of native dead that littered the plain. Forty caciques gathered to sue for peace with the strangers, and Cortés took the opportunity to bear even harder on the technological leverage his side enjoyed. He told the assembled native leaders that for their treacherous and obstinate behavior they ought all to be put to death. And, he added with a wink to his men, his iron machines and savage "deer" were still very angry. At this moment his concerted demonstration began with the ignition of the fuse to the largest cannon. It exploded, remembered Díaz, "with such a thunderclap as was wanted, and the ball went buzzing over the hills, and as it was mid-day and very still it made a great noise, and the caciques were terrified on hearing it." Then followed the demonstration with the "deer," a stallion being brought into proximity with a mare that had recently foaled. The effect of this was even better than that of the iron machine, but now Cortés soothed the caciques with a promise of protection, provided peace could be speedily concluded. Plainly the leaders had much to think about as they returned to their people that evening.

On the next day the first treaty of the mainland was concluded in very much the same fashion as that sealed by Columbus at La Navidad. The thoroughly cowed caciques brought presents of quilted cotton and twenty women, among whom was one of noble bearing and obvious intelligence who was to serve the Spaniards as chief interpreter in the ensuing months of conquest.* Doña Marina, they christened her, and the cool captain general, always with an eye to the control of his own forces as well as the natives, handsomely gave her away to Alonzo Hernandez Puertocarrero. Later, when Puertocarrero went back to Spain, Doña Marina lived with Cortés and bore him a son.

The peace deputation brought something else as well—gold work: a couple of coronets, some masks, and some cunningly fashioned little animals. Cortés asked them where they had procured these, and they pointed westward toward the setting sun. "Culúa," they said. "Mexico." No one present was quite sure what that meant except that it was a place where they would surely have to go.

And there was a final condition: Cortés through Aguilar described the particulars of the True Faith and commanded the Tabascans and their allies to give up their idolatry. Two carpenters constructed and erected a very tall cross, and as it was now Palm Sunday, Cortés compelled the caciques and their relations to attend Christian services in the main square of what had been their town. With the defeated natives looking on in bewilderment, the conquerors marched in solemn procession to the cross, decked it with flowers, and kissed its new wood. They sailed early Monday morning.

Through four days they hugged the coast northward and on Holy Thursday made port in the shelter of a barrier reef. As they floated at anchor looking across the sun-bright waters to the low sand hills

* Las Casas makes the sarcastic observation that her linguistic abilities were hardly needed since all the Spaniards ever said to the natives was, "Take this, gimme that."

topped with salt grass, two large canoes of natives came to meet them. Doña Marina recognized the language of their occupants as that of Mexico. They told Cortés then that they were ambassadors of their great lord Moctezuma who had dispatched them to discover what manner of men these might be.* There was a plentiful display of gold among them, and they readily told Cortés that it came from their great lord. "Send me some of it," Cortés replied, "because I and my companions suffer from a disease of the heart which can be cured only with gold." (Words, as Freud observed, even in the vagaries of wit, are always faithful to some reality.) The remark was noted by both sides, further courtesies were exchanged, and the ambassadors departed. On Good Friday, 1519, the Christians went ashore to the sand hills they now called Vera Cruz and set up their cannons.

As he surveyed his smallish retinue on those hills, with mosquitoes stinging the horses and men and the tips of the high mountains far inland just visible, Cortés could not have known what he was on the verge of. But there was never any doubt in his mind that this expedition would turn its face away from the open sea and march onward toward those magnificent peaks until some great new thing was discovered. He had now passed under his own thumbs and fingertips evidence of golden plunder that lay somewhere within the unimaginable spaces he gazed toward, and even if they were to encounter a huge host guarding the gold, these men had learned the truth of the Spanish proverb forged in centuries of such conflicts: "The more Moors, the greater the spoils."

Here is what the leader had at his command:

> 508 soldiers
> about 100 shipmasters, pilots, sailors
> 32 crossbowmen
> 13 musketeers
> 4 falconets
> some brass guns
> much powder and ball
> 11 ships, large and small, plus a supply launch
> 16 horses

* I have tried to steer a consistent course through these difficult orthographical seas but have found it impossible. Generally, I have yielded to common practice in the spelling of names. Thus "Columbus," which probably ought to be rendered "Colón," after the Spanish fashion he himself adopted—or at least "Colombo." So too with "Coronado," who ought to be referred to as "Vásquez." But more obvious corruptions like "Montezuma" for "Motecuhzoma" or "Moteçuçuma" or "Moctezuma" have been discarded.

Díaz characterized these last wonderfully by gender, disposition, and utility: "The Captain Cortés: a vicious dark chestnut. . . ." "Alonzo Hernandez Puertocarrero: a gray mare, a very good charger which Cortés bought for him with his gold buttons." "Juan de Escalante: a light chestnut horse with white stockings, not much good." "Baena: a settler at Trinidad: a dappled horse almost black, no good for anything."

And this is what he was facing: the Aztecs.

Lévi-Strauss has called them an "open wound" on the flank of the Americas. There was surely plenty of blood, and anyone trying to understand the meaning of these confrontations in the New World must be concerned with the scale of violence practiced by these people of whom the resolute captain general knew nothing.

Until the middle of our fourteenth century no one else knew much about them either. Maybe there was not much to know, because before this time the Aztecs were but one of the many wandering, hungry hunter tribes that troubled the centers of aboriginal settlement in the Valley of Mexico and Central America—another instance of the old antagonism between settled peoples and the nomadic atavisms who periodically appeared like blights on the landscape.

In the Valley of Mexico the Aztecs were the last such nomadic group to break in upon the settlements; last because here the old process of invasion, conquest, assimilation, and evolution was about to be shockingly disrupted by the white men from the east. Like other such groups throughout the world, the Aztecs appear to have gone through the transformation from the worship of feminine deities to the worship of aggressive masculine ones. In her superb study of Aztec thought and history, Laurette Séjourné makes this case and shows that it was under the aegis of the newly-supreme war god, Huitzilopochtli, that these savage strangers shouldered their way into the valley and eventually to the shores of Lake Texcoco. Here Huitzilopochtli's message to his people became the starkly simple dictate of conquest: willpower and strength of arm must win for you an empire out of the lands of others.

Yet, at first the Aztecs had been more persecuted than persecuting, and it is probable that they were driven away from one city-state after another on their way to the lake shores. In 1325 they chose a swampy bit of land in the lake that the neighboring tribes had considered unfit for settlement and erected there a few wretched huts and an altar to Huitzilopochtli. Little more than a century later this ragged group of outcasts had achieved complete independence from their proud and established neighbors. In the next century they emerged as the tyrants of the valley, extending their rule of fear and tribute from coast to coast and as far into Central America as Guatemala.

As Séjourné and the distinguished student of Nahuatl culture Miguel Leon-Portilla argue, this phenomenal career must be understood ultimately in religious terms. These scholars claim that in their nomadic trek through the valley the Aztecs had been deeply impressed by the beauty of those features of Olmec and Toltec cultures that still survived among the living Nahuatl peoples. They themselves spoke the Nahuatl language, and gradually they set about the business of attaching themselves to honored traditions that originally had not included them. They sedulously practiced exogamous marriage, relating themselves in this way to the established cultures they would now rule, using their wives as tutors to their children. They chose as their first king a noble of Toltec origin from the lakeside city-state of Culhuacan and put him to stud with selected Aztec women. His numerous descendants formed the nucleus of Aztec aristocracy. They established too in their central city of Tenochtitlán (Mexico City) and elsewhere groups of scholars whose task was to master, preserve, and transmit the religious forms of the Toltecs.

Leon-Portilla writes that the major figure in this process of cultural assimilation was the royal counselor to the kings Moctezuma I and Axayacatl, a singularly power-oriented man named Tlacaelel. Twice (1440, 1469) on the deaths of rulers he refused the throne, preferring to dictate his designs of aggression and assimilation from beside it. It was he, says Leon-Portilla, who convinced the Aztec rulers that they had a historical mission to subdue their neighbors, who identified this mission with the dictates of the war god Huitzilopochtli, and who managed the elevation of this nomadic deity to the leading position in the reformulated Nahuatl pantheon. It was Tlacaelel who devised and presided over the destruction of the ancient narratives and the reworking of the history of the Aztec people so that it now revealed the ineluctable and divinely ordained march to their present position as lords of the valley and beyond. At this point the prophecy of Huitzilopochtli had been fulfilled: "The four corners of the world shall ye conquer, win, and subject to yourselves . . . it shall cost you sweat, work, and pure blood. . . ."

Most of the pure blood—but not all—was now being drawn from the captive bodies of prisoners of war or slaves to feed the insatiable appetite of the Sun. In the emerged Aztec religion the power of the Sun would fail without this constant flow of blood and the world would come to an end, for each individual was a dismembered particle of the Sun and so must return to it. Thus, Séjourné writes, the Aztecs held it a sacred duty to wage unceasing war against their neighbors and relations to furnish the blood that would keep the cosmic wheel oiled. With one city-state, in fact, a state of constant warfare was artificially perpetuated since it was clear to all that the Aztecs could have overwhelmed the Tlaxcalans had they wished to end things with them. Instead, it better suited their purposes to war periodically with Tlaxcala and so furnish bodies, blood, and trials for their own young warriors. Nothing else could be so sacred, and in the name of this terrible duty all else was sacrificed: personal liberty and life, the honor of treaties and alliances, and the hearts of tens of thousands. The Aztec religious calendar was an endless round of sacrifice—so many this day, so many on that—and the inner walls of their magnificent temples were like giant scabs of the blackened blood of unnumbered victims. Death kept watch over this life and marked its passage with awful regularity.

Through the work of scholars as well as through our knowledge of what was now about to happen as Cortés turned his face toward Tenochtitlán, we can see that the history of the Aztecs reflects this people's inability to absorb the spiritual essence of the ancient cultures they so much admired. By the time the Spaniards made contact, the Aztecs had succeeded in mastering many of the ancient technics and traditions of that Olmec/Toltec past; so much, in fact, that both they and their conquerors believed that the empire, its art, architecture, designs, and myths, were indigenous. Yet the evidence is that there is little warrant in the Olmec or even Toltec past for warfare and sacrifice on such a scale. The Aztecs had only superficially assimilated the ancient religious traditions and then perverted and betrayed them in the service of their temporal designs. They imitated the exterior of what they so rightly admired and set their wise men to the study of the spiritual past, yet behind the façade of reverence for the principle of life their roving war god, Huitzilopochtli, still reigned supreme.

"No society is perfect," writes Lévi-Strauss. "Each has within itself, by nature, an impurity incompatible with the norms to which it lays claim. . . ." Everyone who has confronted the phenomenon of Aztec culture has been struck by its incredible contradiction: great beauty and utter cruelty. I remember once a group of us Westerners staring in silent fascination at the dramatically highlighted exhibit of an Aztec sacrificial knife in a Paris museum. It was beautiful: beautiful in its balance and symmetry, in its jade handle with precious stone insets . . . and in its wide and tapered obsidian blade with which the priests had once hacked the palpitating hearts out of their victims.

The "resolution" of the contradiction lies in the betrayal and perversion of the spiritual traditions of the past. Their residual power conferred beauty on the productions of these ruthless usurpers, but there was terror and tyranny within. Séjourné illustrates this betrayal best in the process by which the Toltec god Quetzalcoatl was gradually supplanted by Huitzilopochtli.

Quetzalcoatl appears as the major figure in the religious thought of ancient Meso-America, a combination god and culture hero. Creator of man out of his own blood, he also provides the staff of sustenance, maize, and like many other such figures he saves mankind when that sustenance is maliciously hidden away by other beings. He is the original artificer, showing humans how to work in textiles, feathers, and stone, and the originator of science, teaching his charges how to chart the course of the stars and record this in calendrical form.

Quetzalcoatl's legend insists most strongly on his spiritual purity and on his intense desire to maintain this condition. Yet there exist in this mythic world demonic powers angered by Quetzalcoatl's purity, and they devise a scheme whereby he is made aware for the first time that he possesses a body. Horrified by this revelation, he gets drunk and has sexual intercourse with a beautiful maiden. In penance for this fall he abdicates his throne and kingdom, going away to the shores of the divine water, and there, in all his feathered finery, he sets himself ablaze. Out of his ashes all manner of rare birds appear, but dawn itself does not. During four black, dawnless days Quetzalcoatl undertakes that mythic journey to the land of the dead. When he appears again it is as a bright star, the Lord of Dawn. And someday, says his legend, he will return to claim his kingdom.

What this narrative seems to speak of is that familiar theme in the world's mythologies—the resolution of opposites, body and spirit. Quetzalcoatl is first all spirit, then all body, but both are necessary to his drama of transfiguration. Séjourné remarks on the myth's suggestion that only through a reconciliation of supposed opposites can man rediscover the unity that should be his estate. This, of course, is the holistic wisdom of myth: how to honor the body and yet know it as a part of the divinely created world. Or, as Séjourné puts it, it is only through matter that the spirit world may be visited.

The legend is a long way from the tyrannical rule of the Aztecs. And yet there was some vestigial reverence for it in Tenochtitlán, where Quetzalcoatl was invoked, adored, prayed to even in the midst of the daily round of sacrifices. Though the god had been displaced by the vicious Huitzilopochtli (so lacking in wisdom other than military), all the externals of Aztec culture paid a kind of subversive tribute to his power. For it was not Huitzilopochtli who had authored all this beauty, but the Lord of Dawn. It was as if the Aztecs had cannibalized the Toltec god just as the Spaniards said they cannibalized the flesh of the sacrificed.

What we seem to be confronted with here is the spectacle of a culture that had given itself over to a historical mission in a world that knew no such thing as linear history and was still devoted to the cycle of myth. At the same time, this culture was powerfully influenced by the ancient examples of myth it encountered in the Valley of Mexico. The Aztecs may have seen themselves as marching toward their destiny over roads paved with the dismembered corpses of their sacrificial victims, but they paid their own sort of tribute to the mythic past they imitated in style and structure. In fact, they were still in some ways captive to that violated past, as Cortés and his forces were now to reveal unwittingly.

This cultural condition may help to explain the striking contradiction of Aztec culture—its beauty and its bloodiness, its magnificent technical achievements and its crucial spiritual failures. It would help to explain also the spectacularly rapid conquest of the valley's established city-states, for as the Aztecs were to find out, those still in the grip of myth are never military matches for those devoted to a historical mission. Always in such confrontations the iron tip of destiny penetrates the sacred round.

The internal conflict between historical mission and assumed mythic heritage became evident only after the Aztecs consolidated their rule over millions of subject peoples. Then, in the reign of Moctezuma II, a series of bad omens began to oppress the minds of leaders and commoners alike. Ten years before Cortés established his beachhead at Tabasco, the Aztecs had recorded the appearance of something like a flaming ear of corn in the heavens—or as it was later remembered, "a wound in the sky." After this the omens appeared with terrifying regularity. The temple of Huitzilopochtli spontaneously exploded in flames and burned to the ground. Another temple was badly damaged by a lightning bolt. On a sunny day fire suddenly streamed eastward through the sky, dropping showers of red-hot coals. In the lake of the capital the waters unaccountably lashed themselves into a frenzy and hurled themselves against the houses, wrecking them. Night after night the people heard a weeping woman pass through the city's hushed streets exclaiming, "My children, we must flee far away from this city!" Or at other times, "My children, where shall I take you?" It was said that this was Cihuacoatl, the ancient earth goddess. One day fishermen caught in the lake a strange cranelike bird the color of ashes. It bore a mirror in the crown of its head, and when it was brought to Moctezuma he looked into the mirror and saw the night sky, though it was now broad noon. He looked away aghast, then looked again. This time he saw a distant plain and moving across it people in ranks. As they hastened toward him, they warred among themselves and rode on the backs of great deerlike creatures. The ruler called for his wise men and asked them to explain this vision, but when they looked into the mirror they could see nothing. Soon thereafter monstrous two-headed men appeared in the streets of Tenochtitlán, but when they were taken to Moctezuma they vanished.

Moctezuma was awed and depressed by all this, especially since the omens were repeatedly connected to failures of his own rule. He alternated between severe self-imposed penances for his transgressions and cruel reprisals against those who brought him news of the omens or else failed to interpret them satisfactorily. He ordered the execution of the noble Tzompantecuetli and his sons because this man had refused to return a good answer to Moctezuma's question of how best to decorate the temple of Huitzilopochtli. Instead, he had responded that Huitzilopochtli would not long be their god but that another was coming who would be lord of all.

Now Moctezuma sought to evade the terrific responsibility that seemed his: the dissolution of the empire. He wanted to die, and he sent emissaries to the Lord of the Land of the Dead bearing gifts of flayed human skins. But the messengers returned from the temple with another negative augury: the Lord Huemac had refused them. Even death was closed off, so Moctezuma ran away from his palace and hid on a small island in the lake where he was discovered by one of his ministers. Then this man whose face none dared look upon except by permission, whose feet were never allowed to touch the earth, was led in shame back to the throne, his prison. By the time the Spaniards anchored off Tabasco, a resolution of a sort had evolved out of despair and the ruler was determined to see what would come.

What came was a report from the seacoast of moving mountains and strange men fishing off them, translated by Leon-Portilla as follows:

Our lord and king, it is true that strange people have come to the shores of the great sea. They were fishing from a small boat, some with rods and others with a net. They fished until late and then they went back to their two great towers and climbed up into them. There were about fifteen of these people, some with blue jackets, others with red, others with black or green, and still others with jackets of a soiled color, very ugly, like our *ichtil-matli* [maguey fiber shoulder cloak]. There were also a few without jackets. On their heads they wore red kerchiefs, or bonnets of a fine scarlet color, and some wore large round hats like small *comales* [flat pottery dish], which must have been sunshades. They have very light skin, much lighter than ours. They all have long beards, and their hair comes only to their ears.

Moctezuma was silent, but he was thinking. His thoughts were in themselves the single most impressive testament to the latent power of the violated myths of the valley. He was thinking that perhaps these strange white beings from the east were the returning god Quetzalcoatl and his retinue. Perhaps this was the meaning of the series of omens—the comet streaking eastward, the vision in the bird's mirror, even the cryptic remark of the murdered Tzompantecuetli. Quetzalcoatl had promised to return and to claim his own, and perhaps now this promise was being fulfilled. From this point on, the curious vacillations of the Aztec ruler in the face of the relentless advance of the whites can be explained by the conflict in Aztec culture between its present position and the mythic past. This was a conflict that Cortés only dimly understood, but in so far as he did so, he exploited it.

The history of the West is replete with ironies great and small, but surely in the history of the New World there can be none greater than this: that it should have been this ruthless and calculating soldier of fortune who would be mistaken for Quetzalcoatl—this emissary of a civilization that had gone so far beyond its own myths as to deliver him here where he might impersonate a native god!

Cortés was a sincere Christian, though somewhat less devout than the Christ-haunted Columbus, and he wore on his person various religious icons, including one of John the Baptist (an appropriate one here in another kind of wilderness). Time and again in his march inland toward the Aztec kingdom he proved his sincerity by his willingness to jeopardize the military and diplomatic successes so precariously won for the instant conversion of the natives. In his first surviving dispatch to his king, Charles V, in July 1519, Cortés stressed the urgent necessity of converting those whom he styled minions of the devil and inveterate sodomites. "It is our hope," he wrote, that His Holiness the Pope "will approve the punishment of the wicked and rebellious, as enemies of our Holy Catholic Faith, after they have first been properly admonished. This will inspire fear in those who may be reluctant in receiving knowledge of the Truth." Cortés and his men had been witnesses to the Inquisition and intimately connected to the wars against the Moors. Behind them were the Crusades and the strangely compelling figure of Raymond Lully. All of these examples showed the way to deal with differences. In a contest between light and darkness, there could be no temporizing. Tolerance or, worse, partial acceptance of differences was the death of belief. In a curious way Christianity had come to impersonate the perverse practices of the Aztecs with Jesus fed and strengthened on the sacrificed blood of unbelievers.

So at Vera Cruz in the first perilous days of trying to establish amicable relationships with the natives, particularly with their Aztec overlords who were there observing on Moctezuma's orders, Cortés ordered his resident friar to give them all a lecture on the True Faith. It was so thorough, remembered Díaz, "that no good theologian could have bettered it." And then Cortés followed this up with a diatribe against the "cursed" practices of the natives—surely a dangerous gamble. On the following morning it seemed he had lost it: not a single Indian remained in camp, for they had all departed during the night in horror at the sacrilegious language of the captain general.

Again, at the town of Cempoal to which he had been welcomed and whose people he had just enlisted in a fragile alliance with other hill towns, Cortés was willing to risk it all for the natives' immediate abandonment of all of their religious rituals. The fat cacique of the town had come out bearing gifts for the white men, including eight maidens. But the leader told him that they could not accept any of the gifts until the townspeople had given up their idols, the practice of human sacrifice, and all that "kept them in darkness." If they would consent to Christian instruction, Cortés promised to make them lords over provinces now under the Aztecs. If not, he promised mortal enmity. The caciques answered that it did not seem good to them to give up their gods and practices, at which Cortés flew into such a fury that he lectured his men to the effect that even if it should cost them their lives to the last man, those idols must come down *this very day*.

The lines were drawn, and when fifty Christians moved toward one of the lofty temples, the caciques ordered their warriors into battle readiness. Cortés through Doña Marina threatened again, adding this time that since these natives had already betrayed Moctezuma, they could expect his vengeance as well as that of the Christians. When the caciques heard this they wept and said that the matter was now out of their hands, and that if the Christians wished to desecrate their idols, they were powerless to prevent it. Here Díaz recalled:

> The words were hardly out of their mouths before more than fifty of us sol-
> diers had clambered up [to the temple] and had thrown down their idols
> which came rolling down the steps shattered to pieces.
>
> .    .    .
>
> When they saw their idols broken to pieces the caciques and the priests who
> were with them wept and covered their eyes, and in the Totonac tongue
> they prayed their gods to pardon them. . . .

Cortés ordered the shattered remains to be taken out of sight and burned while the caciques restrained their warriors only with great difficulty.

Cortés was to act in similar mortal sincerity with his Tlaxcalan and Cholulan allies. Indeed, once he had attained entrance to Tenochtitlán itself, he insulted the Aztecs' gods on his very first visit to their temples—in the presence of the great Moctezuma and surrounded by a swarm of Aztec warriors. So whatever else may be said of this man, his sense of religious mission can hardly be doubted.

That this conviction was allied with the prospect of enormous material gain cannot be doubted either, but such was the state of the faith that—as in the Crusades—its bearers saw no real contradiction between the two. Everyone knew that some sort of emolument was a necessary inducement to extending Christianity into the hazardous wildernesses of the world, and if a certain cynicism lurked about the edges of this admission, it was mitigated by the belief that the light of the faith had to be carried into darkness by men of low degree to whom gain was an essential hope.

Thus the morale of the expedition was immeasurably enhanced on that day at Vera Cruz when Moctezuma's ambassadors returned to Cortés bearing rich gifts. They so inflamed the desires of the Christians to get to Tenochtitlán that not even the most brutal hardships and severe reverses could dampen their ardor. Nor would it be an exaggeration to say that what passed that day into Spanish hands inflamed the cupidity of a whole civilization. For Cortés's second dispatch to Charles V, in which he described the splendors of the Aztec empire, was quickly translated into Latin, French, Italian, German, and English. Here at last was solid proof of the golden rumors of the lands beyond.

And even though these riches did not come from the Oriental treasure houses Polo had described in such tantalizing detail, still in this crucial respect the strange New World was turning out to be true to expectations. Indeed, it might even be more fortunate that these riches came from such savage sources since they could be taken with greater right. Here is a partial inventory of the Aztec ambassadors' gifts as remembered many years after by Bernal Díaz:

> The first article presented was a wheel like a sun, as big as a cartwheel . . . which those who afterwards weighed it said was worth more than ten thousand dollars. Then another wheel was presented of greater size made of silver of great brilliancy in imitation of the moon . . . and the chief brought back the helmet full of fine grains of gold, just as they are got out of the mines, and this was worth three thousand dollars. This gold in the helmet was worth more to us than if it had contained twenty thousand dollars, because it showed us that there were good mines there. Then were brought twenty golden ducks, beautifully worked and very natural looking. . . . Then there were presented crests of gold and plumes of rich green feathers, and others of silver, and fans of the same materials, and deer copied in hollow gold and many other things that I cannot remember. . . .

Significantly, the bulk of this treasure appears to have come from the temple of Quetzalcoatl, and Moctezuma is said to have instructed his messengers that "Our lord Quetzalcoatl has arrived. These jewels," he said, "that you are presenting to him on my behalf are all the priestly ornaments that belong to him."

In delivering such treasures the Aztec lords showed a fatal want of understanding of their visitors. Perhaps they all believed with Moctezuma that the Lord of Dawn would be contented with these dazzling offerings. Or perhaps they thought that these were mere men like themselves, who would now be satisfied and go back to wherever they had come from. The Aztecs could not have known that they were confronted here with the vanguard of a civilization that could quite literally stop nowhere and for whom such mute and tangible items as those now spread out on cotton quilts had come to serve as substitutes for lost, interior riches. Though Cortés smiled courteously when the ambassadors told him that a visit to Tenochtitlán was out of the question, in his mind the matter was settled. He later wrote of this moment, "I decided to go and see him [Moctezuma], wherever he might be, and I vowed that I would have him prisoner or dead or subject to Your Majesty." He was to have him all three ways before it was ended.

It would be an intolerable cliché to write of this moment that "the die was cast" were it not for the fact that Cortés himself resorted to it. He secretly ordered holes bored in all the ships (the work of worms, he said) so that there could be no turning back. He convoked a meeting of the caciques of the hill towns now allied with him to ensure peace at Vera Cruz, and then he gave his men a speech on the duties that lay before them as Christians. He drew many comparisons with the heroic deeds of Caesar and the Romans, and when he had finished, all agreed that the die had indeed been cast for good fortune. In the middle of August 1519 the Spaniards set out for Tenochtitlán with their Indian allies and bearers.

Along the march they added indispensable allies from the city-states of Tlaxcala and Cholula. The former, as has been remarked, were in an artificially maintained state of warfare with the Aztecs, but they were on friendly terms with the Spaniards' allies from Cempoal. They were not friendly to the Spaniards, however, and in two major engagements they came close to so crippling Cortés's forces that he would have been forced to turn back or die there in Tlaxcalan territory. Peace was made just as the battered invaders were wondering whether they would get out of the engagement alive, to say nothing of continuing the march on Tenochtitlán.

In pledging themselves to Cortés the Tlaxcalans excused their late hostility by claiming they had thought the Spaniards friends of Moctezuma. Whether this was only an excuse is not clear, but it is evident that once the two armies began talking, Cortés was able to exploit the great hatred the Tlaxcalans bore their Aztec persecutors. Madariaga is right to call this the turning point in the conquest, for the Tlaxcalans were to provide the crucial manpower for the final taking of Tenochtitlán. But Madariaga and others are wrong in styling this takeover a "conquest," for it never could have occurred had not the Aztecs made themselves so deeply offensive to allies and enemies alike that on balance the newcomers seemed the saviors they claimed to be. Séjourné is much more accurate in suggesting that the "conquest" be regarded as a religious revolt against the Aztec interlopers who had so swiftly established their fearsome hegemony.* This was a revolt of myth against historical destiny, and it is another irony that it should have been fo-

---

* After peace had been made between the Spaniards and the Tlaxcalans the latter revealed that they had been told that one day a host would come from the direction of the sunrise to rule over them. It thus appears that here too the prophetic expectations of the return of Quetzalcoatl played into the designs of the Christians.

mented by those who would soon put an end to myth in all the regions they marched through.

The details of the winning over of the Cholulans remain obscured. The issue is the Spaniards' massacre of natives at Cholula in which perhaps as many as three thousand fell in two hours. What incited this slaughter? Treachery discovered in time by Cortés? Lies told him by Tlaxcalan allies seeking revenge? Or simple military expediency? We shall never know, but it is certain that in the hush that followed this carnage the waiting Moctezuma knew that his last barrier against these advancing creatures had been removed. He had tried everything: presents to propitiate them; excuses that he could not receive them because it was not the right time, or because he was ill, or because he was too poor to host such august visitors (but the presents had belied this last excuse); military plots; and finally magical incantations. Now the massacre at Cholula had opened the road to the heart of the city. As the Aztecs remembered it:

> When the massacre at Cholula was complete, the strangers set out again toward the City of Mexico. They came in battle array, as conquerors, and the dust rose in whirlwinds on the roads. Their spears glinted in the sun, and their pennons fluttered like bats. They made a loud clamor as they marched, for their coats of mail and their weapons clashed and rattled. Some of them were dressed in glistening iron from head to foot; they terrified everyone who saw them.
>
> Their dogs came with them, running ahead of the column. They raised their muzzles high; they lifted their muzzles to the wind. They raced on before with saliva dripping from their jaws.

There was nothing left for Moctezuma but to meet them. As the Europeans moved along the broad causeway to the city, the size of their undertaking began gradually to dawn on them, and they wondered if they were not living out something from the pages of a Spanish chivalric romance. It was so huge and beautiful and unexpected here in this savage land. Prepared on the islands, they had not expected the towns they had met on their march inland. But the towns had not prepared them for this. Now they saw all of it—towered, lake-set, white in the sun, the causeways, side streets, and canals choked with anxious dark people gazing toward the tiny band and their Indian allies. The Europeans saw too the great, the magnificent lord of it, Moctezuma, advancing toward them along the causeway, feathered and jeweled like the incarnation of Quetzalcoatl. His shaded litter stopped some yards from this farthest

Western penetration, and the king alighted on robes flung down to cushion his sandaled feet against the earth. The captain general moved forward to embrace the king, but Aztec attendants prevented his presumption: Cortés had to be content with giving Moctezuma a necklace of glass beads, only slightly more valuable than the usual trading truck his people had learned to give away in such encounters. Then they spoke, through interpreters, across distances: the white man of his sworn mission from his king, the native ruler in the captive syllables of that mythic past from which he could not quite free himself. Face to face with the impersonator of that past, Moctezuma described himself and his people as but temporary interlopers in realms that had always awaited the return of the true lord. And so, he concluded, whether you are He Himself or ambassadors of Him, all that you see here is yours by right. Then Moctezuma led the Spaniards through the crowds to their palatial lodgings, the treasure house of the kings.

Five days later Cortés stood with Moctezuma atop the great temple that overlooked the humming capital. The king pointed out the four quarters of the city and their attractions, the marketplace far beneath them and the bowered canals. Behind them were two blood-spattered sanctuaries, one devoted to that Huitzilopochtli in whose name so much of this had been accomplished. Cortés suggested to his friar that this might be a good time to ask the king whether a church might not be constructed just here. When the friar, with somewhat better sense, replied that he thought the time not ripe, Cortés overruled him and spoke insultingly to Moctezuma of these devilish idols, wondering aloud how so great a ruler could be so credulous. Then, Madariaga quotes him as remarking to a group of his men:

> "What do you think, gentlemen, of this great favor which God has granted us? After having given us so many victories over so many dangers, He has brought us to this place, from which we can see so many big cities. Truly, my heart tells me that from here many kingdoms and dominions will be conquered, for here is the capital where the devil has his main seat; and once this city has been subdued and mastered, the rest will be easy to conquer."

What else *could* have happened other than the enactment of this prophecy? Though Moctezuma was quickly disabused of any lingering thoughts that Cortés might be Quetzalcoatl, it would have been too late to turn to this god in any case: futile against the Christians, and impossible in a landscape rendered rebellious by generations of tyranny. He and his people would have to resist as what they had become.

Appropriately, war began during the feast in honor of Huitzilo-pochtli. The captain general was away from the capital squashing a force sent from Cuba by Governor Velásquez to arrest him for disobe-dience. While Cortés was suborning and subduing these forces led by the inept Pánfilo de Narvaez, his deputy in Tenochtitlán, Pedro de Al-varado, treacherously and rashly attacked the Aztec celebrants in the midst of their rituals. An Aztec account of this tells how, just

> when the dance was loveliest and when song was linked to song, the Span-iards were seized with an urge to kill the celebrants. They all ran forward, armed as if for battle. They closed the entrances and passageways, all the gates of the patio. . . . They posted guards so that no one could escape, and then rushed into the Sacred Patio to slaughter the celebrants.

They cut off the arms of the ritual drummer and then his head, which rolled across the floor. Others were slashed in the stomach and tripped on their own dangling intestines as they tried to run away. The Span-iards pursued the survivors into the ceremonial houses and there con-tinued their work. "They ran everywhere and searched everywhere," says the account. "They invaded every room, hunting and killing." Then the whole city turned against them, besieging them in their palace within whose walls lay the manacled Moctezuma, now only a cipher.

By forced marches Cortés arrived back in Tenochtitlán on June 24, his troops caked white with dust and sweat. On the 25th the Aztecs stormed the palace and within five days had forced the Christians and their allies into costly retreat. In the treasure room of the palace Cortés supervised the hasty loading of as much gold and jewelry as horses and Indians could bear and then set out into the murderous night. At the bridges the Aztecs caught them and many of the Christians died there, plunging through the gaps of the wrecked bridges to the waters below, where they were too weighted with plunder to swim to safety.

But the following April the relentless Spaniards were back at the city's gates, this time for good. They had brought overland the fittings and timbers for brigantines that would control the lake while their in-fantry laid siege along the causeways and streets. They burst the pipes of the city's water supply, (a maneuver perfected during the sieges of the Crusades) and cut off all supplies of food into the city. Soon there was no fresh water within, only stagnant pools or brine. The people ate lizards, swallows, mice, and even the salt grass that grew on the margins of the lake. Smallpox worked in concert with the invaders. The disease had been introduced to the mainland by a black slave of the Narvaez forces, and as the natives had never known it, they died in great numbers. Here as elsewhere the Old World was this New World's plague, the very organisms tolerated in one sphere fatal in another.

They fought on through May, through June, and over the bridges through July, until it was street by street in the heart of the city, then house by house until the survivor Aztecs were crowded atop each other in one of the city's quarters. Cortés several times suggested peace to them, but they would not surrender, and so, as he wrote to Charles V in his third dispatch, "I came to these conclusions: first, that we would recover little if any of the treasure that had been taken from us; and, second, that they would force us to destroy them totally. This last caused me the greater sorrow, for it weighed on my soul."

The siege was completed on August 13, 1521. After ninety-three days a terrific silence fell upon the shattered, charred hulk of the great metropolis, its streets clogged with putrescent corpses. Cortés and his officers stood talking under a colored canopy on top of one of the remaining houses while a bow shot away were the huddled, cloaked figures of the last warriors. Then the captured king, Cuauhtemoc, was brought before the captain general. Moctezuma had died in that first siege, either killed by his own people, as the Spaniards claimed, or assassinated by Toledo blades, as the Aztecs said. His successor, Cuitláhuac, had died a victim of the plague. Now there was only this slim young man who is reported to have said to Cortés, "I have done everything in my power to save my kingdom from your hands. Since fortune has been against me, I now beg you to take my life. This would put an end to the kingship of Mexico, and it would be just and right, for you have already destroyed my city and killed my people."

But Cortés only smiled and patted the young man on the head and motioned him to take a seat beside him under the canopy. It was late in the day, and it had begun to rain. . . .

The war was not quite over. There remained the obligatory period of looting and raping and some further killing by the Indian allies who now took vengeance with impunity for the generations of Aztec cruelty. But the Aztec accounts stress the brutal greed of the Christians as they picked through the city's ruins and felt through the survivors' ragged clothing for hidden nuggets or gems. They felt through the hair of the women, fondled their breasts, thrust horny hands under the hems of their dresses. Those they found especially attractive they set aside for more leisurely attentions. All this was capped by a riotous victory banquet Cortés gave for his army. There was great drunkenness on Castilian wine and other forms of debauchery. It was not quite Jerusalem after its glorious retaking by the Crusaders, but for the New World it would more than suffice.

Now it was over, and the slow process of invention could proceed in the husk of Tenochtitlán as it had on the islands: with native labor. During the nine-month period when plans for the final taking of the city had gestated, the first step had been the branding of Indians who had at one time or another "unlawfully" resisted the invaders. The Christians burned "$\mathcal{G}$" for "Guerra" into them, and continued the practice in the sanguinary aftermath of victory. Some conquerors did not even bother to brand in their haste but merely slapped painted signatures on their chattels. No matter. The markings, however affixed, were to be indelible.

As Cortés oversaw the process by which "New Spain" emerged out of old Mexico, enslavement was regularized. He wrote to Charles V that he had regretfully been "forced" to compensate his Christian troops for their valiant services with gifts of Indian laborers. "There was nothing else I could do," he wrote, "not only for the maintenance of the Spaniards, but to assure good treatment of the Indians." This regularized apportionment, the same system in force on the islands, had in fact been predicted by the Aztecs themselves. In the hellish days of the siege amidst the sounds of burning and crashing buildings, the defenders had taunted their native adversaries by telling them that no matter who won the war, the natives now destroying the city would have to rebuild it. "It pleased God that this . . . should turn out to be true," wrote Cortés, "for they are indeed the ones who have to do this work." Later he euphemistically described the extension of slavery throughout New Spain as "putting those Indians under the imperial yoke of Your Majesty." Those who had suffered under native masters were now betrayed by their alien deliverers.

Much of the treasure the Spaniards had been forced to leave behind on the so-called *Noche Triste* of June 30, 1520, was not recovered, though the young king who had received the patronizing head pat had been tortured to reveal its whereabouts. But the exploitation of the natives and their lands compensated for these lost bars of gold and silver. Cortés instituted the mining of humbler but useful metals like tin and iron ore and exacted metals, cocoa, and cotton from the Aztecs' tribute provinces, using the royal income books as his guides to locations and appropriate returns. And surely there were other lands adjacent to New Spain that would be found rich in precious metals. He wrote to the king, assuring him that once the peoples of these lands were pacified, "our settlers say they will take over the mines and reduce the people to slavery."

Meanwhile, it was necessary to govern his own men, an irksome task for one of such martial energies, especially since the conquerors showed no more disposition to stay on these beaten lands than they had in their island phase. Like Columbus before him, Cortés now had to enforce unpopular rules, and the settlers, he confessed, "are not very pleased with some of them, in particular with those which bind them to strike root in the land; for all, or most, of them intend to deal with these lands as they did with the Islands first populated, namely to exhaust them, to destroy them, then to leave them."

After all, why not? Rootlessness had driven them here to begin it all, and even so sagacious a man as Cortés could not legislate against a disposition sunk beneath the reach of any laws. Madariaga remarks that Cortés was captive to his own conquest. So were they all, for as the captain general had once written in approbation of their heroic efforts, "we were only doing what we had to as Christians."

# The Lost Colony

At the end of Paul Green's play about Raleigh's "Lost Colony" at Roanoke, the scared and starving colonists are shown in the act of abandoning their settlement on the island in the hope that they will find temporary succor with some friendly Indians on the nearby island of Croatoan. John Borden, a colonist elevated to command through a series of disasters, reasons to the huddled group that their position here has become untenable and that the best chance for the survival of the colony lies at Croatoan; if they can hold out there until Governor White returns from England with supplies, then the grand venture can continue, and their sufferings will not have been in vain. The burden of his argument is that the colonists have a sacred trust to keep the settlement alive. It is a tiny outpost, to be sure, on a dark and dangerous seacoast, but it is the very first English settlement and a light in that darkness that will guide the many who shall come after them. If that light should go out, they will have proved faithless and their terrific sacrifices will be robbed of meaning. "We'll never yield," Borden yells over the confused murmur of the colonists. "We'll carry on the fight—on Croatoan—in the wilderness—wherever God sends us—and to the last man."

Then, in relief against an American night sky, the colonists are shown marching off into the "vast unknown," their voices trailing away in a somber hymn. The only light left onstage is a shaft that plays on the flag that flutters above the now-deserted chapel. The lights of the amphitheater come slowly on, the play is over, and another of the hundreds of audiences who have seen this summertime production files musingly out into the carefully tended woodlands that stand on the very site of the colony.

This is Fort Raleigh, a designated Historic Site, maintained by the National Park Service on North Carolina's Outer Banks. Members of the audience who wish to can walk the "Thomas Harriot Nature Trail" that leads to the shore of Roanoke Sound. There the surf slams almost against the very back of the stage, giving an impressively realistic sense of the sea-girt fastness of the old outpost.

In the daylight you can walk the beaches and look across Croatoan Sound to the low mainland with its scattered clumps of trees and marsh grasses lining a flat horizon. The nature walk back takes you through a maze of loblolly pines oozing their aromatic gum and live oaks writhing low over the loose, sandy soil that is traced by vines, creepers, and a species of the yucca plant. Nowhere on this northwest tip of the island can you escape the sound of the surf pounding muffled through the woods. And here, perhaps under your very feet, is the site of the village the English built and then abandoned under circumstances probably very much like those Paul Green has imagined in his play. But no chronicles of that failed colony have survived, and to our day the precise location of the village has, perhaps perversely, remained undiscovered. So too has that of the Indian village supposed to have been close by it.

Between the amphitheater and the visitor center is the site of the fort, located by archeologists in the late 1940s. Its earthworks have been restored, partly on the basis of the local evidence and partly on that of a similar fort the commander Ralph Lane built on Puerto Rico. It looks now like a lopsided star listing eastward, and the earthworks have been gentled by the passage of time. The center of the fort is emptiness, and standing there looking at it with the sound of the surf and the random rustle of leaves and branches in your ears, it is not hard to imagine the brooding sense of calamity that must have sat on Governor White and his search party that August day in 1590 when they surveyed the tenantless ruins. There is still an air of mystery and tragedy at Roanoke, the air that Green wanted to get into his play, the air that obscurely lures American tourists here in the summers. How terrible it is, we feel, to have vanished so utterly as to leave behind only the pathetic talismans preserved in the glass cases of the visitor center: a twisted nail, some shot, a pipe stem.

There is indeed tragedy here at Roanoke Island, but it is not what Green and his audiences have supposed. It is not the tragedy of the failure and disappearance of the colony, for the long view discloses this to have been but a momentary setback in the successful campaign of English colonization. The tragedy here is of far greater dimensions, encompassing both those vanished Europeans and the equally vanished natives who once called these sandy spits home. It is the tragedy of misperception, of fearful ignorance, of arrogance and baseless reprisal. Most poignantly, it is the tragedy of a failure of vision and of that missed opportunity for spiritual growth and renewal that comes but once to a civilization. We have seen how on island and mainland the Spanish had been unequal to their great finding. How melancholy but understandable to find that the English experience with the New World is of a piece with that of their archrival. The place names the English gave to America were different, but the map was the same.

So much of the historiography of exploration has been concerned with the rivalries of the several nation-states that only grudgingly and over the reasoned objections of specialists do we even entertain the proposition suggested above and adumbrated at the outset of this essay: that the states of the West had vastly more in common than they had dividing them. Of this preoccupation with international rivalry, Francis Parkman's histories of the contest for North America can serve as examples. Parkman presents the contest in the form of a mortal, gigantic chess match played out between European superpowers on the scarcely sensed board of a wilderness continent. Yet what comes through his pages is something other than intended, for juxtaposed against the lands they schemed about, the European rivals appear about as indistinguishable as Parkman found those tribes whose lives were pawns in the game. In reading his histories it is easy to forget whether it was the French or the English in this or that action because in fact they were both engaged in the same thing, in the same ways, possessed of the same religion, the same symbols, the same technology. Even in enmity the states of the West would still sell each other the military hardware that would make overseas conquests frighteningly easy.

Surveying Europe at the beginning of the age of exploration, A. R. Hall observes that beneath the "inevitable gradations in civilization . . . lay an essential cultural unity, undisturbed by nationalism or religious schism." The intellectual life of all Europe, he continues, was then as homogeneous as that of the modern nation-state. From the point of view of what I attempt in these pages, this is the essential matter. Thus I find the perspective of the brilliant Brazilian historian Darcy Ribeiro most useful, for he writes of the Americas from the point of view of the colonized, not the colonizer. To face *outward* from the Americas toward the alien crafts of the invaders is a salutary alteration of vision far more significant than the blurred international distinctions. The grand phenomenon is that so strongly phrased by Ribeiro:

> The history of man in these last centuries is principally the history of the expansion of Western Europe, which, constituting the nucleus of a new civilizational process, launched itself on all peoples in successive waves of violence, cupidity, and oppression. In this movement the whole world was shaken up and rearranged according to European designs and in conformity with European interests. Each people, even each human being, was affected and caught up in the European economic system or in the ideals of wealth, power, justice, or health inspired by it.

Never, he concludes, in the known history of the world has any other civilization shown itself so "vigorous in its expansionist energy or so contradictory in its motivations. . . ."

Ribeiro sees the first wave of this expansion as a mercantilist invasion made possible by significant advances in ocean navigation and in firearms. As we have noted, the West invented few of the technics and implements of either. What it did do was to gather to itself the inventions of others and extend them to fuller realizations and ever more ingenious applications. Why it did this lies beyond Ribeiro's concerns but not ours. With the spectacle before us of the invasion of a new world by a whole civilization, we are led as by a dowser's wand beneath politics, technics, and economics to the world of the spirit. And the spirit of the West is to be found in the history of that religion to which all Europe subscribed.

Christianity, as we have seen, had a unique orientation to the world, an orientation that emphasized the capacity of rational thought to render Christians lords of all earthly creation. In the age of exploration Christians of all nationalities and persuasions were united in a conception of the earth as a divinely created *thing,* there for the enjoyment, instruction, and profit of man. Though the nearest derivation of this view seems to be Augustine, who viewed the world as of no intrinsic interest, its ultimate derivation is Old Testament scripture as rendered through Christian exegetes. There, in the deeply incised record of a new monotheism turning away from the worship of the natural world toward the adoration of a god so otherworldly that his name could not even be written down, is the beginning of the superimposed sacred history.

Max Weber, for one, traces the West's gradual, inexorable elimina-

tion of the magical or numinous from the world to the influence of the
Old Testament, and he finds enormous entailments to Christianity's developed view of the world as neutral and even empty of all spirit life.
To Weber this view resulted in the conception of the world as an open
field for such human activity as might be pleasing to a god infinitely removed from it. Here human ingenuity and restless creativity might
enjoy almost limitless freedom, governed only by the increasingly qualified stricture that such behavior not work unnecessary hardships on
fellow Christians. Such a nonsacramental world, bereft of spirit, its
gods and sacred groves and megaliths reduced to euhemeristic ciphers,
or else banished to devilish realms, could pose no resistance to those in-
tensive investigations of nature that ultimately resulted in the West's
celebrated ability to expand.

A. R. Hall points out in his study of the scientific revolution
(1500–1800) that it was precisely this world view that progressively
eroded the authority of Hellenistic science. Christianity's conception of
the world as the mere material artifact of God, he writes, allowed for
increasingly "objective" considerations of that material artifact, rather
as if the maker had gone away and left it behind. Such considerations
inevitably conflicted with the received wisdom of the ancients so that
after a time this wisdom came to be honored in name only, and at last
was discarded. For us, this process is perhaps most strikingly illustrated
in the field of cartography where the old Ptolemaic conceptions that for
centuries had been unassailable now increasingly came under challenge
by the observations of Crusaders, voyagers, and casual travelers. For a
while cartographers wrenched and stretched Ptolemy this way and that
to fit over the newly expanded globe; then he was paid the hollow
honor of lip service (as in the famous 1448 map of Andreas Walsperger
that claims to be founded on Ptolemy but is in fact largely freed of his
errors); and finally he was openly flouted.

But our interest in this goes beyond the process itself and its opening
up of the globe to Christian activity. Our deeper interest lies in the fact
that the erosion of authority was also at work on the Christian religion
itself and for exactly the same reason. For just as the Christian world
view had permitted challenges to the scientific wisdom of the ancients,
so now with incremental power that view unwittingly permitted, indeed encouraged, covert disregard of the authority of religion through
nontraditional investigations of nature that would turn up no "objective" evidence for the existence of spirit life. If the world is wholly susceptible to understanding through rational investigation, then the exis-

tence of spirits must be denied, and without belief in spirits there can be no true religious conviction. When in the late sixteenth century the English astronomer Thomas Digges extended the heavens of science beyond the borders of the theological ones, he was pointing the way that the West would travel. This was a way abetted by religious authorities, as in their persecution and execution of the mystic pantheist Giordano Bruno in 1600 for his heretical insistence that spirit life pervaded all of nature and welled from its many sources. And writing of Puritan theologians of the next century, Perry Miller notes that they welcomed the scientific investigations of their time only to realize too late that they had let the "wooden horse of rationalism into the Trojan citadel of theology."

Unquestionably, it was a high achievement to see gradually past the ancient intellectual encumbrances, and Hall's study of the scientific revolution, or that of Marie Boas, gives the modern reader a vivid sense of the courageousness of Western thinkers engaged in their often solitary and dangerous tasks. It is doubtful that any of them could have guessed that his work was preparing the way for a mechanistic conception of nature and man that would have the practical effect of denying the operation of God in a world He putatively had created. This attitude, of course, lies far in the future, and to have uttered it even after a century of work by Galileo, Harvey, and Descartes would have seemed blasphemous and, to a great many Europeans, foolish. For the majority there were still spirits about, but these were mostly malignant ones, such as the fallen angels who had become the nature dieties of the pagans. But the forms and observances of a great religion die slowly, and long after its positive and life-enhancing aspects have fallen into practical disuse, its hag-ridden residues may be keenly felt. So, in the middle of the seventeenth century, one can find Christians who seemed utterly consumed by the life of the spirit and as convinced of the operations of the devil as they were of divine providence. And yet these same Christians were denying life to much of the world and acting upon that world as if it were a passive configuration of matter devoid of its own interior life, laws, and spirit and existing only to be "civilized" for gain.

The great question of the new science of the European Renaissance was: What lives and what does not? Francis Bacon in his classic document of the new science, *The Dignity and Advancement of Learning* (1605), showed how far the West had traveled from the mythological world view in which everything has life and is infused with spiritual significance by remarking that myths had served long enough as Pillars of Hercules beyond which human speculation dare not venture. Now was the time to sail beyond myths by seeing them accurately for what they had always been—allegorical poems veiling the secrets of nature. Bacon then demonstrated the Christian scientific mind at work as he dissected various mythic narratives to show that they were really parables instructing humans to develop a wholly rational understanding of the natural world.

Such an understanding did produce its intended result—a greatly increased knowledge of the workings of the natural world—and in consequence the life of humans in the Christian West was remarkably eased and improved. Yet there were losses too, for the spiritual life of the West, already seriously enfeebled by the centuries of relentlessly advancing Christian history, suffered another wound. Lewis Mumford observes that the greatly increased fund of scientific knowledge was accompanied by a "deformation of experience as a whole." The instruments of science, he writes, "were helpless in the realm of qualities. The qualitative was reduced to the subjective: the subjective was dismissed as unreal, and the unseen and immeasurable as non-existent." How fatal such an attitude is toward the life of religion is obvious. What has not been so obvious is the manifold effects of this attitude on the behavior of the West in the centuries of its great outreach.

As for the natives of the wildernesses who still dwelled in the numinous universe where the unseen has at least as much reality as the visible, we have already seen something of their fate. Now at the close of the first century of massive exploration others of them were to find themselves in the path of driven white men from the east. These were the English, belatedly aware of what William Carlos Williams has called the "fat of the new bounty." The Spanish had opened English eyes and now the latecomers were bound outward to get their fair share.

Of course, the Spanish were fiercely determined that no one else should have some of that bounty, and the English were aware of the lengths to which their predecessors would go in protecting their finds. With almost a century's head start on the English, Spanish hegemony had begun to spread up the eastern coast of a huge continent, the interiors of which the luckless Soto and Coronado had painfully felt. They had established outposts on what they called "the Island of Flowers" (Florida) and in Georgia, and they had made forays as far north as the "Bay of Santa Maria" (Chesapeake Bay). So when the French touched at Florida in the early 1560s, the Spanish reacted swiftly and with brutal decisiveness. The massacre of Huguenot colonists at Fort Caroline was intended as a grim warning to all others to keep out. In the dawn hours of September 20, 1564, a Spanish force under Pedro Menéndez de Avilés attacked the fort, and within minutes the ground inside the stockade was slippery with French blood. Some of the colonists were lucky enough to be aboard ships anchored offshore, and when they refused Spanish terms, the victors tore out the eyes of the heaped corpses and flung them at the ships in baffled rage. The moral of this could not have been lost on the observant English.

Still, their great privateers had gotten a taste of wilderness riches as they had plundered Spanish shipping and Spanish ports. Though in much of Anglo-American historiography it is the Spanish who are stigmatized as "get-rich-quick" adventurers, the truth is that the English, the French, and the Dutch were too. As always where some people scheme to get rich quick, there will be others who scheme to rob them of their riches. For thirty years Sir Francis Drake and Sir John Hawkins had been doing just that, and when the English began their voyaging again after the long post-Cabot lapse, they did so in order to scout out an American base of operations against Spanish colonial activities. They had not entirely given up the hope that somewhere north of Spanish holdings another Mexico or Peru might be discovered. That glistening lesson had been deeply absorbed by the English as well as by the rest of Europe, and though a half century had passed since the last golden strike in Peru, still one could always hope. Who could say what might be found? But meanwhile, there was the lucrative business of privateering. Obviously, it could only become more lucrative with a local base of operations, and Drake's return from a raiding voyage in the fall of 1580, laden with Spanish plunder, hastened forward the plans of Sir Humphrey Gilbert and others to establish themselves on the northward coast of the continent.

Gilbert's letters patent empowered him to search out remote "heathen and barbarous" lands not occupied by Christian princes. The locations of these are properly vague, but the cultural assumption is plain: such remote lands belong by right of the True Faith to such Christian princes as should first discover them and not to the native inhabitants. The assumption was never stated outright since on the face of it the claim might have seemed unjust. For a full century, while the English established themselves in Virginia and New England, a more benign-seeming version was offered: the Christians had as much right to the wilderness lands as did the natives, indeed a greater right if the latter proved unwilling to share with them.

Another assumption of these letters patent is worth attention, and that is the recognition of the legitimate rights of Europe's cognate rulers to conquest and exploitation even if these rulers were at war with each other, as would soon be the case with England and Spain. In allowing Gilbert to go to the New World, Queen Elizabeth was claiming the right of her subjects to act upon that New World after the fashion permitted by her fellow rulers. For in the mind of the West, the New World was there to be taken.

Gilbert was thought singularly qualified for the taking, at least partly because he was half-brother to the Queen's favorite, Walter Raleigh. More practically, he had rendered distinguished service in another colonizing venture, this in Ireland. Ireland was a sort of apprenticeship for many who would be engaged in New World activities, the Irish natives serving the English as the Moors and West Africans had once served the Portuguese and Spanish. Irish campaigns were schoolings in savagery and in them the "wild Irish" forfeited their lives and lands by reason of cultural differences.

Gilbert had behaved with exemplary dispatch and authority in quelling a Munster uprising in the fall of 1569, driving his forces, relentless as a harrow, through the ravaged land. Immediate and unconditional surrender or massacre was his iron rule of campaign, and it is said that he once forced capitulating Irish lords to walk to his tent between two rows of the severed heads of their countrymen. He remarked of these lords that he thought his dog's ears too good for the words of the highest among them.

However much his Irish victories might have recommended him to Elizabeth and her advisors, the results were disappointing: Gilbert seems to have been a choleric and inept bungler in his New World assignments. A first abortive voyage in 1578–79 sank him so deeply in debt that he complained he had been forced to sell the clothes off his wife's back. These losses made him rash and desperate to recoup quickly with another expedition to discover unfound wealth, thereby illustrating the truth of his half-brother's wry remark that the "needie are alwaies aduenturous."

With five vessels, one of which enraged him by putting back, he blew out in June 1583, to found a colony/base. He touched at Newfoundland, claimed it for England, and left there a shipload of colonists too sick to do anything except wait for health or death. In August he started back, losing another of his flotilla as they cleared the jagged coast. His last ill-considered act was his refusal to transfer to the larger and safer of the two ships that now remained to him, and on the night of September 9 the lanterns of his craft disappeared forever.

One battered ship might seem small return to us, but the English fire for exploring was stoked now, and nothing short of unmitigated disaster could put it out. Indeed, as we shall shortly see, even the disaster at Roanoke resulted in only a temporary banking. But as we attempt to probe the mental world of these English with our time-blunted instruments, we can see that this voyage of Gilbert's did in fact yield something of value, a remarkable document written by Sir George Peckham called *A true reporte of the Late discoveries*. Published shortly after the return of Gilbert's lone ship, it is one of the very earliest real-estate advertisements in our language.

Peckham had been one of Gilbert's backers, and as a Catholic suffering financial persecution at home he had an additional interest in seeing a colony planted where his co-religionists might hope to experience greater toleration. He himself had not been out with Gilbert but had compiled his report from information supplied by one who had. Yet, in a curious way, his words are more revealing of the English variety of the European mind after a century of exploration than if he had been an adventurer himself. This is because Peckham was less interested in the actual details of the voyage (bleak as they were) than in laying a theoretical groundwork for future planting. His advertising scheme was to assure potential investors and recruits that exploration was not only a profitable thing but a just and righteous one as well.

The question that Peckham sets himself is whether Gilbert's late voyage and the others that must surely follow are just in the eyes of God, and in putting the question he gives a good example of his method:

> By occasion of this historie, I drewe my selfe into a more deeper consideration, of thys late undertaken voyage, whether it were as well pleasing to almightie God, as profitable to men? as lawfull as it seemed honourable. As well gratefull to the Savages, as gainfull to the Christians. And upon mature deliberation, I founde the action to bee honest and profitable, and therefore allowable. . . .

Thus seen, exploration and colonization became Christian callings. The country newly discovered by the coasting Gilbert is inhabited only by savages without knowledge of the true God. Having become aware of this, how could a Christian nation turn away from its spiritual duty? And how could God disapprove of their planting there? "Is it not," Peckham asks,

> to be lamented, that these poore Pagans, so long living in ignoraunce and Idolatry, and in sorte, thirsting after christianitye (as may appeare by the relation of such as have travailed in those partes), that our heartes are so hardened, that fewe or none can be found which wil put to theyr helping hands, and applie themselves to the relieving of the miserable and wretched estate of these sillie soules? *

Plainly this is a duty that must be discharged with full and immediate vigor and not merely with distant good wishes. God, says Peckham, did not mean for such souls to languish amidst an uncleared wilderness, creeping about in its fearsome shadows, pursuing the profligate existence of nomadic hunters. For God did "create lande, to the end that it shold by Culture and husbandrie, yeeld things necessary for man's life." This means that the high business of bringing Christianity to the natives must entail the introduction to them of all manner of civilized pursuits, chief among these being agriculture.

All of this, obviously, necessitates English settlement and English working of the soil. Drawing sustenance from the Old Testament in the manner that was already obsessive among radical Protestants of his time, Peckham argues that only by living amidst the savages could Christians hope to cleanse these souls of their "horrible Idolatry" and their equally horrible practice of "sacrificing of humaine creatures." For, he reminds his readers, if after the death of Joshua even the children of Israel lapsed into idolatry, how much more precarious was the state of the New World savages surrounded by wild nature?!

Inevitably, since this was God's business, the calling would result in material profit to English investors and relief to the needy at home. So Peckham prefaces his report with a garland of poems attributed to those most conversant with the economic realities and potentialities of colonization—Drake, Hawkins, Martin Frobisher. The major theme sounded in these otherwise unmemorable lines is the holy exchange of Christianity for New World goods. Captain John Chester said it best:

> Marke well this booke when you to reade beginne,
> And finde you shall great secretes hid therein.

* When Peckham wrote, "sillie" (silly) meant something like "simple" or "ignorant."

For with your selves you may imagine thus,
That God hath left this honour unto us.
The journey knowne, the passage quicklie runne,
The land full rich, the people easilie wunne.
Whose gaines shalbe the knowledge of our faith,
And ours such ritches as the country hath.


Put this way at the outset of the tract, Christians were bound to assent to the rightness of the enterprise, but as if to perform exegesis on his own scripture, Peckham returns to the question later in his text. The Gospel should, of course, be free to all, he writes, for such was the example of the Apostles. But then he invokes his era's favorite apostolic authority, Paul, glossing him in 2 Corinthians 9: "If we have sowen unto you heavenlie thinges, doo you thinke it much that we should reape your carnall thinges?" After all, he concludes, the workman is worthy of his hire, and however great the financial returns, they are inconsiderable weighed in the scales with Christianity and civilization—which are the same thing.

This idea of the vigorous pursuit of a calling as a religious duty and of the justice of the profit to be derived therefrom will be familiar to many through acquaintance with the Puritan heritage, in which it is given its most precise articulation as well as its most rigorous application. But interestingly, the idea is espoused here by a Catholic, for, like exploration, this idea was not confined to a single sect but was generally shared.

Weber finds this idea at the heart of Calvinism, and in his controversial work, *The Protestant Ethic and the Rise of Capitalism,* he argues that it was a major inspiration for the development of finance capitalism. And yet for reasons Weber himself suggests, the idea has a far more extensive root system than Calvinism or even Protestantism. Here again, Darcy Ribeiro's analysis seems more perceptive. He distinguishes between the Iberian colonizers and their rivals and successors merely on the basis of efficiency of exploitation. His view assumes that what the French and English were trying to do in the New World was essentially what Portugal and Spain had showed them was possible, first in Africa and then in the Americas. The difference was that the latecomers had the advantages of example. As England began to launch into the colonizing business in 1583, it was clear to those involved that Portugal and Spain had not really known what to do with the fabulous wealth they had found. In the literature of the time there are numerous smug references to the fact that Portugal and Spain were really only serving as conduits for all that wealth into the rest of Europe and that they were keeping precious little of it within their own economies. A Venetian observer commented that the New World gold came to Spain "as rain does on a roof—it poures on her and it flows away."

Their successors thought they could do better through a more efficient form of mercantilism, one that de-emphasized mining in favor of exporting surplus goods and humans to the colonies and importing raw materials in short supply at home and among neighboring states. This involved a shift in economic policy but no shift in attitude toward the New World or its native populations. The same God who had rewarded the Spanish and the Portuguese would now reward the English for the same reason: their Christian mission to the heathen. Columbus's letter to his sovereigns on his return from his first voyage could have been written by some of the adventurers now interested in the English plantings, for religion and profit both lit up the Admiral's lines. There would be some difficulties in this exchange of faith for goods, to be sure, but none could doubt the ultimate issue.

With a perspicacity that now appears almost frightening, Peckham outlines these difficulties and the manner in which they could be overcome. There are, he wrote, two types of planting. The first is one in which the Christians ingratiate themselves with the savages by peaceful and gradual means, giving them gaudy trifles, defending them against the raids of supposed cannibal neighbors, teaching them the faith, and reaping the resulting economic benefits. The second type comes into being when the savages, ungrateful for all this, turn against the Christians in an unlawful attempt to dislodge them from their justly won portion of the New World. When this should happen

> there is no barre (as I judge) but in stoute assemblies, the Christians may issue out, and by strong hande pursue theyr enemies, subdue them, take possession of theyr Townes, Cities or Villages (and in avoyding murtherous tyranny) to use the Lawe of Armes, as in like case among all nations at thys day is used. . . .

What Peckham really means is that there is only one type of planting: the second is envisioned as the consequence of the inevitable failure of the first. Savages, being what they are, must resist civilization and Christianity. This is the logic of the infamous *Requerimiento,* here doing service in another nation-state of the West under another title. Drawing again on the Old Testament, Peckham clinches his point by reminding his readers that the children of Israel in their time experienced difficulties in planting themselves in a heathen and resistant land, and as if with that ancient cadence in his ear, he exhorts his countrymen to "plant, possesse, and subdue."

It was Raleigh who succeeded to the planting rights of Gilbert, and the wording of his patent is the same as that of his half-brother's. There is much talk in the document about relations with other Christian princes, but there is no mention of the natives, their rights, or the missionary work to be performed among them. The truth is, as David Beers Quinn has pointed out, that for Raleigh the venture was almost exclusively concerned with finding and establishing a privateering base. Only after it had become clear through the voyages of 1584 and 1585–86 that the tricky sea stretch of the Outer Banks was militarily unsatisfactory did Raleigh begin seriously to project along the lines of a permanent settlement.

Despite the fact that he was an individual of far greater reach than that half-brother whose picked bones now rested God knew where, as a colonizer Raleigh had serious faults. Like others he would associate with his "Virginia" scheme, he was always more attracted to the handsome, quick strike than to the slower and unheroic route to wealth through colonial trade and settlement. At the end of his checkered career, long imprisoned in the Tower, he secured a last reprieve only by promising James I to find the fabled mines of Manoa and so finance a gigantic campaign to smash the colonial power of Spain forever. He failed to find what he searched for along the Orinoco, which the sacrificed Arawaks had so long ago migrated down, and thus he became himself a sacrifice in the murky ritual of international power play. By that time (1618) the English were in fact planted in Virginia whereas they had failed at the Outer Banks under his aegis.

At the end of April 1584, Raleigh had commissioned Philip Amadas and Arthur Barlowe captains of an exploratory voyage under the guidance of the pirate pilot Simon Fernandes. Following the blazes of the Spanish, the fleet went south through the Canaries and then west.* On July 4 they sighted the North American mainland and nine days later were at a place on the Outer Banks the natives called "Hatorask" (Hatteras).

---

* In the very earliest days of European traffic to the Americas the primitive sailing directions had been simply, "south till the butter melts, then west."

Barlowe has left a partial record of this touching of what was soon to become the site of the first English colony. Almost nothing is now known of him, but either he had the poetic gift—perhaps conferred as a casual blessing of sea life—or those magnificent verdant and vine-heavy banks temporarily overtook him and loaned his pen something of themselves. Whatever the reason, Barlowe's "Discourse" is a wonderfully fleet and clean glimpse into a portion of the New World as yet unshadowed by anything not native to it.

The new land announced itself far offshore with a strong, sweet smell, as it had to those of Columbus's crews almost a century before. It was, Barlowe recalled, "as if we had bene in the midst of some delicate garden." In the next sentence he has the expedition taking possession of all they could see and beyond. What they could see was a low, sandy bit of land, heavily wooded with tall cedars. At closer view they were astonished at the bursting clusters of ripe grapes, so plentiful everywhere that the "very beating, and surge of the Sea ouerflowed them. . . ." The profusion moved the man to exclaim that though he had been to many places in Europe where the grape abounded, here he found "such difference, as were incredible to be written."

Seemingly the first humans in this splendor, they climbed the dunes and saw sea on one side, sound on the other, and as if to break into all this with something of their own, they fired off their harquebuses, whereat "such a flocke of Cranes (the most part white) arose vnder vs, with such a crye redoubled by many Ecchoes, as if an armie of men had shouted all together." For two days this Adamic solitude continued while they took note of the "goodly woods, full of Deere, Conies, Hares, and Fowle, even in the middest of Summer, in incredible aboundance."

We would give much to know what musings this world inspired in these men during those two days as they swung at anchor and tramped the beaches and margins of the woods; whether amidst all this seaspangled freshness, the sounds of wind, water, birds, and branches, they could think of anything more than their mission here; whether to Barlowe or to any of them the landscape nodded ever so slightly. Or were they merely taking inventory?

On the third day they saw a canoe with three natives coming to the northeast tip of Hatorask. It landed "four harquebushot" from the ships, and one of the natives got out and strode the beach in obvious invitation. The English rowed in to pick him up and take him out to the ships, where he was given meat, wine, and presents. When after several hours he left them, he went a short way off in his canoe where he "fell to fishing, and in lesse than halfe an houre, he had laden his boate so deepe, as it could swimme. . . ." This catch he divided among the strangers' ships and then departed. Such was their introduction to the savages.

The day following, the natives returned in some numbers, including the brother of the local chief, since the leader himself, known as "Wingina," was recovering from battle wounds at a town on the mainland. In Wingina's stead, Granganimeo treated with the strangers in the same friendly fashion as they had known the day previous. All communications, of course, had to be made in the language of gesture, but such was the openness of the natives that it was clear the white strangers were welcome. Granganimeo did his best to "shewe we were all one, smiling, and making shewe the best he could, of all loue, and familiaritie."

Here trade relations began, and as they did, something else became clear: these friendly natives had had some prior experience with whites. In exchange for hides they wanted weapons, indicating they knew the uses of firearms and the enormous leverage these would give them in battles with their enemies. Thus far had the technology of the Old World begun to alter the immemorial realities of the New. From the natives' gesticulations the English thought they were being told of Spanish coasting reconnaissances and of two Spanish shipwrecks.

The point is an important one since it tells us that these natives had obviously guessed something of the material advantages that might accrue from intercourse with their white guests. In other words, they had intelligent reasons for being friendly. In view of subsequent relations on the Outer Banks and then in nearby Virginia, and in view especially of insistent English allegations of Indian treachery, it is well to know that in these first meetings and for a good time thereafter there is no evidence of treachery. There is only evidence of friendliness and desire for further cooperation. This judgment derives from the numerous references in Barlowe to native hospitality and consideration, and it is buttressed by the fact that two natives volunteered to return to England with the ships when the English made ready to return after six weeks of amicable trading. Their names were Manteo and Wanchese, and the histories of these failed ambassadors have much to tell us of the real tragedy at Roanoke.

·   ·   ·

What Amadas and Barlowe reported of this first voyage was more than enough to keep Raleigh and others interested. Raleigh mounted a campaign to widen the circle of investors and there was talk of "Virginia" as a trading station as well as a naval base. Sir Richard Grenville came in and agreed to head up a larger expedition to return and plant. In all this the two native ambassadors were utilized as propagandists, and as part of the program they were tutored in English by a young scientist in Raleigh's household employ, Thomas Harriot (or Hariot). In the process Harriot learned enough of the Algonquian dialect to serve as a valuable interpreter on the impending voyage, and his subsequent book on that experience has been a mine for modern ethnographers trying to piece together the truths of a vanished culture.

Now they were ready to return and plant: a first fleet of seven ships, six hundred Englishmen (half of these soldiers), and the two natives stood to sea in April 1585. The ships split up on the passage out as they chased likely Spanish prizes or traded with their putative Spanish enemies in the islands. At last, near the middle of July they were reunited at the Outer Banks. By that time the first comers were already in negotiations with the natives for the lease of some land on Roanoke Island, which was a part of Wingina's territory. No doubt the favorable reports of Manteo and Wanchese helped sway the chief, who ended by allowing the strangers to use a tract on the northwestern tip of the island, quite near a native village. Before Grenville headed the fleet back to England for additional supplies and armaments, construction was under way for the first English building in the New World. It was a fort.

The man Grenville left in charge of the remaining 107-man garrison was Ralph Lane. The choice was a logical one but in retrospect singularly unhappy. For what happened under Lane's command between August 1585 and June 1586, when Drake cruised by and took what was by then a rabble back to civilization, had the effect of confirming for the English the darkest view of the New World and its inhabitants. As for the natives, what happened to them during these months was enough to disabuse them of any illusions about mutually beneficial cooperation.

At first Lane paused to muse on the pleasing spectacle of this new world, as Barlowe had in 1584. The land he described as "the goodliest and most pleasing territorie of the world (for the soile is of a huge vnknowen greatnesse, and very wel peopled and towned, though sauagelie) and the climate so wholesome, that we have not had one sicke, since we touched land here," and the people he found naturally courteous and very desirous of European goods. But that moment passed and he and the others quickly began to give offense and to term the natives' reactions to such behavior treason. But never, even in that first brief moment, were they so bemused as to forget themselves. Never did they consider these natives their equals. How could they, as they looked on the comparative nakedness of the Indians, their public acknowledgment of the body and its needs, their devilish, nature-bound rites, and their inexplicable toleration of the uncleared lands on all sides of their settlements? How could humans be content with this? Even more, how could Christians be content with the spectacle of it? Not even Manteo and Wanchese could have suspected that in those few August days they and their people had been silently relegated to the status of vassals by the armored and businesslike guests who were so dependent on them. But when they began to protest against the arrogant behavior of Lane and his men, then the natives began to get a new sense of what had seemed familiar territory: these whites evidently considered these coastal lands theirs. Here was that inevitable condition of friction that Sir George Peckham had predicted would lead to the "second type of planting."

Even before Grenville had left there had been trouble over a trifling matter. In July he had taken a party to probe across Pamlico Sound and up several of the mainland watercourses. In an anonymous journal of that trip we find these revealing entries:

> The 12. we came to the Towne of Pomeioke.
> The 13. we passed by water to Aquascococke.
> The 15. we came to Secotan and were well intertayned there of the Sauages.
> The 16. we returned thence, and one of our boates with the Admirall was sent to Aquascococke to demaund a siluer cup which one of the Sauages had stolen from vs, and not receiuing it according to his promise, we burnt, and spoyled their corne, and Towne, all the people being fledde.

What if the Indian did take the cup—what an incommensurate reprisal! And what such a reprisal tells us about that magazine of arrogance, discontent, and fear that lay within those armored breasts!

Lane was the one to ignite it. A veteran of the Irish wars, he was first and last a military man. In his letters and in his discourse on the garrison he commanded, he reveals himself efficient after a narrow military model, brutal, insensitive, and self-sorry. If we are to judge by the remarks of Barlowe and those of Harriot, who was under Lane during these crucial months, the man was more than ordinarily disposed to misunderstand and mistreat the natives, more than ordinarily equipped to poison that sandy soil for further Christian and native congress.

Lane's orders to his rebellious "subjects" drove them to resentment, sullen withdrawal, and at last outright hostility. But he was only a pro-tem commander of a wilderness outpost, poised between his own unruly forces and what all of them viewed as an untamed and threatening environment. If his orders, actions, and reactions seem to us terribly misguided, for him they had centuries of sanction.

Thus by early spring of 1586 Lane was thinking along lines laid out by other wilderness commanders: he was planning a campaign against the local tribes and then an overland invasion of the territory of a chief on the Chesepiuc (Chesapeake) Bay who was rumored to be fabulously wealthy in pearls. Already he had summarily taken prisoner the chief of the nearby Chawanoac people, and throughout the succeeding months he would keep that chief's son hostage in an attempt to neutralize his presumed enemy. We have no evidence—even in Lane's often paranoid account—that Chief Menatonon had ever meant the whites harm. There is also little doubt that Lane's tactics did neutralize him—at a cost.

From September 1585 to March 1586 the commander kept spotty records, so we cannot follow well the chronology of the steadily deteriorating relations with the natives. But well before his notations allow us a more detailed sense of things, he had driven the hospitable Wingina into enmity. No doubt the chief was beginning to get a different view of the situation he had committed himself to. These guests were building permanent structures and had sent ships back to bring in others of their kind. This could only mean the whites would require more territory, and they probably meant to carve it out of his. Perhaps he now thought differently of the irenic offices of Manteo and Wanchese and wished he had not offered the use of a portion of Roanoke.

But even more prominent in his considerations must have been the daily offense of arrogance and contempt the whites offered the chief and his subjects. Before many weeks had elapsed, Wingina could hardly have failed to understand that Lane considered him and his lands at the disposal of the Christians. Nor could he have missed the trend of Lane's actions, which was pretty plainly to divide and conquer, so that soon the natives would be completely dispossessed. Lane had probably even tried to impress Wingina in his campaign against the Chawanoacs and Mangoaks a full fifty miles inland.

Increasingly Wingina felt the burden of supplying his dependent overlords with fish, game, and plant foods, the more so because this freely assumed task had been transmogrified into tribute owed. Nor could he have been unmoved by Lane's policy of swift and brutal reprisal for the most minor incidents of friction. Even Harriot in his promotional book on the colony was moved by this policy to remark that "some of our companie towardes the ende of the yeare, shewed themselues too fierce, in slaying some of the people, in some towns, vpon causes that on our part, might easily enough haue bene borne with all. . . ."

As the one responsible for the visitation of this plague on the countryside, Wingina probably felt compromised and therefore obligated to take some sort of action. Lane notes that on the death of his brother Granganimeo, Wingina assumed a new name, Pemisapan. This is usually the outward sign to all who will read it that an inward change has occurred. Hereafter in Lane's lines the change was read direly, and Pemisapan appears in the warpaint of the treacherous agent of discord, fomenting bloody plots in unreachable swamps, biding his time till he might put together an alliance of tribes to sweep down on the fort and spill the Christians off the edge of the continent.

Since Christian history has cut his tongue out, others must try to speak for him and note that there was nothing treacherous in Pemisapan's behavior. Treachery implies a breach of trust, but where none has been ventured, none can be breached. Pemisapan was instead pressed by the designs of the English into that fated and fatal role that Alvin Josephy has called the "patriot chief"—a role in which there is neither exit nor encore. John White, the extraordinary draftsman of the expedition and later, governor of the 1587 colony, has left us a drawing of what must be Pemisapan. Silent, reproving, his head held slightly askance, the chief gives us a long stare from within the imprisoning confines of a primitive ethnographic text: here is one of the natives.

In the course of the swiftly widening gulf between Pemisapan and Lane, the chief went to the Chawanoacs and Mangoaks to inform them of Lane's intentions, and he convoked a great meeting with these people to plan a common course against a common threat. Lane says he took a force to invade this meeting and break it up. He calls it Pemisapan's "conspiracy." Whatever the intent of the meeting, the conspiracy was one of Lane's own devising, and it is a measure of how far he had driven things that even the ambassador Wanchese was now to be counted as one of Pemisapan's principal aides. While Manteo still clung to his position as the white man's Indian (perhaps his only hope), his companion in foreign education resumed his place among his own people. It is clear that his experience with the strangers had been repugnant to Wanchese. Sitting through the long gray mornings in Raleigh's study while Harriot ran them through a child's vocabulary of an alien world, walking the noisome streets of London dressed up in taffeta and greeted as a freak by passers-by—this had not been what Wanchese had envisioned when he had volunteered to go with Barlowe and Amadas to England. And then to see these same strangers back in his own world busily changing it into theirs must have been too much. Wanchese had no further illusions about the role of the cultural go-between. Better to side with the patriot chief and let matters fall as they must. There is no more revealing aspect of New World settlement and conflict than that of the "renegade," over whose visage whites drew the implacable mask of the Bad Indian, obscuring for us its real features.

We have only Lane's word that Pemisapan in his "conspiracy" ever planned offensive action against the garrison. Lane mentions poisonings and plots and raiding parties, but the truth is that Pemisapan never did attack. What he did do was to advise the region's native leaders to follow his example and leave the English literally alone. Lane indignantly records that

> The King was aduised and of himselfe disposed, as a ready meane to have as-suredly brought vs to ruine in the moneth of March, 1586, himselfe also with all his Sauages to haue runne away from vs, and to haue left his ground in the Island vnsowed, which if he had done there had bene no possibilitie in common reason (but by the immediate hande of God) that we could haue bene preserued from staruing out of hand. For at that time wee had no weares for fishe, neither could our men skill of the making of them, neither had wee one grayne of corne for seede to put into the ground.

Lane had told the departing Grenville that he and his garrison could live off the land until supplies were fetched back in the spring of 1586. What "live off the land" meant, of course, was living off the gifts of the natives, gifts that somehow turned into tribute. Both Lane and Harriot attest to the fact that the garrison was almost entirely dependent on the natives, that the whites could not—or would not try to—learn the native arts of fishing, hunting, and agriculture. Why should they when the savages provided for them? Here as in so many other instances there seems little difference between these English and their Spanish counterparts, many of whom instantly assumed the airs of the *hidalgo* as soon as they set foot in the New World with its abundance of enslaved laborers.

Pemisapan's strategy began to work. The supply ships were very tardy, the garrison could not support itself, and Lane had to begin dispatching small groups to Croatoan and the mainland to live as they could. All were ordered to keep their weather eyes out for signs of sail. Lane's construction of Pemisapan's strategy is that the enforced dispersal was a device enabling the savages to deal with the whites piecemeal. Perhaps. But it seems more likely that Pemisapan hoped to discourage the whites into abandoning the site forever. That would be a more effective strategy than any series of raids.

It is revealing of the situation that Lane and his men had provoked that it was they who indeed attacked. The commander's reasoning was that since matters stood so it was best to have it out now on chosen terms. This is the logic of a military man, and it was efficacious—in the short run. In the long view, the attack was what finally rendered the English presence here untenable on any basis other than the extermination of all the natives of the region. The English were not yet strong enough for that.

Lane planned a night attack on Pemisapan and his headmen at the town of Dasemunkepeuc immediately across Croatoan Sound from Roanoke, but this miscarried when members of the advance guard fell on two native canoeists and decapitated them in full view of the town. A cry went up, the surprise was spoiled, and the plan was abandoned in favor of a more frontal approach. Now Lane sent word to Pemisapan at Dasemunkepeuc that he was coming to complain to him about recent native outrages. "Hereupon," he writes,

> the King did abide my comming to him, and finding my selfe amidst 7. or 8. of his principal Weroances * & followers (not regarding any of the common sort) I gaue the watchword agreed vpon (which was Christ our victory,) and immediately those of his chiefe men, and himselfe, had by the mercie of God for our deliuerance, that which they had purposed for vs.

The chief, bored through by a musket ball, fell to the ground, but his wound was not immediately disabling, and amidst the tumultuous shoutings of Christian victory he leaped up, running full stride for the woods. Here Lane's "Irish boy" drilled him again with a sidearm, and another Irish servant, one Nugent, was sent to track him down among

* Subchiefs or head men.

the trees. A short time later Nugent emerged from the greenery with Pemisapan's clotted head in his hand. Utterly unaware of the irony of his situation, this tamed Irishman presented his English master with the trophy, grinning perhaps, for approval, himself a trophy. The chief's mutilated corpse, seed for dragon's teeth, was taken to the burial house of the chiefs, fellow to his tribal predecessors and to those other patriot chiefs to come—King Philip, Pontiac, Tecumseh, Sitting Bull—those whose vision was sufficient to comprehend the future in the present.

For Lane and his garrison this siege solved the immediate problem. But it is here that mythic paradigm and historical fact come together and one feels the full and terrible force of the entire misadventure. For in cutting off Pemisapan, the questing whites had destroyed the crucially necessary guardian of the gates to a new world without whose assistance further passage would be hopeless. Lane and his men and the civilization that sponsored them had not been equal to the challenge of the New World as they had met it here on the Outer Banks, and enclosing themselves within the fortress of their world they had thought to make this strange surrounding world attend upon them. When this failed, they struck out, believing that force equaled conviction and could make an entrance. But the murder of the patriot chief had destroyed whatever marginal possibilities may have remained for gaining the assistance of the natives, and without that assistance there could be no colony.

Not only had the whites failed to learn the requisite skills for survival here, the delicate and precise adjustments the natives had made to their lands and waters, but they had been contemptuous of such skills, such living, and for this there was no remedy. When they had witnessed the half-naked people dancing in their Green Corn ceremonies, they had thought "Devil," failing to see that such ceremonies formed an integral part of those very survival skills they themselves lacked. For what was being celebrated in the Green Corn ceremony was the natives' deep earth sense that in one way or another all things are interdependent. So in John White's painting of the Green Corn Dance, while the symbols of myth flourish in profusion—leaves, gourds, virgins—the basic symbol is the circle in which all things are joined and inseparable. But the circle belongs to that archaic world view into which these Christian outsiders had stumbled in their pursuit of history. White could depict the external likeness of the mythic ceremony but was excluded from its interior significance. Had any of the English been able to enter that circle, perhaps the desolate resolution might have been avoided. As it

was, on this first day of June 1586, as Lane and his men stood triumphant in the routed and disordered town, the fate of what was to be
known as the "Lost Colony" had already been decided. More: here, ir-
revocably, the English portion of New World history locked into place.
For the enmity incurred here on the Outer Banks precluded cooperation
and congress not merely for those who stood with that severed head
smoking into the ground at their feet but also for those who would
come out here the next year. And the disappearance of the colonists of
1587 would lurk in the English mind ever afterward as a dread cautionary tale of the perils of the wild world and of the unregenerate savagery of its inhabitants.

A week after this triumph at Dasemunkepeuc, the much reduced
Pyrrhic victors spied the sails of what they hoped were Grenville's
supply ships. They were instead those of Drake, cruising the waters at
the tag end of a grand piratical scheme against the Spanish colonies.
Wasting away in a wilderness of their own making, Lane's garrison
eagerly seized Drake's offer to transport them back to civilization,
spurning his alternative offer of sufficient supplies to hold out until
Grenville's arrival. They knew better than Drake could have how hopeless their situation here had become.

In panicked fashion they scooped up from the fort what came quick-
est to hand and made their way to the boats, spilling out maps, books,
journals, and drawings along the sands and into the waters of what
Lane now called "that miserable road." So great was their haste that
they did not scruple to abandon three of their company who were off
foraging somewhere.

When Grenville finally did arrive (badly delayed by chases after Spanish prizes), he found the fort a shambles and everyone gone, even Manteo. He organized several search parties, and one of them was able to
catch a native and learn from him that the men had all gone back in
ships. Now, perhaps frightened for themselves in this empty landscape,
they too determined to go back. Grenville left behind a corporal's guard
of fifteen to hold the fort until the colony could be restocked. Pathetic
hostages, these men were not the lost colony but the marooned one.

Naturally there was some disappointment at how the Roanoke venture had turned out. The colony had failed to plant itself; the fort was
practically dismantled and tenuously held; and the site was clearly un-
suitable as a privateering base. But it is the peculiar genius of the northern capitalistic eye to see potential profit even in the midst of apparent

waste. Harriot had been able to save his manuscript during that messy departure, and it contained an explicit inventory of items the English would be able to merchandize. Indeed, reduced thus to its bare economic potential, Raleigh's Virginia looked far better on Harriot's pages than it had looked to those who had been out there. For here the Christian tendency to strip the landscape psychically gave speculators at home the illusion that this wilderness had in fact been pruned back and laid open for exploitation.

It was within this view that plans went ahead for further colonization, not at Roanoke where things had run so foul, but at a site somewhere on the Chesepiuc Bay. This new planting would be essentially a trading factory with privateering operations only a small part of it. Raleigh blithely handed out paper grants of the unknown land, 500 acres to each colonist, and on May 8, 1587, 150 of those who would one day be known as the Lost Colony set out from England.

Their governor was the artist John White, who had knowledge of the terrain through his service with Lane. A far better draftsman than administrator, he was befuddled and bullyragged almost from the start by the piratical Simon Fernandes, who once again served as pilot. From mid-passage to the unscheduled terminus at Roanoke in late July it was Fernandes who called the turns, not White. They had stopped at the island on their way to the designated site to the north only to establish contact with the men at the fort and to deposit Manteo, who was to serve as their man on the Outer Banks. However, Fernandes was eager to discharge his homely cargo of colonists and their stores in order to have time to prey on the Spanish ships that were due to pass through the Azores in August or September. Incredibly, he was successful in outmaneuvering Governor White and dumped the colonists at Roanoke in a flat contravention of plans.

Like the Columbus of the second voyage, White and his colonists found no welcome at the site of their outpost. In the long shadows of the ending of that first day they found the bones of one of the fifteen soldiers; there were no signs of the rest. The next day they came to the site of the fort, but it was empty and indeed not even a fort; the natives had razed its works, and in place of soldiers, barracks, and ordnance there were grazing deer and "Melons of diuers sortes" growing within the enclosure and gripping the disused houses in viney embraces. White immediately set his men to repairing the effects of this silent assault on civilization, but he did so "without hope of euer seeing any of the fifteene men liuing."

Four days the colonists worked at repairs; on the fifth the natives showed themselves. A colonist was hip deep in the sound hunting crabs when a band of Pemisapan's people, over from Dasemunkepeuc, arose out of a bed of tall reeds, filled the unlucky man full of arrows, and beat his head to bits. Two days later White sent a deputation with Manteo to Croatoan in the forlorn hope that some sort of alliance might be patched together through the ambassador's offices.

This mission was relatively successful. After recognizing their kinsman among those they had learned to fear, the Croatoans offered their friendship once again. But they could do nothing to conciliate a countryside now in fact as hostile as the whites had always imagined it. From them the English learned the identity of the ambushers of the crabber, the same men who had killed at least eleven of the soldiers at the fort sometime the previous year. Wanchese had been in both actions.

White had wanted Manteo and his people to convoke a peace conference with the hostiles, but when this appeared impossible he determined, like Lane, to take the offensive. In a bungled night raid on Dasemunkepeuc he succeeded only in mistakenly killing a few Croatoan allies. The hostiles had known his plans and had fled. There was nothing left to do but to crown Manteo puppet lord of Roanoke, Croatoan, and Dasamunkepeuc and hope that he would furnish warnings of enemy plans.

It quickly became clear to the new colonists that in this climate they could not long hold out. They were, of course, no better equipped to support themselves than their predecessors had been, and they could plainly expect nothing but armed attack from all but Manteo's people. One month from the time the ships had arrived at the Outer Banks the entire company came to White begging that ships be immediately dispatched back to England for necessary supplies. After much wrangling about who should go, it was agreed that it should be the governor himself, and on August 27, 1587, White weighed anchor, leaving behind 110 colonists, including his daughter, son-in-law, and Virginia Dare, said to be the first English child born in the New World. None of these people was ever seen again by whites.

It is not necessary to give details as to why White and Raleigh so long failed to relieve the stranded colonists other than to say that open war with Spain, the threat of the Armada, and Drake's designs against Spanish ports relegated the investment on the Outer Banks to insignificant status. When it became clear that no privateering operation could base itself there, the colony dipped still lower on official horizons. Even Raleigh seems to have lost much of his enthusiasm. He, White, and Grenville all made efforts of one level or another to pry loose the ships needed to return, but it was not until late March 1590 that a fleet led by White got clearance.

Astonishingly, they spent much time in their crossing chasing likely prizes, and it was not until the middle of August, nearly three years after the colonists had been left on Roanoke, that White arrived with the supplies supposed to be so urgently needed. Before White left the colony in 1587, there had been talk of moving the site from Roanoke to a less hostile climate some fifty miles inland. The colonists had told White then that they would leave word of their destination carved in a conspicuous place where he could find it on his return. Perhaps now he was merely stopping at Roanoke to find that message and to make certain that some were not still holding on at the fort.

The commander was greatly heartened on the evening of August 15 as they lay off Hatorask when he saw a "great smoke rise in the Isle Roanoak neere the place where I left our Colony in the yeere 1587. . . ." In the morning White led two boats up the sound to the island but was distracted by another column of smoke rising from the high dunes of Hatorask. They spent the day investigating this latter signal but mysteriously could find nothing. Another day was wasted.

On the day following, one of the ships capsized as it attempted the dangerous passage over the bar into the sound and the captain and first mate were lost along with five others. Now White had all he could do to persuade the rest to go on with the search. At sunset two boats with White again at the sheets made their way toward the island, but in the darkness that swiftly closed down upon them they overshot their intended landing place. Now, however, they could see through the black trunks of the forest the light of a great fire. We "let fall our Grapnel," White writes, "neere the shore & sounded with a trumpet a Call & afterwardes many familiar English tunes of Songs, and called to them friendly; but we had no answere, we therefore landed at day-breake, and comming to the fire, we found the grasse & sundry rotten trees burning about the place."

With gathering grimness White now urged his reluctant companions through the woods to the site of the fort, noticing along the way the prints of animals and savage men. As they clambered up a sandy embankment they saw before them a tree upon which "were curiously

carued these faire Romane letters CRO: which letters we knew to sig-
nifie the place, where I should find the planters seated, according to a
secret token agreed vpon. . . ." Missing from the inscription was the
Maltese Cross that had been the code for distress. Yet that "CRO"
looked as if the carver had been interrupted in the midst of his message.
By what, White could surely guess.

They went on to the site of the fort. Here

> we found the houses taken downe, and the place very strongly enclosed with
> a high palisado of great trees, with cortynes and flankers very Fort-like, and
> one of the chief trees or postes at the right side of the entrance had the barke
> taken off, and 5. foote from the ground in fayre Capitall letters was grauen
> CROATOAN without any crosse or signe of distresse. . . .

Inside the palisade, however, there was the same disconcerting evidence of interrupted and hasty activity as in the "CRO" of the first tree:
pigs of iron and lead, shot, and other heavy items strewn about and
overgrown. But the small pieces of ordnance were gone. It looked as if
the colonists had had time to take away some of their possessions but
not all. A few yards from the palisade sailors found five chests that had
been buried and then exhumed by native hands, and White looked
ruefully over his maps and charts spotted with rain and mildew, his pictures torn from their frames, his once bright and serviceable armor
eaten through with rust. Here truly the horrid hand of the wilderness
had reached out and grasped them to itself.

As if to reinforce the sense of this, the searchers were suddenly besieged by such heavy and unseasonable weather as to make both the an-
chorage off Hatorask and the searching of Croatoan extremely hazardous. The crews, already unenthusiastic about the search, demanded to
leave this cursed place where winter seemed to be setting in on the very
heels of summer. Bowing to their mood and to that of the weather,
White abandoned the search on August 28, and they set out first for
Trinidad and then the Azores. From this point on White's journal is full
of talk about the weather and privateering, and we find no more in it of
those he had set out to relieve.

His last word on Roanoke is in a packet he sent the great historian
Richard Hakluyt three years later from Ireland. In it he enclosed his
journal of the last voyage. Of this, he writes Hakluyt, you may find
"the successe of my fift & last voiage to Virginia, which was no less vn-
fortunately ended then frowardly begun, and as lucklesse to many, as

sinister to my selfe." Here, he concludes, he is finally committing the "reliefe of my discomfortable company of planters in Virginia, to the merciful help of the Almighty. . . ." And with these words and wishes White disappears like his planters, the place and day of his death unknown to us.

Yet, obscure as was the fate of the commander, that of his colony was to become the subject of intense speculation. For several years it was assumed that the colonists were still alive somewhere on the mainland, a belief possibly encouraged by those who stood to profit by continued interest in planting. But gradually it came to be accepted that they had really vanished, victims of the dark New World and its redhanded natives. This may explain why there was a lull in colonizing efforts for a decade and why it was a full two decades before the English would once again attempt permanent settlement a few miles northeast of the Outer Banks. By that time the Lost Colony was well on its way to becoming one of those white legends that lit the American landscape with a strange and awesome glow, pointing out to the settling whites the ways that they must go and the perils that crouched along those ways.

Only when the New World had been pruned and leveled so that none need fear becoming lost in it, and its remnant natives had been disarmed and corralled, could Europeans begin to rest easy in their possession. And it was at that moment in our history that we began to experience the spiritual enormity of what had been done here.

# Things of Darkness

The English had learned one thing from Roanoke. Planting would take some time because it would be attended by the same sort of difficulties encountered by their Iberian rivals. The New World, rich as it might be in promise, was obdurate, and it would yield grudgingly and at cost. In 1610, thirty years after interest had been kindled, the thinking of the English had come this far. Our window into their state of thought is not, however, a writer centrally engaged in colonizing operations but a stay-at-home, a thinker. And though the setting of Shakespeare's *The Tempest* is ostensibly an uninhabited island somewhere in the Mediterranean, the true psychological locale is that unmade continent the English were even now beginning to clear at Jamestown. In the fiction and the lines, even in the stage directions of this play, we are privileged witnesses to a great mind at work on the experiences of civilizing a wilderness.

Shakespeare had been inspired by certain "Bermuda pamphlets" that had circulated in 1610 from the survivors of a shipwreck in those islands. There is a direct reference in his text to the "Bermoothes," and it is known that Shakespeare was friendly with several members of the Virginia Company, so it is likely that he had seen the company reports. But in a truer sense *The Tempest* was not inspired by any single topical incident but rather by the far more compelling drama of a whole civilization extending itself outward into the wilderness. What would happen out there? Could civilization stand the wilderness? Would the bonds that held it together remain in force beyond it? Or would the spectacle, the brute impact, of the wilderness and its wild people be too much and tempt Christian men to lawless and awful behavior? If, as many were claiming in those days, this was the greatest event since the coming of Christ, it was so because the wild lands and peoples represented such a profound challenge to the existence and order of Christian civilization. Here is the real drama that lies beneath the surface fiction of Shakespeare's play.

That fiction, as its many witnesses know, concerns the machinations of a white magician, Prospero, exiled to the island by his brother, Antonio, who has usurped him as Duke of Milan. Prospero stirs a boiling tempest to bring to the island a ship carrying Antonio and Alonso, King of Naples, with that king's brother, Sebastian, and his son, Ferdinand. Alonso has been an accessory after the fact to Antonio's treachery, and they are now all returning to Italy from Tunis where the king's daughter, Claribel, has just been married. Prospero's purpose is to put all of them under his temporary control so that he can correct the great wrong done him and thus restore the right and natural order of things.

In the magical tempest the ship seems to split and the survivors are cast up on the apparently deserted shores where Prospero will work upon them. But before this resolution of order can be accomplished, the wild setting beyond the reach of law and order must tempt and find out the weaker natures among the survivors. Disorder must deepen before order can be restored. Thus Antonio and Sebastian plot regicide; two of the king's servants plot the murder of Prospero; and Ferdinand the royal son must be reduced to slavery in his suit of the magician's daughter so that his constancy may be fully tested. Eventually, Prospero's magic so manages things that all is benignly resolved: the regicidal plot is foiled; the two seditious servants are thwarted and punished; Ferdinand is promised his bride, the marvelous Miranda; and her father, restored to his rightful place, forgives his brother. Now order is restored, all are content in their proper stations, and the entire company is ready to return to civilization.

The play moves by a series of paradoxes, that much beloved device of writers of the time. The basic one is that great, unexpected good and renewal may somehow issue out of compounded disaster and destruction. Here the restitution of order and the marriage of the royal lovers strengthen and renew the states of Milan and Naples out of the unlikely matrix of political intrigue, savage weather, shipwreck, and isolation on an uninhabited island where ugly designs threaten even greater ill. Thus old Gonzalo, the faithful counselor, in summing up the business of the plot:

> Was Milan thrust from Milan that his issue
> Should become kings of Naples? O, rejoice
> Beyond a common joy, and set it down
> With gold on lasting pillars. In one voyage
> Did Claribel her husband find at Tunis,
> And Ferdinand her brother found a wife
> Where he himself was lost; Prospero his dukedom
> In a poor isle; and all of us ourselves
> Where no man was his own.

This paradox of unforessen good issuing from such an unlikely setting was probably most immediately suggested to the artist by the Bermuda pamphlets, for in them the survivors vividly recounted the terrors of a tropical tempest, of shipwreck, and of finding themselves castaways on islands long shunned by Christian mariners as devilish places, the haunts of cannibals. The pamphlets also told of internal threats to survival from disaffected crew members who seemed to have succumbed to the temptations of the wilds. But what the company actually found in the Bermudas was more paradisal than devilish. There were no cannibals, either, and they all survived the internal threat.

But this particular event, stirring to the imagination as it so clearly was, was merely a local illustration of that far grander phenomenon in which the English were settling along the shores of the New World, facing up to and facing down the attendant hardships and terrors of taming the savage land to civility. And in this process, the latest in a swelling tide of English power and prestige, Shakespeare saw the same paradox revealing gradually its same happy resolution. True, the Roanoke colony had vanished, and herein the hopes of many investors had been smashed. Sir Humphrey Gilbert had also vanished, into a watery grave; and Raleigh, that jewel, rotted in terminal disgrace. But at this moment of bleakness, bright news; in the midst of wilderness darkness, some shining lights. The successful planting of the Jamestown colony and the discovery of the paradisal "Bermoothes" made it possible to hope that three decades of loss and heartbreak had really been but the necessary sacrificial trials before the breakthrough into delight, order, and profit. Assessing this same situation, a contemporary of Shakespeare's was moved to remark that such great happinesses were routinely attended with "manifold difficulties, crosses and disasters, being such as are appointed by the highest providence." It now appeared as if that providence was ready to underwrite the establishment of the Christian order that was its highest and truest earthly manifestation.

In his suggestive little book *The Elizabethan World Picture* E. M. W. Tillyard shows that in Shakespeare's age order was so dominant a value that almost every other one was conceived of as subordinate to it. Without order there could be no true worship of God, no society, no profit, no civilization. Essentially what order meant was a political state of the West functioning in its appointed fashion, a condition in which each being knew its place in the vast, God-ordained hierarchy that stretched from the Creator to the inchworm. This happiest of all conditions was variously imagined—as a chain, a dance, a set of correspondences—but its opposite, chaos, had a far greater number of forms, for it was everywhere. Elizabethans, says Tillyard, were obsessed by the fear of chaos. It perpetually gnawed and threatened at the edges of their world like the barbarian hordes of an earlier age. Thus in Shakespeare's plays so often the order of the state, a family, or an individual totters or is shattered, but at the conclusions of the plays order is restored. In *Othello* the order of all three—state, family, and individual—breaks down on the island outpost of Cyprus, a process that continues unchecked right through the suicide of the hero, which is the ultimate expression of chaos. But the villainous Iago himself is not violently murdered in the last scene, as perhaps we might wish in satisfaction of our own desires for instant retribution. Instead, the state here claims its high prerogative to regulate violence, and Iago is marched off to an orderly execution.

Certainly one does not have to search very far for explanations of this fear of chaos, for eruptions within the bodies of the states of the West were as frequent as pox. Shakespeare knew his history well, as his plots and sources demonstrate, and he also knew that deeper history of the flagitious heart of man. Beneath these was that sense so powerfully urged in Old Testament writ of dark chaos rolled back to the edges of lucid creation, where it crouches waiting to engulf the cleared spaces of human endeavor as the desert sands of the Near East seemed waiting always to sweep again over cultivated lands too long neglected. Order was the cleared and cultivated field or the carefully tended garden, nature made better by man, and it was the obligation of the Christian to bring this order to places lacking it. Only then might one hope for prosperity.

Prospero succeeds in this task in the play, fighting a two-front battle against the inherent wildness of the isolate island and against those castaways who are here tempted to outlawry. In fact, however, the island is not quite "uninhabited" despite the fact that the folio stage directions tell us that it is, for in those same directions we find a certain "Caliban" under the Names of the Actors: "Caliban, a savage and deformed slave." This is neither oversight nor contradiction. Instead, it is a reflection of the fact that Christians do not recognize beings of the wilds as kindred souls. Caliban is one of these, and in the fiction of this play he embodies the wildness, the chaos, that perpetually threaten order.

Doubtless, Shakespeare had the Indians in mind here; indeed, there are several references to them in the text. But while it is the Indian who is uppermost in the artist's mind, there is an older model for Caliban. This is the European folk tradition of the Wild Man, a tradition that Richard Bernheimer traces to that substratum of oral tradition that underlies so much of the Old Testament. There certain shaggy creatures were said to roam through equally shaggy zones where they lived a scarcely imaginable existence outside the bounds of God's concerns. Occasionally these creatures are shown as the inheritors of the ruins of civilizations blasted by God's wrath where they howl to each other in what the divine curse had rendered a wilderness, and it was in this context that they entered the traditions of the West. Bernheimer says that Jerome in his making of the Vulgate Bible was evidently thinking of these creatures when he translated certain Old Testament passages. So the voice of the old Desert Father, wasting away before the consuming fires of the flesh, echoes down the centuries, identifying the shaggy creature with the earth, with sex, blood, sin.

There is also in this figure some element of the ancient nature gods, though as filtered through Christian culture this influence is restricted to the negative imputation of an ungovernable animal appetite, Priapus rampant. The figure that emerges in the folklore of the West's Middle Ages is a hairy, half-naked creature who lives a mode of existence somewhere between those of the beasts and men in that no-man's-land of the European outback. He is without knowledge of God, and this makes him insane and a slave to his impulses, especially his sexual ones. Typically, he desires to satisfy these by raiding some village and carrying off to his woody lair a fainting white maiden. In medieval tradition this brings on the white knight, the exemplar of order, whose subscription to the chivalric code of desperately controlled worship of the white maiden is the perfect antidote to the lawless lust of his adversary.

Yet despite his loathsome appearance and even more loathsome behavior, this Wild Man has some positive characteristics, existing like other stray signs we have noted as the distress signals of a divided and deeply repressive civilization. For one thing, it was said that the Wild

Man knew certain secrets of nature that had been lost to men in the process of civilization. If you could capture him, you might be able to learn what these were and in this way reattain some portion of the lost human heritage. Again, it was told that if you could capture the Wild Man alive and break him to civilization, he would then become a uniquely valuable and even heroic member of it. Apparently something of his life in the woods had conferred on him a mysterious virtue available nowhere else.

But despite these positive characteristics, and indeed just because they suggest the perilous attractiveness of this figure, the burden of this tradition is the hunting down and conquest of the Wild Man by the forces of civilization. On the breath of narrative, in the fabric of tapestry, graven in stone, wood, or metal, and in the fictive play of folk pageants, the Wild Man is compulsively stripped of his wildness. As he is and would be, he is too great a threat to us, for as Hayden White has pointed out, the Wild Man is not only without in his wilderness habitat; he is also the unreconstructed beast within the jungle of the single body, and there, like chaos, he is always lurking and looking for a way out. So as the Wild Man he must cease to exist, must either be civilized or sacrificed to civilization—which amounts to the same thing. Bernheimer describes springtime folk pageants in which the stirring up, hunting down, and killing of the Wild Man symbolizes the banishment of winter that makes way for the advent of spring in the waiting fields. In this can we not see a sad perversion of rites of renewal? For here it is the natural human, the old earth figure, who is identified with death while the civilized killers are imagined to be on the side of life. We must now know that such sacrifices as that imagined in this tradition cannot be as they are supposed. They cannot be signs of renewal but only of duration, and that is surely something different.

Obviously, the experiences of the Christians in the New World would reinforce for them the truth dramatized by the tradition of the Wild Man. He could hardly be allowed to remain his brutish and unrepentant self since as such he directly opposed the imperative extension of order. How indeed could civilization hope to gain entrance to the New World, to colonize and profit, without civilizing its wild men? It was an impossibility. The wild men's claims to ownership of these lands were dwarfed into nullity by the greater need to order the claimants and their lands. We might note that in ancient Israelite culture wild animals were thought to have had *their* rights, but these were conceived of as distinctly subordinate to the needs of humans. The wild animals had a right to their wild nature, but wherever man appeared in the land, the animals must submit to domestication or else to destruction.

In *The Tempest* Caliban is shown to be the rightful owner of the island to which the white magician and his daughter are banished. At first, relations between native and newcomers are cordial enough, and Caliban shows Prospero how to get along in this alien land while Prospero in exchange begins that process of tutelage that he hopes will redeem the savage for civilization. But, being what he is said to be, the savage can do nothing else than rebel against this arrangement. True to his lustful nature, he tries to rape Miranda. To Prospero—and presumably to Shakespeare's audience—this ghastly attempt is the fit symbol of savagery and savage obduracy. In light of it we are meant to feel that no reprisal could be too harsh, and Prospero's subsequent usurpation of the island and his enslavement of Caliban are made to seem perfectly appropriate. Indeed, these acts are measures of Prospero's grand powers since he must accomplish them as preliminaries to the rectification of his Italian political affairs.

Prospero's addresses to his slave as "earth," "filth," and hag-born "seed" forcefully remind the audience of the creature's subhuman origins, while also serving to explain Caliban's dispossession, current status, and future. Thus in the grand resolution of the play Caliban has no part, or better, his part is over. Scourged to his cell, he is made to mutter contritely, but it is clear that he will have no part in this "brave new world." He is instead fated to slave out his days to an order renewed and reaffirmed here, a necessary and even annoying sacrifice to that order.

As in *Cymbaline* and *The Winter's Tale*, there are dark tinges in this late comedy of Shakespeare's, and on an aesthetic level these lend a wonderful depth of tone that by comparison sometimes makes the early comedies seem frivolous. But beyond aesthetic considerations, these dark tinges tell us much about that larger drama of the New World that would be written by those coming after the artist. A great artist, Ezra Pound once remarked, is an antenna of his race, and in this fiction we are witness to the mind of the white Westerner, the new magician of the globe, imagining a future for the New World Wild Man that is really no future at all, and imagining as well a future in which that New World ceases to be itself and becomes instead an indistinguishable appendage of the Old. Caliban is made to meditate rape, but Christians were meditating things far more final.

·  ·  ·

Shakespeare was no Protestant radical, but he was fully a man of their world, and however much the scripture-haunted outcasts to New England would have disapproved of his art, surely they would have intuitively understood his drama of the New World and approved its resolution. For all their distrust of literary drama, the Puritans also saw human existence as a terrific drama in which God and the Devil were joined in struggle toward a divinely appointed resolution. That the resolution was divinely appointed, foredoomed, and foreknown did not lessen one jot the tension of the drama, for uncounted millions had leagued themselves with the Cosmic Loser and millions more would yet do so. And at every moment the Devil and his misbegotten minions threatened the sanity and sanctity of the few who fought for God. Here in the New World wilderness, amidst wolves, bleak woods, swamps, and cruel Indians, the drama was more fearsome than it had ever been in those lands from which the Puritans had chosen exile. Here the drama was so stripped to essentials that all could readily sense its presence in the very midst of the workaday round. Equipped thus, the Puritans sought an entrance here.

Winter was their season. One feels this truth in trying to assess the history they were now to enact, its curious, indelible stamp on North American culture. And where they landed was a match for their temper, these exiles who believed so deeply in their own drama that they would take untraveled routes to a northeastern coast, avoiding more hospitable ones, warmer latitudes. They knew that sliver of New World history other whites had made, knew of the Spaniards to the south, of the savages of Raleigh's Roanoke, and of the Virginia patent. They would settle a bit beyond any of these.

In Holland the Puritans had been trapped between their position as aliens and the tendency of their youngest generation to stray off into the customs of that country. Regarding this, the elders could foresee the slow death of the sect. But where to go? Holland was a center of information about the New World, and as the exiles cast their eyes in that direction, they felt their very bowels "grate within them and make the weak to quake and tremble." For to these there was no meliorating image of the New World but rather an almost unrelievedly grim one of a savage terrain filled with savage beasts and bestial men, cannibals really, who delighted in rending and eating their victims while yet they lived.

That they made the choice to come here after all tells us much about their temper. After sixty-five days out they made Cape Harbor (the present Provincetown) and gazed across November's gray waters to a land wooded to the very brink. Many years later their first historian, William Bradford, recalled that first glimpse with a vividness that is the brand of a powerful impression. It was already winter in these parts, he recalled, and they who know New England winters know them as sharp and violent. "Besides," he adds,

> what could they see but a hideous and desolate wilderness, full of wild beasts and wild men. . . . Neither could they, as it were, go up to the top of Pisgah to view from this wilderness a more goodly country to feed their hopes; for which way soever they turned their eyes (save upward to the heavens) they could have little solace or content in respect of any outward objects. For summer being done, all things stand upon them with a weather-beaten face, and the whole country, full of woods and thickets, represented a wild and savage hue.

They made their reconnaissances in a shallop as winter shut down on the land, the spray from the bay waves turning their clothes to iron as they jounced shoreward. They dug in the frozen ground with their swords and turned up native graves and then several bushels of cached corn. Bradford got his leg caught in a native deer trap. They saw some men and a dog, but these ran from them. Camping at the edge of the woods, they heard the wilderness lash and howl about them, not knowing whether it was wind and trees or wolves or men. And one morning there were quite definitely men, tall, tawny, shooting arrows at the English muskets, dodging behind cover.

From their landfall on November 11 until after Christmas they were at this depressing business, nibbling at the locked and frozen edge of an inhospitable coast: winter, primitive travelers' hardships, Indians, fear. Bradford's wife went overboard the *Mayflower* into winter's waves; their first birth, from Goodwife Allerton, was stillborn; and the closest they had come to the natives was those graves into which their ignorant eyes had peered. Within three months half the company had died.

In the middle of March the wasted company was suddenly penetrated by the bold appearance of a lone Indian. Tall and straight with long hair falling down his naked back, he stood among them discoursing in their own tongue. His name was Samoset; he was not a native of these parts but of those farther eastward where the continent yawned seaward. None of his people were left but him: sickness brought by strange ships. The wind was rising, and they gave him a cloak to cover himself as he talked on about the country, its tribal divisions, locations, and leaders. He left, promising to bring a great sachem and another who could speak the white tongue better than himself.

In four or five days Massasoit, the Wampanoag sachem, arrived in his native state, attended by Squanto, the white-tongue speaker who, like Samoset, was the last of *his* tribe. Through Squanto the whites concluded a treaty with the sachem, brazening out their otherwise hopeless and indefensible situation into articles that specified a good deal more what the Indians should perform than what obligations accrued to themselves.

Here began the utilitarian legend of the Good Indian: faithful, a little doglike in his patience and kindness, teaching the whites how to plant their corn and where to take fish, serving as their interpreter as they penetrated "unknown places for their profit." A very prerapine Caliban.

They reaped their first harvest, fired off a salvo of gratitude, and Massasoit and ninety of his people came among them to feast and to give out venison for three days. This was the first Thanksgiving.

Through the starving time, through all the cruel deprivations, they must often have asked themselves what they were doing here. And the answer to fall back on was to be found in that great Reformation of which this perilous and forlorn little migration was to them the latest and most significant act. For within the monstrous cathedral of European Christendom another revitalization movement had been sparked off by the rapt internal attentions of a German heresiarch. The idea was to *go back*, to return to the primitive vigor of the early Church, to that glimmering moment before the building of the cathedral and the institution of its far-flung offices when the Church existed only within the hearts of the convinced. They wanted to strip away the layers of appliqué, the pomp and vestments, to come again to what even they could call the primitive, naked sources. Perhaps, tucked away inside this noble effort, was the deathless wish for revelation in their time and lives, for a new age of miracles in which the indwelling spirit might surge through and lift them as it had the Apostles gathered in the house at Jerusalem in the wake of the Ascension.

If so, the Reformation was a failure, and it is lamentable to reflect on how much of Protestantism was merely a reaction to Catholicism, having so little blood and fiber of its own. Luther, hounded by his opposition, was as viciously set against the millennial sects and their attempts to chant and storm their ways through to revelation as were his persecutors. The movement's other great figure, Calvin, was as authoritarian as any caricatured pope, expounding the meaning of the scriptures in a light, careful voice from his Geneva pulpit while students from all over Europe took notes and then submitted them to the Master for inspection and correction.

Still, there were those who heroically or tragically—which is it?—persisted in trying to go back, and of these they who took the struggle to this northeastern coast were the most earnest. They constituted, as their voluble historian Cotton Mather said, the very marrow of that generation of godly men that from the very beginning of the Reformation had been most desirous of closing with what they took to be the law of Christ and the spirit of primitive Christianity. That first age was the *"Golden Age,"* Mather wrote, and to "return unto *That*, will make a Man a *Protestant*, and, I may add, a *Puritan*." So here it was once again in a somewhat different guise: Christianity's commitment to its own history, dooming even these radical dissenters to the defeat of their much desired ends.

Whatever else we might say about these people, we must sincerely admire their energy and the strength of their commitment. They came, "forsaking," as one of their historians wrote, "their fathers' houses and the pleasant heritage of their ancestors," to what even after several decades they still styled the "howling wilderness," to hack out of a stony ground a purer form of Protestantism. In such a light and from our vantage point we cannot but feel the tragedy of their foredoomed failure, for the closest they could imagine themselves back to the primitive vigor of their faith's "Golden Age" was the written records of it, texts that did not even exist for the followers of Christ, did not exist as recognized scripture until the fourth century.

And not even these texts could be gotten at unmediated but instead had to be filtered through the iron sluices of Calvin, Augustine, and Paul, and then endlessly intellectualized by their ministers to those sitting in the darkness and hardness of their little frontier churches, the painful anatomy of the sermons ending only when all the life juices had been pressed out and nothing more was to be said. The Puritans would take their notes on these sermons, too, so that they might be chewed over at home as they chewed their family meals. Hard fare, for with their sermons they got their Calvin and through him their Augustine and Paul, and so had reiterated over and again the desolateness and worthlessness of the human condition in consequence of Adam's sin.

They heard the pitifully few changes that could now be rung on the historic Christian theme of being strangers and sojourners in a world of sin and tribulation, and these fell on their ears in the wilderness with a doubled force. They heard with somewhat less clarity their ministers attempt to explain this: how God could have created this world and still not be of it; how all creation had been perfect as it had issued from the Maker's hand and that visible nature was a sign of the divine handiwork, yet to seek God in nature or to worship divinity through any aspect of it would be the terrible sin of idolatry. Drummed to church in martial array, they had drummed home to them that those who worshiped the earth, the earthy, belonged to the vast company of the damned, while those who steadfastly kept their eyes heavenward might hope at least to belong to the City of God. Between these lay a huge and impassable gulf like that separating body and soul, Dives and Lazarus, or Prospero and his "filth."

Hard fare, indeed, and it is a touch-test of just how hard that some considerable attention was paid by seventeenth-century Puritan ministers to softening and sugarcoating those uncompromising stony lumps that Calvin offered in place of the wafer. Human beings, the clergy reasoned, being the poor weak creatures they are, need something attainable to hope, something less terrible to contemplate than Calvin's vision of earthly life, so removed from heaven, so close to the awful precipice of a judgment from which there could be no escape. So these would-be purifiers developed the theology of the Covenant of Grace whereby God would undertake to save some sinners on the condition of their belief in Christ. And even if it would never be entirely clear whether one had been so arbitrarily elected for salvation, still it was something to hope for. Meanwhile they would stick close to those letters in the sacred book, letters that would tell them who they were and through what they moved in their earthly passage.

The letters, as rendered through their ministers and the greater authorities who bulked unseen behind the pulpits, told of something more: of history, that not-endless but lengthy chain of actual events that stretched from that first divine breath of creation, moving over the lifeless waters, through the covenant with Abraham, to the New Dispensation of Christ, and on down into the wan light of these very hours spent within the fortress walls of the New England churches. Unbroken, unbreakable, the chain's links were also inescapable and if human thought had been capable of omniscience, it could have seen that the links had all been forged of old. All that had happened and all that would happen had been foreknown in an instant at the outset. Here in the letters of the texts this awesome truth was partially preserved for human understanding.

The method of discerning how ancient events in history had forecast succeeding ones was called "typology." A "type," as the eminent Puritan divine Samuel Mather explained, is some outward or sensible thing, some event in history, ordained by God and recorded in the Old Testament to foreshadow something about Christ in the New. Christ's actions in the New Testament fulfill and thus abrogate the events of the Old Testament and announce the New Dispensation. These New Testament actions are therefore styled "antitypes," and the study of typology revealed to these English that God's workings were logical, progressive, continuous, and linear. All was according to plan, and history was the witness of this.

Not surprisingly, it had been Paul who had first discerned the relationship between events recorded in the Old Testament and those of the ministry of Christ. But it was Augustine and more especially Calvin who firmly established the method of typology. Augustine provided the patristic authority for the method with his blunt comment that in the Old Testament the New lies hidden and with his own exegeses of the texts in *The City of God*. In Calvin's assiduous hands the ancient practice became the science of the inescapable. In typology the historical tendency of the faith reached its nadir.

Characteristically, the action of the antitypes on the types was felt to be confined to those events recorded in the texts. Though history still continued and would until God had had enough of it, still the great events in history were presumed to have ended with the last words of the New Testament. But here in New England there were many who would see their own history as the fulfillment of the Old Testament type of the Israelites in the wilderness. Were they not children of God too? Were they not also traveling through a horrid "desert," beset on all sides? And though there was no longer a Promised Land, since in their theology all the world had been made into a moral wilderness through sin, there were yet resisting heathens to be conquered here like the Canaanites and Amalekites of old. And did not God have a great plan for these children too? Perhaps if they conducted themselves out here so that their influence would gradually radiate throughout Christendom, they could hasten the ultimate fulfillment of all the divine promises.

Of course, neither Calvin nor Augustine would have sanctioned this interpretation of the English errand into the wilderness, and many sober students of typology like Samuel Mather specifically argued against it. But the temptation was very strong since such an interpretation not only made glorious sense of this terrific endeavor but also elevated it onto the stage of that cosmic drama of the godly against the Devil-driven worshipers of the earth.

By conceiving of themselves as "this little *Israel* now going *into* a *Wilderness*," as Cotton Mather put it in his history of their movement, the Puritans were hoping to have their own special role in Christian history, to be thus reattached to a drama that had become progressively remote. To become reattached to the primal source was a disguised way of trying to escape the "terror of history." And yet the way in which they attempted this escape precluded real success, for the typological construction of Christian history, even so extended to include them, could not issue into anything truly regenerative, and these radical Protestants, even at their most convinced, must often have harbored the strange, somnabulistic sense of merely mimicking actions original to the mighty days of the ancients. Of themselves they could initiate nothing of significance.

Indeed, we now know they did not. Their brief and bitter hour here—less than a century, really, and little enough to make their papist adversaries smile within their long view—is but a sad, microcosmic recapitulation of the history that Christianity had already enacted in the Old World. Securely tied to that history, they extended it into another wilderness, repeating helplessly its negative achievements: the suppression of dissenters; the search for and destruction of alien enemies; and the costly self-repression that would finally sunder their very sect itself. To have been genuine reformers, they would have had to accept the New World. But nothing in their history told them how this might be done.

The Puritans' way with dissenters was a short one, conditioned, as was everything else they did, by the umbrageous presence of the wilderness that loomed and towered on all sides. Huddled in their villages, they felt it and so felt the necessity of tribal solidarity, that internal dissent was an intolerable confederate of that wilderness. So they did not temporize. We know well the names of their most disturbing dissenters—Anne Hutchinson, Roger Williams, Thomas Morton (whose heinous sins we shall confront later)—as well as the sectarian affiliation of those nameless others, the Quakers. But it is surprising how remarkably the dissenters' offenses guide us to their persecutors' own interior misgivings.

In the matter of Anne Hutchinson, for example, the root of the discord that divided her from the community that eventually sent her into exile and death was nothing less than the means and meaning of revelation. Even in the modified Calvinism preached in seventeenth-century New England, revelation was not something humans could come by on their own. Like other fundamental religious matters it had to be almost endlessly mediated, first by the Bible, then by Augustinian and Calvinistic exegeses of the text, and finally by the vitalizing preaching of the ministers. Direct divine revelation belonged only to the age of miracles. Now God spoke through the scriptures, and His ministers lined these out to their congregations, instructing them in the proper responses.

Here, then, was this person—and a woman at that!—who spoke of experiences of direct infusions of the Holy Spirit, who babbled dangerously of intense, private, wordless revelations. Such an opinion obviously ignored the careful, legalistic steps by which Puritans had said one might become justified. But behind this was that old, troubled question of the control of revelation that went back to a much earlier period of Christian history. In these "American desarts" the need to severely control revelation was the more deeply felt. Again, it was the woods, the swamps, and the Indians. Who could tell to what wild freedoms these things might tempt the sojourners? Even Roger Williams, so tolerant of the Indians and receptive to the landscape of New England, was sensible of this temptation and feared to gaze upon the natives at their woodland rites. Doubtless they all recalled vividly the temptations unsuccessfully encountered by the wandering Israelites and also the punishments that so swiftly followed. Those like Hutchinson would have to be effectively silenced.

At her trial Hutchinson admitted to her opinion. In a sudden moment she flamed out at Governor Winthrop, sitting in his moldy judgment, and defiantly insisted upon her immediate revelations—though in Cotton Mather's account she recanted and then withdrew the recantation, trapped between interior confusions and palpable deceits. And, Mather smugly concludes, she was exiled more for the lie than for the original offense and so was justly slain by the savages. But the real truth is that she was the scapegoat after the Old Testament type, sent away from the community bearing less her own sin than a temptation many felt too nearly. As Governor Winthrop correctly observed, "such bottomless revelations" could not "stand with the peace of any state. . . ."

Still less could those revelations of the Quakers, which to the ortho-

dox seemed even more bottomless, and utterly lacking in civility besides. In 1656 the majority levied a fine of forty shillings an hour on any
of its citizens so foolish as to harbor these pretenders to inner light. As
for the interdicted themselves, they were subject to the loss of an ear if
they ever returned to Massachusetts after initial punishment; both ears
for a subsequent offense; and for a third aggravation to have the tongue
bored through with a hot iron. One William Brend was so severely
whipped by authorities that his flesh hung about him in blackened,
jellylike bags and he was left for dead.

Yet still they would chant danger to the theocracy, and Increase
Mather recorded with his family's characteristic mixture of malice and
credulity the end of the wagging tongue of one of them. A certain Har-
ris, missing for some days, was found

> with three holes like stabs in his throat, and no tongue in his head, nor the
> least sign thereof, but all was clear to his neck-bone within, his mouth close
> shut, one of his eyes hanging down upon his cheek out of his head, the other
> sunk so deep in his head that it at first seemed quite out, but was whole
> there. And Mr. Joshua Hobart, who was one of them to view his dead body,
> told "that there was no sign of any tongue left in his mouth."

Belief in Jesus may have been what was necessary for justification, but
manifestly this was not His work but that of the Old Testament God.

Nor were they always content to leave it to Him: on a fall day in
1659 William Robinson and Marmaduke Stevenson were hanged on
Boston Common while drummers hammered home the sentence above
the din of the vindictive and thirsty crowd. Mary Dyer was hanged the
next year and William Leddra in 1661, after which executions of
Quakers ceased by royal decree, to be superseded by public floggings
and humiliations.

With Roger Williams the issue was equally fundamental and more
immediate. His whole life as he both chose and was compelled to lead it
presented the ruling majority with a direct question: What did it mean
to be a Christian? We know, through rumor, about his seventeenth-
century liberalism as well as his sense of the limitations of civil authority. What is somewhat less well known is that the question of Christianity for him came to turn on the whites' relations with the Indians.
As for Las Casas before him, the spectacle of the whites' callous appro-
priation of native lands and equally callous treatment of the dispossessed called into doubt the validity of the whites' claims to spiritual superiority. Not that Williams ever doubted that Christianity itself was

superior to the Indians' religion; to him the Indians were damned, lost souls awaiting Christ's saving message. They erred, he thought, in their polytheism and in their idolatrous worship of nature, including animals. But the question of the real meaning of Christian conduct occurred to him because the Bay Colony had forced upon him the opportunity of viewing matters from the Indian vantage. Living amidst them in Rhode Island, he entered their round, learning their dialects and customs with a thoroughness and sensitivity unmatched in his time. In noting their generous and affectionate ways, their natural piety, and their ability to regulate satisfactorily their internal relations, he came to wonder whether they were not potentially better Christians than those who so arrogantly wore that mantle.

That these whites should pretend to the God-given right to the native lands, expelling under guns and various other more flimsy devices the rightful inhabitants, came to seem to the outcast a monstrous perversion of what Christianity ought to be. Francis Jennings quotes a letter of his decrying the whites' insatiable hunger for ever huger portions of Indian land, "land in this wilderness, as if men were in as great a necessity and danger for want of great portions of land, as poor, hungry, thirsty seamen have, after a sick and stormy passage." This, he remarked tellingly, "is one of the gods of New England, which the living and most high Eternal will destroy and famish."

For Williams, then, the white claim to the lands was unjustified and blasphemous on its face, invalidating the claimers' pretensions to Christianity. And if not Christians, then what were they and what was their mission here? He saw the frail and largely unsupported English attempts at missionizing for what they were, a mask for territorial acquisition and exploitation. From his Narragansett Bay exile Williams could see the face behind that mask and describe its features in print that reached all the way back to England.

Whatever other disagreements he had with the Bay Colony, it was this matter that cut deepest, as—again—Cotton Mather reveals in the attention he gives it in his ecclesiastical history of New England. Calling him a hot-headed man with windmills in his brain who preached furiously against the colony's charter, he remarks that Williams did so "on an Insignificant Pretence of *Wrong* thereby done unto the *Indians,* which were the *Natives* of the Country," a country that all knew had been given the whites by the English crown. For Mather knew, as did John Cotton and the others who crossed pens with Williams, that to question the moral foundation of the charter was to question all. And not just for the English, either. The entire Western invasion of the wildernesses of the globe was predicated on the God-given right that this cantankerous man now questioned. Perhaps it was the savage surroundings to which they had banished him that had put such notions in his head. Whatever, it was with considerable satisfaction that they beheld him in his last years forced to close ranks with them in King Philip's War. There is a touching image of him at the outset of those hostilities, thinking to calm a group of armed warriors with his mild manner and pacific reputation, only to be turned away by one of their old men who told him to go back to his own while he still had the chance. As he had long ago foreseen, the whites' policies had at last rendered the tribes as furious in fact as they had long been in folklore.

All this was not enough: however silenced, sentenced, and banished the dissenters, there remained the threat and the challenge of the external enemies, the heathens. The cosmic drama the exiles felt themselves involved in required external enemies, and as we ponder the words of their favorite scriptural passages, the history of the faith in that world they had geographically left, and their circumstances here in New England, we cannot be surprised that they found enemies and attacked them. Their historical precedent was the Israelites coming up out of the wilderness and forcing their way through idolatrous enemies into a land they would make into their god's earthly kingdom. In the New Testament their man was Paul, whose sense of the cosmic drama had been profoundly strengthened by his personal history as both persecutor and persecuted. And lastly, they had Augustine. The vision of God's people warring against those of the earth is the foundation, the walls, buttresses, and joists of his vast edifice, *The City of God*. Without such a vision, it would surely disintegrate into jumbled fragments of dogmatic assertions and idiosyncrasies. Such a vision founded on dualisms and assumed oppositions *needs* them, for without oppositions the vision is robbed of meaning and becomes absurd.

In the first exile in Holland the Pilgrims had gotten their reports of the New World and had been terrified. But they had also recognized it as the wilderness of old and its inhabitants as those damned ones who as in the long ago would have to make way for the children of God. So too for those who would follow the first wave a decade later: they expected opposition, even slaughter, and came armored for it. When the first colonists received news in 1623 of the great Virginia uprising of the year previous, their response cannot have been too much different from the perverse satisfaction evinced by those colonists who had themselves borne the terrible blow. The Virginian Edward Waterhouse had written in the wake of the uprising that it was well finally to have it out and go on with the necessary business of clearing and ordering the land:

> Our hands which before were tied with gentlenesse and faire usage, are now set at liberty by the treacherous violence of the Sausages. . . . So that we, who hitherto have had possession of no more ground than their waste, and our purchase at a valuable consideration to their owne contentment gained; may now by right of Warre, and Law of Nations, invade the Country, and destroy them who sought to destroy us. . . . Now their cleared grounds in all their villages (which are situate in the fruitfullest places of the land) shall be inhabited by us, whereas before the grubbing of woods was the greatest labour.

It was, in other words, turning out precisely as Sir George Peckham had anticipated.

In poking about the records of the New England past, one feels just this sense of armed anticipation, a fierceness, even a desperate courage, waiting its time to break out. One feels they were only waiting out the good offices of such as Squanto and Massasoit, never doubting in the security of their theology and its texts that strife must come. Some of them would intensify the sense of this conflict's inevitability by their descriptions of North America as the last stronghold of the Devil and the Indians as *his* peculiar and chosen people. Few except Williams were equipped to imagine that these same Indians might be their key to the interior wonders of the New World. Perhaps none were capable of seeing Jacob's struggle with the angel as a type of their situation here. Johannes Pedersen reads that narrative as the resistance of the guardian of the gates to Canaan against the alien adventurer seeking entrance there. If he is right, then it is significant that Jacob does not attempt to destroy the angel but rather to gain his blessing. This would be more in keeping with the directions of myth than with those of history, but it was by the latter that the English sought to find their way inland. For them myth was the hideous error of the ancient past that they were here to oppose and finally destroy. They got their first opportunity against the Pequot.

As Francis Jennings has revealed with admirable clarity, the English used for this war and their subsequent claims to Pequot lands in the upper Connecticut Valley a palpable pretext: the Indian killings of two white traders. Neither trader had actually been killed by the Pequot. One was held anathema by the Bay Colony, and the other had been subjected to a humiliating breech-thumping by a guard of musketeers as he was being expelled from the colony for multiple offenses. It is true that he was later rehabilitated and entrusted by Massachusetts with considerable trading responsibilities, but neither of these men was significant as a cause of hostilities.

As a war it was not much, but its single significant action revealed that the imperative war against a handy enemy was what truly informed the whites' behavior and was in the long view a more important motive than politics or economics. At dawn of May 26, 1637—a short enough time since the great migration of 1630—a body of Connecticut militiamen with sizable Narraganset Indian reinforcements attacked a Pequot village on the Mystic River. The aim was neither conquest nor negotiated capitulation. It was the utter extermination spoken of in the old texts. The whites formed two concentric circles with themselves in the inner one and set fire to the village. Those who were not shot or burnt within were cut down without by the English and their Indian allies of the outer ring.

In words that remind us chillingly of the Crusaders in Jerusalem, one of the combatants described the heaps of bodies, many those of women and children, "so thick, in some places, that you could hardly pass along. . . ." Sometimes, he continues, "Scripture declareth women and children must perish with their parents. Sometimes the case alters; but we will not dispute it now. We had sufficient light from the word of God for our proceedings." In that light, amidst the burning and mutilated bodies, the smoldering mats and personal effects of as many as five hundred Pequot, other matters became visible too. Here the Narraganset saw for the first time that religious rage of their current allies that had lain waiting its moment of profession. And ritual warriors as they were, they cried out in horror at a magnitude of slaughter such as they had neither seen before nor contemplated. This was not war as they had known it. This was something more terrifying, more total, and at the same time curiously less focused.

It was as if here for the first time since landing on these weather-beaten shores the whites had been able to *get at* this threatening wilderness; as if here on the Mystic River they had found out its lair, had carefully surrounded it, and had then subjected it to a purifying fire. Almost forty years were to pass before another such signal opportunity presented itself, and when it did, the Narraganset were to see the truth of this May morning. It had not been the Pequot who had been the target. The target had been that old aboriginal world of which they had been part and that had so lately been mysteriously transmogrified into the "howling wilderness." The Narraganset were forced to see this truth because in what was called "King Philip's War" it was they who became the target of annihilation.

The ostensible target was the "insolent" and "treacherous" Wampanoag, that very tribe whose help had been so crucial to the survival of the English during the first years of settlement. Trouble between the whites and the Wampanoag had actually preceded the arrival of these English, for as had happened all along the continent's coasts the tribes had been treacherously handled by the cruising Europeans. Kidnappings, perfidious shipboard slaughterings, and epidemic disease were these natives' introductions to the civilization of the West—which makes Massasoit's friendship the more remarkable to muse upon. And before the old sachem died he had incurred the enmity of the whites by his increasingly intransigent insistence that English colonization should not include the spiritual corruption of his people. In the last years of his life, witnessing the enlargement of white settlement and the decay of aboriginal purities, he protested the English missionary efforts, which typically commenced with the blasphemous slander of the natives' religious beliefs and practices.

Massasoit died in the winter of 1660–61 and was gathered to his fathers, being succeeded by one of his three sons, Wamsutta. By the summer of the year following, Wamsutta was dead, victim of a disease mysteriously contracted while in the protective custody of an armed detail sent out of Plymouth to haul him in for his indiscriminate sales of land to unapproved purchasers. His successor was Metacom, or Metacomet, who in an effort to appease the English had taken the name Philip. From the time of his ascension until January 1675, Philip and his people were in constant friction with their English overlords.

The surface issue was land—who should own it and under what circumstances. Now after a half century of sales and what the English were pleased to call "quiet possessions," matters looked different to this son than they had to Massasoit. But his complaints against the "quiet possessions" were contemptuously brushed aside by the English as patent fabrications. So now he began to hearken to the wind-blown words of the old Narraganset sachem Miantonomo, who, brooding on the fate of the Pequot, had urged pan-tribal solidarity. No one had heeded

Miantonomo except the English, who had arranged his execution by the Mohegan.

Now, squeezed by the conflicting and overlapping land claims of Massachusetts, Rhode Island, and Connecticut, Philip saw that the ultimate aim of the English was not land but the extirpation of the Indians. He was rumored in these days of 1673–74 to be going to and fro like some tawny Satan among the tribes, stirring them to resistance and outright rebellion. Probably the rumors had substance. Probably Philip was during this time donning the fatal mantle of the patriot chief. Nevertheless, in January 1675, a converted Harvard-trained Massachusetts Indian, John Sausaman, or Sassamon, who had at one time been Philip's amanuensis, visited Governor Winslow's house at Marshfield to report Philip's hatching conspiracy. Sausaman, at least a double agent, never got home. His hat and gun were soon found on the surface of a frozen pond and his body beneath the ice. He was buried, but as the sap began to rise, so did the rumor of foul play. Sausaman was exhumed and pronounced the victim of assassins. A witness was then produced who fixed the blame on two of Philip's men. In this way the conspiracy was confirmed, and one of the convicted Indians told the English that it was to break forth with the leaves of summer.

As usual, the whites initiated hostilities even as a group of Rhode Island Quakers met with the supposedly ferocious Philip and heard accommodating words. But these peacemakers, hateful to those in Massachusetts, were shortly informed that their northern neighbors would have no terms whatever and would presently bring Philip to heel by force. Accordingly, Plymouth Colony arranged the evacuation of those whites nearest Philip's seat at Pokanoket (now Bristol), Rhode Island, and prepared to move against him. In this lull some Indians of that neighborhood began pilfering the now-deserted houses, and one of them was killed while doing so. When an Indian deputation called at a garrison to complain of this, it was told that it was a matter of no consequence whatever to the English that an Indian had been killed. The Indians left straightaway, and the day following the white killer and six others were found dead. So began this most disastrous and decisive of the "Wars of the Lord."

In his official record of the war the Puritan historian, the Reverend William Hubbard, set his title page with appropriate scriptural references intended as guides through the narrative that followed. Their usefulness is terribly clear:

And the Lord said unto Moses, write this for a Memoriall in a Book, and rehearse it in the ears of Joshua; for I will utterly put out the Remembrance of Amelek from under heaven.

(Exod. 17:14.)

Wherefore it is said in the book of the Warrs of the Lord, what he did in the red sea, and in the Brooks of Arnon.

(Numb. 21:14.)

As cold waters to a thirsty soul, so is good news from a far Country.

(Prov. 25:25.)

Not trusting entirely to allusion, Hubbard clinches the connections for his readers. Historical recording, he remarks, which had been for the children of Israel both a guide through the wilderness and a memorial of their experiences in it, could be the same for the "remaining Israel of God in these Ends of the Earth. . . ."

So the historian, alert to all signs of God's hand in this history, records that as the whites moved against Philip at Pokanoket on June 26, 1675, they witnessed an eclipse of the moon, and some among them averred that they had seen in the very center of the moon "an unusual black Spot, not a little resembling the Scalp of an *Indian*. . . ." As they drew closer, there were other ominous signs. Philip and his people were gone but the advancing whites could be sure that they had been here from the smoking homesteads they began to encounter. They went on. In a field near a burned-out house they found a gutted Bible, its leaves profaned by the touch of earth, and farther on, the heads and hands of slain English mounted on pikes alongside the road. But they could not close with their enemies, meeting only many Indian dogs that seemed to have lost their masters.

It was not until the middle of July that they came upon Philip and a sizable group of his followers, after trailing him through "many desert Places, inaccessible Woods, and unknown Paths, which no Geographer's Hand ever measured, scarce any Vulture's Eye had ever seen. . . ." It was near his original seat, and when their first volley killed seven or eight Indians, the rest retired deeper into the swamp, leaving behind a hundred green bark wigwams. That night Philip escaped westward and so prolonged into a very costly war what might have ended here as a skirmish. Thereafter the heaviest engagements were in the flatlands of the Connecticut Valley.

Near the site of their council meeting place some miles east of Deerfield the Indians surprised Captain Richard Beers and his detachment of thirty-six. Twenty of them were killed, their heads fixed on pikes, and one was dreadfully slung up into a tree by a chain hooked through his jaws. In the middle of September the tribes struck at Deerfield itself where they ambushed a company of Essex County men and killed most of them, suffering heavy losses themselves. The next day they signaled their continuing defiance by hanging captured white garments in the bushes on the far side of the river where they fluttered maddeningly in the breeze. In October they hit at Springfield and Hatfield. But now an early winter began to close down on New England, depriving the tribes of food sources and stripping from them the sheltering foliage of woods and swamps. They began to retire back into the country of the Narraganset, leaving their battered persecutors to ponder the inscrutable ways of the Old Testament God who had thus far exposed His children to a fearsome scourging.

The Indian retreat also left the English with the opportunity to reconsider aspects of their strategy, and it was during this time that the fate of the Narraganset was settled. This tribe had been pretty effectively neutralized by English intimidations, and they had even been conscripted once again as nominal allies. In a "treaty" the whites forced on them the Narraganset had been prohibited from giving aid and comfort to any of Philip's followers and had been promised bounties for any enemy heads they should bring in. As a pledge of faith the Narraganset placed several hostages in English hands. But few heads had been produced, and it now appeared that the Narraganset were indeed supplying sanctuary to the combatants of the river valley engagements.

Scruples were aired and blown away on winter winds: here was a chance to deal a blow that would strike exemplary terror into all who might be thinking of joining with Philip come spring; in the longer view, a blow at the Narraganset and their guests would serve the purpose of clearing that neck of the woods of natives. So plans were laid and troops mustered to march against the huddled and wintering Indians who, if driven from their shelters, might be expected to lose hundreds more to the very sharpness of the season's edge.

At Wickford on Narragansett Bay the troops impressed an Indian guide who directed them southwest across the flatlands toward the Narraganset fort. They marched through a bitter and snowy evening and bivouacked in the open fields under a steady fall of flakes. Some hands and feet froze that night, but at dawn the stiffened men were roused and moved off again to the edge of a great gray swamp in what is now West Kingston. Here, so their guide told them, they should find Indians enough before another nightfall.

For once nature seemed in its cruelty to abet their designs: the swamp was frozen and the troops could travel atop the ice, passing through the tangled, slanted, and spindly trees and brushing aside the craggy branches of the low shrubs until they had penetrated to the very heart of the wilderness and to the Narraganset fort. But here none knew how to gain entrance without sustaining ruinous losses until the Old Testament God once more came to their assistance, showing them a corner way in. "Wherefore," Hubbard tells us, "the good Providence of Almighty God is the more to be acknowledged, who as he led *Israel* sometime by the Pillar of Fire, and the Cloud of his Presence a right Way through the Wilderness; so now did he direct our Forces" to the right side of the fort where they could raise their own pillar of fire, both of shot and flames. Once again the wilderness howled with the doleful cries of the women, children, and warriors within.

Did these Indians, as they beheld the English suddenly among them, have time to think back on the Pequot and on old Miantonomo? As they and their fort went up in flames into the darkening sky, did they know the truth of that sachem's prophecy that soon there would be no Indians or any land left to them unless the tribes united? Now it was too late, too late as this massacre told them, even for Philip. Those not killed on the spot were driven out of the swamp before night onto the frozen fields, where many of them died, perhaps a thousand in all, including many women and children. In the swamp, in the glow of a fatal fire, the whites rejoiced in their victory over this part of darkness. The savages had been about to have their dinner when we surprised them, Hubbard writes, when they and their food "fryed together. . . ." Richard Slotkin quotes a contemporary poem of that celebration that strengthens the ghastly image:

> Had we been *Canibals* here might we feast
> On brave Westphalian gammons ready drest.

This was the practical end of King Philip's War, though both sides would fight on through spring and into the first of summer with serious losses. But what happened here had its desired effect on any potential pan-tribal alliance. Before winter had relaxed, some groups were requesting peace—which the English steadfastly refused, knowing they could eventually dictate all the terms—and the flimsy stitchings of Philip's confederacy came apart as news of the Narraganset's fate spread. By early spring the war had degenerated into a hunt for the Wild Man and his shrunken pack of followers and relations.

During these days as winter grudgingly loosened its grip and a sullen spring settled on the devastated land, Philip and his group flitted about, striking here and there in desperate and random fashion, burning out settlers, destroying flocks and herds, and stuffing themselves hastily with what they could get from the ruins they made. A white woman, Mrs. Mary Rowlandson, who was taken hostage in a February raid on Lancaster, Massachusetts, has provided us with our solitary inside glimpse of Philip and his harried people as they ran, dodged, forded streams, and starved through what had once been familiar territory, carrying on their backs, in their arms, or on commandeered horses their old, their infants, and their wounded. With them Mrs. Rowlandson drank horse's leg broth and ate the still-frozen ears of last year's corn. With the Wampanoag women and their infants she forded the ice-flecked and swollen streams. Old Testament references litter the pages of Mrs. Rowlandson's narrative like boulders in a New England field: drenched and demoralized after a hasty fording, the captive quotes the Psalmist on those storied ancient ones: *"By the Rivers of Babylon, there we sate down: yea, we wept when we remembered Zion."*

By July 1676, when the English killed about a hundred Wampanoag near Pokanoket, the only remaining matter was Philip himself, imaged in white literature as a rebel against God, an infernal monster. On August 12 they ran this Wild Man down in the swamps of Pokanoket where, so Hubbard says, God had caused him to return for the execution of divine punishment. This time there was to be no escape, even though after first jumping their quarry the hunters became mired to their middles in treacherous bogs. Waiting at one stand, a turncoat Wampanoag and a Plymouth man intercepted the fleeing Philip, but in the swamp dankness the Englishman's gun flashed harmlessly in its pan, and it was left to the Indian with his big-boled, two-barreled weapon to shoot Philip through the heart.

So, concludes Cotton Mather in his account of the war, on that very spot where Philip contrived his conspiracy, this "Agag" (the Amelakites, again) was cut down and then cut up in quarters, his head that had brewed this serpent's broth being sent over to Plymouth where it providentially arrived on the very day the people were keeping a solemn thanksgiving to God. And God, writes Mather, "sent 'em in the Head of a *Leviathan* for a *Thanksgiving-Feast.*" This was the second Thanksgiving. Mather records that visiting Plymouth twenty-four years later he beheld the whitened skull of the rebel fiend. Not satisfied with victory, nor even with death and extermination, this driven man tells us how he tore off a jaw of that blasphemous Leviathan, perhaps to keep it in a study crowded with other mementoes of God's providential handiwork.

Such as it now was, the land was indisputably theirs. There would be other wars, and the French would harass them on the north and west, but New England by any name belonged to the English. They had cleared much of the landscape, and Hubbard notes with satisfaction the relations of several who had traveled from Massachusetts well into Connecticut without once encountering an Indian or even the rumor of one who had recently passed that way. But the cost of this had been frightful, not only to the outlying settlements but to the larger interior ones too; Boston itself had felt Philip. Larzer Ziff tells us that one in eleven families had been burned out and one in eleven of the militia killed. Large areas of croplands had been left untended or else had been destroyed. Food stocks had been seriously depleted, poverty was extensive, and vagabonds now appeared in a land where once any sign of shiftlessness had been suspicious.

Other calamities now visited the New Englanders in the clustered shapes of huge fires in Boston, a smallpox epidemic, and another one of shipwrecks. In 1679 the general health of the colonies had been so weakened by the purging of the Indians that an official inquiry by the ministers led to the confession that God had a controversy with New England in consequence of its decline from its primitive purity. Sad as this was to them, it may be sadder to us, announcing as it does that old Christian theme of declension. In less than a century those who had pledged and sacrificed so much in the name of purity were now voicing the familiar lamentations.

The lamentations were perhaps, as Perry Miller persuasively argues, ritual ones. Yet the ritual itself is more significant than he allows—though he grants that something had gone out of the Protestant movement as early as 1600. Certainly the dream of grand renewal had died here in New England. And kindred dreams had so often died in the Christian past that people had learned the appropriate responses. What lived on were the vestiges and phantasms of that dream—a dry, legalistic theology, churches, a clergy, and nightmare notions that shadowed the bustling brightness of these days when a new commercial culture arose out of the failed theocratic one. For even as New England succumbed to commerce in the last years of the seventeenth century and correspondingly altered its conception of its errand into the New World, those nightmare notions of the Satanic wilderness they had cleared still haunted them.

For one final time they all felt the old demonism of the continent. In 1692 the Devil danced in the woods outside of Salem and the long-silent voices of the repressed clamored sorcery. Fittingly, it was two West Indian slaves who gave shape to this last visitation with their archaic incantations from out of that darker continent, violated so long ago by Prince Henry's Portuguese. Their magic drew children hungry for some of it in their own lives, but there were unsuspected others with similar needs, and shortly the thing was out of hand. Suddenly again the landscape was malignly alive, prompting that lightning rod Cotton Mather to his most hysterical pronouncements. "The *New-Englanders* are," he writes of this time

> A People of God settled in those which were once the *Devil's* Territories; and it may easily be supposed that the *Devil* was exceedingly disturbed, when he perceived such a People here accomplishing the Promise of old made unto our Blessed Jesus, *That He should have the Utmost parts of the Earth for his Possession.*

So it was natural that "never were more *Satanical Devices* used for the Unsetling of any People under the Sun" than those already employed against New England's saints and those now so horribly visible as the witchcraft contagion spread beyond Salem Village to other towns until it reached into the very quarters of the colonial governor himself.

Devils were everywhere, as the Indians had so lately been, and Mather and his ministerial associates flung scripture and demonology at them as if these reverend men were themselves possessed. But in the end it was neither scripture nor demonology but New England practicality that calmed the haunted ones. Simply, too many had been accused, too many convicted, and already too many had died. This was not Spain of the Inquisition, nor even New England of the first years where the wilderness had pressed on all sides, but a large cleared space whose ports carried on an intercourse with a world that had banished such contagions to the crowded realm of delusion. New England had come near to being destroyed in King Philip's War; it would not now commit suicide after a hard period of economic recovery had brought it to a new level of prosperity.

Cotton Mather might fulminate about this world as the "Devil's *Field*" in every nook of which devils were encamped, and he could express amazement when, in the year after the contagion had been broken, his call for further communications of remarkable providences went almost unanswered. But his audience had other ears now. Malign spirits were departing from this new New England before the bright blade of progress. The spirits of aboriginal America would go westward into the sinking sun, with the skulking tribes themselves, out into the ragged edges of woodland, and beyond to the unimaginable prairies. In the clearing they had made, the white people faced each other and confessed their delusion. Old Samuel Sewall, standing in his pew at Boston's South Church, hoped publicly that his judicial errors would not pollute the land with the innocent blood that had been shed. His fellow parishioners still knew his scriptural reference, of course, even if they had long ago passed beyond a living belief in what it implied:

> . . . ye shall not pollute the land wherein ye are: for blood defileth the land: and the land cannot be cleansed of the blood that is shed therein, but by the blood of him that shed it.
> Defile not therefore the land which ye shall inhabit. . . .

Sewall and the others were probably right to remark this passage, but wrong to think its truth could so easily be avoided. Could they have availed themselves of him, Shakespeare might have been nearer the actual situation where he has his Prospero acknowledge Caliban as in some way his responsibility: "This thing of darkness I acknowledge mine."

PART THREE

# Haunts

# Possession

A mong the thousands of artifacts left to us, the inheritors of the centuries and acts of exploration—stone markers buried deep in jungles, rusted bits of armor accidentally exhumed, moldering slave castles, sailing orders, and cairns in wooded swamps—perhaps none is so absorbing as our legends of whites captured by natives. And none might have as much to tell us of the spiritual stakes involved in exploration.

Such legends began to appear singly like early morning clouds well before the dawning of the Age of Exploration. The great Arab geographer Edrisi, for example, included one in his updating of Ptolemy. Here he told of a fourteenth-century voyage by eight Lisbon relatives who became subsequently known as the Wanderers. They were the first, he tells us, to undertake exploration of the Green Sea of Darkness, sailing into it for eight days until they were wrapped in the fetid mists of mystery. Fearful, they turned southward for twelve days before reaching an island inhabited only by sheep whose flesh was so bitter it could not be eaten. On they sailed another twelve days until they raised another island, this one with houses and fields. Approaching it, they were surrounded by natives of great height and beauty and they were taken captive. After four days in prison they were visited by the king of the island and his Arabic-speaking interpreter, and when they told of their mission to explore the Green Sea, the king laughed in their faces.

"Tell them," he commanded, "that my father once sent slaves to explore that waste, but that after a month's sailing they were still in gloom and so returned without having learned anything."

The Wanderers waited in prison until a homeward wind arose. Then they were blindfolded and put aboard a ship that three days later touched the African west coast, where they were put ashore, still blindfolded, their hands tied. The natives vanished back into the sea's horizon, but the Berbers came to the rescue, and the Wanderers eventually made their way back to Lisbon where they excited so much interest that a street in that city henceforth became known as the "Street of the Wanderers."

More fully reported was the plight of the German Johann Schiltberger, captured at the battle of Nicopolis in 1396 and then lost sight of for thirty-three years except for fleeting glimpses of the luckless man traveling as a captive in the caravans of Bajazet, the great Shah Rukh, and Timur the Lame (Tamerlane).

Of this last potentate the West received thrilling news from Ruy Gonzalez de Clavijo, a Spanish nobleman dispatched to Timur in the spring of 1403. But he arrived at Samarkand under armed guard and there, held firmly under the armpits, gazed on the face of that heavy-lidded, half-blind legend as he played at chess and dispensed summary punishments to offenders, surrounded by elephants and a polyglot population driven here from the four corners of the world Timur had raced through. Clavijo and his little party were suddenly released in 1405 because Timur had died and his headmen feared the Spaniards would discover this and publish it to the world. The news got out anyway and overtook the hurrying Westerners on their way back, and they could already see the loose fabric of Timur's dominion begin to rend in the winds of rebellion.

These were the scattered precursors. As the West extended its probings, the legends, some documented, others not, kept pace, informing and thrilling those who had not yet been out.

Of New World captivity legends, one of the earliest is that of Hans Stade of Hesse, one of those international adventurers so conspicuous in the early history of exploration. German guns and gunners were much in demand in Europe, and Stade was employed by the Portuguese in eastern Brazil in 1548 where he was ambushed and captured by Tupi tribesmen against whom the Portuguese were warring for access to brazilwood, pepper, and cotton. Stade endured ten months in the midst of a people who, so he reports, practiced ritual cannibalism against their enemies, and he describes this practice in considerable detail. He was eventually ransomed and either wrote or dictated a narrative of his adventures after his return to Europe.

On the continent of North America the earliest captivity narrative is that of Cabeza de Vaca. This thirty-eight-year-old veteran of Europe's brutal wars (he had been at Ravenna in 1512 where twenty thousand corpses littered the fields), whose grandfather had participated in the despoilment of the Canaries, was one of four survivors of the fantastically bungled expedition of Pánfilo de Narvaez to Florida in 1528. Here the fat leader had thought to discover another Tenochtitlán to compensate for the city that Cortés had beaten him to.

The expedition landed on Florida's western coast in April. They were shown a few golden nuggets and then plunged feverishly inland after more. Struggling through swamps and palmetto groves, their armor rubbing them raw, they found nothing of value, only small thatched villages and big naked archers who were accurate at two hundred paces with their seven-foot bows. By the beginning of August they had stumbled back out of that "dismal country" onto the beaches of the western panhandle where they improvised barges, their hope now only to escape. Alas for them, very few did. Narvaez, paralyzed by the unsuspected power of this wilderness, could not command, the expedition broke up into flotsam, and the governor was swept out to sea with part of it, calling back heartlessly to Vaca to command himself as he saw best.

What had been six hundred, and then four hundred, now became ninety. Soon they were but sixteen. On an island in Mobile Bay they fell to eating each other and named the place "Malhado," Island of Doom. At last there were only four, and it was these who made their way back along the curving coastline to New Spain. Eight years after Narvaez had touched land the wilderness miraculously opened and the tattered, skin-clad survivors stood again in the presence of their countrymen.

Vaca's story caused a sensation at the Spanish court where he first told it, and it sparked two grand expeditions, which together traversed most of the continent, east to west, but neither Soto nor Coronado found what they sought. When the latter first made contact with the Indians at Há-wi-k'úh he could not penetrate the town's interior nor so much as suspect its mythic dimensions. Where Vaca had led him to anticipate cities of gold, he saw only mud houses (you may still see these golden cities, however, when the sunset strikes the cliffs, but the town itself is rubble).

With the establishment of the whites in North America the captivity narrative became the first form of popular literature, changing its emphasis as the settlers' needs changed, yet retaining always a certain recognizable core that could evoke powerful responses. Until the end of the eighteenth century the narratives emphasize the demonstrable fact that the hand of God could penetrate the darkest and most tangled interiors to rescue His children from the hands of the savages. Whether French Catholic, as was the Jesuit father Isaac Jogues, or English Puritan, as was Mrs. Rowlandson, the protagonists of these narratives never weary of seeing divine providence in even the most ghastly of circumstances. When his thumb is cut off by his Mohawk captors, Father Jogues sees this as an opportunity to praise the Lord for having allowed him to suffer for Him these seven years in the wilderness, and with his whole hand he offers the mutilated member to the Heavenly Father. Thinking back on her ordeal as a captive in King Philip's War, Mrs. Rowlandson is able to make sense of it and of her terrible losses by seeing it as an opportunity to reflect on how lax a Christian she had become.

After clearing the easternmost strip of the continent, whites turned their attention to those pockets and borderlands where aboriginal remnants hung on. Now the captivity narrative served to fan the dimming fires of racial hatred in those whose ignorance of the earlier years of conquest might cause them to harbor notions of tolerance. These narratives showed vividly again and again to the whites of such places as upper New York State and Pennsylvania that there was still a red menace, still work to be done.

Even in the middle of the nineteenth century, when all of America east of the Mississippi was solidly in the possession of whites and such Indians as had been suffered to remain there had been reduced to harmless—if noisome—beggary, the captivity narrative continued in its popularity. For if it no longer spoke to the present realities of the majority of its readers, who gazed with satisfaction on their cities and towns, their broad and ample fields and harmless copses which contained no least whisper of the old menace, it could still remind them of their heroic origins and explain in a satisfactory and dramatic way how all they now enjoyed had been won.

Two introductions to the same narrative, separated by three decades of nineteenth-century American history, illustrate this changed emphasis. In introducing the captivity narrative of Mary Jemison in 1824, James E. Seaver draws our attention to his scrupulous job of editing since "books of this kind are sought and read with avidity, especially by children, and are well calculated to excite their attention [and] inform their understanding. . . ." Consequently, Seaver writes, the style has been kept simple and the content strictly factual lest such young readers derive inaccurate ideas of so crucial a subject. In his 1856 introduction to the Jemison narrative, the eminent specialist in Iroquois culture Lewis Henry Morgan knows he is addressing a different audience. For it the Indian menace is scarcely a reality, so the captivity narrative cannot serve as a manual of immediate instruction. It may still serve, however, as a memorial to the trials of those wilderness generations who had carved this pleasant land into being. "As time wears away," Morgan intones,

> we are apt to forget, in the fullness of our present security, the dangers which surrounded the founders of the original colonies, from the period of the French and Indian war to the close of the Revolution. It is well not to lose our familiarity with these trying scenes lest we become insensible of our ever-continuing debt of gratitude to those who met those dangers manfully, to secure to their descendants the blessings we now enjoy. This narrative, while it brings to light a few of the darkest transactions of our early history, is not without some instruction.

And though Morgan does not say so here (he does in an appendix), his readers were expected to draw some present instruction from the narrative by simply casting their eyes westward where that last and hugest portion of the continent was even now being forced open by the advancing pioneers.

As the frontier moved westward, so did the locale of the captivity narrative, first onto the Great Plains, then into the rugged thrusts of the western mountains, and at last into the forests and watered valleys of the Northwest. But by this time it had dropped all but the flimsiest pretensions to history and was parading itself nakedly as what in fact it had always been: a drama of fear and resistance directed against all of the New World. Handed about within families and from one family to another, sometimes going into more than thirty editions and at other times passing wholly beneath the channel of written literature into oral tradition, the captivity narrative gradually succeeded scripture as the means of understanding why things had developed here as they had.

But while the emphasis changed, the interior of the narrative remained remarkably consistent, even in its late and palpably fictive stage. And, as Richard VanDerBeets has pointed out, that interior is curiously mythic in form, for it describes the pattern of separation, initiation, and return that we have noted as belonging to the myths of heroic questers. Typically, the narrative begins with a scene of domestic tranquility, emphasizing the innocent husbandry of the frontier family: the mother at churn or spindle with her newborn babe in her arms; the older children at play or chores; and the father dropping seed into the fecund fields. Suddenly the wilderness that looms at the edges of the civilized clearing erupts—as it has always threatened to do—the red fiends bursting from their unhallowed sanctuary. The father is instantly slaughtered, the brains of the baby are dashed out against a tree, the cabin is ransacked and burned, and the captives are carried off into worse than Egyptian blackness. Then there is a long, forced flight through the brambles and snags of the wilderness, across rivers of no returning with the panting and often wounded captives torn bloody in their passage. At last they arrive at the heart of darkness, a hellish thing, a clamorous clearing that is a caricature of that left so far behind. Here the captives are subjected to the tortures of these damned souls, perhaps forced to run the gauntlet while savage squaws rain lethal blows on them with studded clubs. Or they are tortured piecemeal like Father Jogues. Even worse, they might forcibly be "adopted" into the tribe, and if women, compelled to marry swarthy, grunting warriors. And finally the "redemption," a significant term almost universally used to describe a captive's return to civilization, whether through ransom, prisoner exchange, or escape.

So the captivity narrative was the perfect scripture for a civilization's sense of its encounter with the wilderness, for in the redemption that rounded it out there was victory. The happy ending was a triumph, an ultimate mastering of everything the wilderness and its natives could throw up in the way of opposition and temptation. And the losses served to strengthen the impression of a grand pattern in which civilization absorbed its defeats, even its lost colonies and failed expeditions, and still went ahead with the work of civilizing.

It is the sense of victory over the wilderness that distinguishes the captivity narrative from that mythic pattern it impersonates, for in the interior of the narrative where there should be initiation there is only resistance. And at the conclusion, the redemption, nothing new is brought back out of that fearsome other world except a strengthened determination to ultimately subdue it, establish dominion over it, and so utterly change it. What really is being dramatized in this tradition is the historic Christian fear of becoming *possessed,* possessed by the wild peoples, yes, but also, more profoundly, by the wilderness and its spirits. We might say that it is the fear of going native.

The dread of being entered into and taken over by malign spirits is not limited to Christian civilization but is in fact widespread and has been found among many cultures living in intimate connection with the natural world. But in Christian civilization possession is wholly negative in imputation, whereas in many native cultures possession is actively sought. Not only are there positive aspects to non-Christian possession trances in which an individual feels himself or herself to have been entered into by divine spirits and so acts in accord with them, but there is also a more general and less spectacular pattern of seeking out direct encounters with the spirit world in which the individual might feel again the primordial surge of at-one-ment and be momentarily transfigured. Individual vision quests, vigils, communal ceremonies, even the recitation of tribal legends and myths, are all means of continually revivifying that necessary sense of belonging to or being merged with an encompassing order of things. In all these situations the individual will is "captured" and submitted to a greater power, and though this must often be a fearful experience, it is nonetheless a desirable one.

To Christians, such behavior had long seemed devilish, and the distinct bias of the Judeo-Christian tradition is that possession is a state to be feared and avoided. True, there are scriptural instances of divine possession in both the Old and New Testaments. Samuel, Saul, David, Elisha, and Deborah in the Old Testament, for example, are spoken of in terms that leave little doubt that they had experienced such possession. So too had Balaam, who would have spoken against the Israelites had not Yahweh entered into him and prevented the cursing. But the heavy emphasis in the Old Testament is on demonic possession, especially that which might overtake the unwary in wild, uninhabited places such as in the desert or about the ruins of blasted civilizations where the shaggy satyrs howled to one another. The Israelites, we recall, had been evilly possessed in the wilderness when they had abandoned themselves in a dervish dance to the golden calf.

This emphasis is taken over into the New Testament where one of the most telling manifestations of Christ's power is his ability to drive out the demons that possess the suffering. Justin Martyr (d. A.D. 165) claimed that Christ's fundamental mission had been to exorcise demons, and A. D. Nock, writing of the early church, claims that Christian exorcism was unique in its time in "that it was an integral part of a religion." So compelling were these precedents that for future generations of Christians they practically defined what possession as a state meant. Against this force the example of the primitive Christians could not prevail even as they were shown encouraging possession, trances, ecstatic speaking, and visitations of the Holy Spirit. We have seen the ways in which mature Christianity severely delimited immediate, ecstatic revelations, including the silent banishment of the Holy Ghost, and we have seen too its treatment of self-proclaimed prophets, of mystics savaged by the Inquisition, and of those like Montanus who claimed to speak not as themselves but as organs of an indwelling spirit. Such individuals were possessed, all right, but malignly so. They had been captured or had let themselves go, had lapsed back into that old ooze of night and nature from which Christian history had redeemed mankind.

This view of possession gained huge authority as Christians began encountering the nature-worshiping natives of the wildernesses, and our documents of exploration speak—in Portuguese, Spanish, French, English—of the horror with which the Christians looked on the spectacles of native religion: the dancing, drumming, idols, and, maybe most of all, the shamans in whose trances and speakings the whites could see nothing but the work of Satan or malicious and meretricious pretense. For like the Israelites at Sinai and the doomed worshipers of Baal, these dark ones seemed plainly to have succumbed to their wild habitats.

No. The thing to do was to *take* possession without becoming possessed: to take secure hold on the lands beyond and yet hold them at a rigidly maintained spiritual distance. It was never to merge, to mingle, to marry. To do so was to become an apostate from Christian history and so be lost in an eternal wilderness.

North Americans are urged to think with a certain cautious fondness on our single great parable of intermingling, the marriage between the white Englishman John Rolfe and the Indian princess Pocahantas. This merging is supposed to symbolize a new and hopeful beginning in the New World. But the records of that arrangement tell us something else: the marriage was not based on any true desire for merging but rather on political expediency and a fear that no marriage could bridge. John Rolfe may, as Perry Miller says, have been horrified to find some genuine affection for Pocahantas overtaking him. God, he knew, had forbidden intermarriage to the Israelites in the interests of tribal solidarity and religious purity, and so Rolfe knew that he must find a means of justifying what otherwise could only be construed as a slide into temptation. So he did, professing and maybe believing at last, that he had entered into this marriage "for the good of this plantation, for the honour of our countrie, for the glory of God, for my owne salvation, and for

conuerting to the true knowledge of God and Iesus Christ, an vnbelieu-
ing creature." On any other terms the marriage would be a wretched
instigation "hatched by him who seeketh and delighteth in man's destruction."

It was a difficult business, this possessing while withholding themselves, and the records show that it was not uniformly accomplished,
though European governments would do their best by sending out
shiploads of white women to keep the white men in the clearings and
out of the woods. For neither the French nor the Spanish was John
Rolfe a paragon, though it was the English themselves who strove
most to avoid his example.

From their earliest days in the New World the English evinced an official desire both stern and shrill to keep the colonists together and to
keep the natives at a safe distance. So they recorded with a sour satisfaction the fate of a splinter colony at Wessagusset, Massachusetts, that
strayed off and kept Indian women but that managed itself so improvi-
dently that it soon came near to starvation and then became "so base"
as actually to serve the Indians for sustenance. One poor, half-starved
fellow was found dead where he had stood in tidal mud in his last
search for shellfish.

In Virginia, Massachusetts, and Connecticut there were stiff penalties
for those who moved ahead of the line of settlement and lived surrounded only by woods, animals, and Indians. Such individuals, it was
reasoned, would soon, like the untended olive, turn wild themselves,
and one of Cotton Mather's most persistent laments was the "Indianiz-
ing" of the whites. As he watched them straggle off into the woods, establish barbarous little outposts, and settle slowly into native ways, he
predicted divine retribution as of old.

What all this tells us is that these newcomers had felt obscurely the
lure of the New World and that some had responded to it out of old
dreams long repressed in the march of Christian history. For the fear of
possession does not make sense by itself, as wilderness travelers Joseph
Conrad and Carl Jung knew so well. Only when the fear is seen as the
obverse of desire does it reveal its full truth. Conrad, witness to the European invasion of the tropics, wrote of those who had succumbed to
the temptation and had become lost forever like Willems in *An Outcast
of the Islands* and Kurtz in *The Heart of Darkness*, who pronounced judgment on his own reversion: "The horror! The horror!"

Jung, so interested in matters of the deep past, has left a record of his
dream life as he penetrated ever farther into the mental zones of alien

places. In these dreams he is menaced by dark strangers and threatened with red-hot curling irons. Of this last, he observes that it is the dream of a white man in Africa who fears that his hair will be kinked, that he will go black. Such threatening dreams revealed to him his own interior conflict. I was prepared, he remarks, to feel a white European's superiority to these people. I was not prepared "for the existence of unconscious forces within myself which would take the part of these strangers with such intensity."

Almost never do our raw wilderness documents speak openly of this lure, of the human need to feel possessed, body and spirit, by a landscape both visible and numinous. Only negatively can they do so, guiding us in their obsessive loathing of those who did succumb. And only this situation can adequately explain the strange persistence and perpetuation through the centuries of the tradition of the willing captive: one who, being captured, refused redemption or ran away after being redeemed, slipping off from the settlements to return to the tribes and the wilderness. They were variously called renegades, white Indians, squawmen, or simply degenerates, and we might imagine that their shadowy careers would have been consigned to a silence beyond obloquy. But it is not so. Instead, their stories have come down to us, loaded, true enough, with all the weight that can be assigned scapegoats, but they *have* survived, and this is significant. Never as popular because never as useful (even as cautionary tales) as the orthodox captivity narratives, their persistence tells us of other visions of New World contact that haunted the whites even in the very midst of their victories.

The stories begin with what was for us the beginning of the Americas, with Columbus. One Miguel Díaz of the second voyage wounded a fellow Christian in an argument and to escape punishment ran away into the woods where he became the consort of the local chieftainess. Like most of his kind, history loses track of him here, but his defection was thought worthy of record in his own time, a strange footnote to conquest.

Then, following like an apparition the traces of further conquests, there are the renegades of Cortés, Narvaez, and Soto. Though, as we have seen, Gonzalo Guerrero was not with Cortés as he pushed off from Cuba for an unknown empire, one would have thought from the diligence with which that driven leader searched for Guerrero that he was indispensable to the success of the expedition. Maybe in some way he was.

With Narvaez there were at least two who preferred to remain with the Indians rather than wander on into unknown dangers, guided only by the hope of coming out once again into Christian territory. Doroteo Teodoro, a Greek, went inland with the Indians and never came out again, though years later Soto's men would catch word out of the thicket that he still lived with his adopted people. And Lope de Oviedo ("our strongest man") turned away from the entreaties of Vaca to escape and stayed on with the natives while Vaca and the others slipped northward in the night.

On the baffled and trackless Soto expedition there were numerous deserters, notably (from a white point of view) Francisco de Guzman, bastard son of a Seville hidalgo; a hidalgo named Mancano; and another named Feryada, a Levantine. And there was an unnamed Christian described by the Portuguese knight who chronicled the expedition:

> . . . the Indians came in peace, and said, that the Christian who remained there would not come. The Governor wrote to him, sending ink and paper, that he might answer. The purport of the letter stated his determination to leave Florida, reminded him of his being a Christian, and that he was unwilling to leave him among heathen; that he would pardon the error he had committed in going to the Indians, should he return; and that if they should wish to detain him, to let the Governor know by writing. The Indian who took the letter came back, bringing no other response than the name and the rubric of the person written on the back, to signify that he was alive.

We have record too of several black slave defectors, notably "Carlos" and "Gomez," both of whom are known to have lived among the Indians for many years. These were presumed to have so little stake in civilization that their actions could more easily be disregarded by their Spanish masters.

Not so easily dealt with, or even entertained, was the possibility that some members of Raleigh's Lost Colony had survived by merging and migrating inland with Manteo's people. Almost two centuries after John White and his party had panted through the sandy seaside woods in much-belated relief of the colonists and had come upon that tree with its interrupted message, the old sentinel trunk was still being pointed out to the curious as evidence of that tragedy. Expeditions were compulsively sent out in 1602, 1608, and 1610 to find some trace of the colonists, even if fatal, but all failed since they could not see the clues. In 1654 friendly Indians showed another expedition evidence that some of the missing colonists had revisited the site of the old fort, but so unthinkable was the alternative of survival through merging that this hint too was left unexamined.

So, for the same reason, were other hints, before and after, including those from John Smith who had knowledge of whites living somewhere inland from Chesapeake Bay, and from a German traveler, John Ledered, who heard in North Carolina of a nation of bearded men living some miles southwest of where he was. Still the hints of survival persisted and in the eighteenth century were so strong that they compelled North Carolina's pioneer naturalist/historian John Lawson to acknowledge them. The Indians of this place tell us, he notes, "that several of their ancestors were white people and could talk in a book as we do; the truth of which is confirmed by the gray eyes being frequently found amongst these Indians and no others." But Lawson could not let the spectacle of miscegenation pass without censure. The evidence of it drove him to remark that the colony had

> miscarried for want of timely supplies from England; or through the treachery of the natives, for we may reasonably suppose that the English were forced to cohabit with them for relief and conservation; and that in the process of time they conformed themselves to the manners of their Indian relations; and thus we see how apt human nature is to degenerate.

It was only in the last decade of the nineteenth century when the menace of merging was dead that a white historian, Stephen B. Weeks, could put all this information together and make the appropriate inference. And this was that the 1587 colonists, being destroyed piecemeal by the vengeful natives of the mainland, took what they could carry and went to Manteo's people on Croatoan. Together these two groups left Croatoan, which was only a seasonal home for the natives, and moved inland by slow stages, away from the hostiles of the Outer Banks area. They were subsequently encountered on the Lumber (Lumbee) River in the mid-eighteenth century when Scots and Huguenot settlers pushed into the area. And their descendants are there to this day in what is now Robeson County. Here one may meet striking mixed bloods who trace themselves back to that intermingling on the Outer Banks and who carry on the names of the Lost Colonists.

Even while the search for the Lost Colonists was vainly going on, others of the English were becoming "lost" under the very eyes of the orthodox, and of these none excited greater opprobrium and more continuing vengeance than Thomas Morton of Massachusetts. With the single exception of a long section detailing the bestiality trial and execution of one Thomas Granger (another horrid example of the temptation to mix in the wilderness), there is no more luminous passage in Governor Bradford's long chronicle of the early years of Massachusetts than that on Morton. This is so because no one transgressed so flagrantly and almost joyously against the powerful taboo that kept the English from mingling with all that surrounded them.

The man had come out from England in 1622. By the summer of 1626 he had usurped his partner's prerogatives in a settlement venture near the present town of Quincy and was master of a heterogeneous group of indentured servants and local Indians. It was not merely that Morton was trading with the Indians and supplying them with guns, powder, and spirits, for by Bradford's admission as well as by other contemporaneous accounts we know that Morton was by no means alone in this practice. Indeed, had all those English been Mortons, there might have been little to fear from selling the natives what would then have been merely hunting arms. What deeply rankled was that Morton was actively encouraging the intermingling of whites and Indians and that in doing this he was accomplishing what was so much feared: the Americanization of the English.

In the spring of 1627 Morton presided over the erection of a Maypole at the place he suggestively and salaciously styled "Mare-Mount" or "Ma-re-Mount." It was too much. A party under the command of the doughty Miles Standish broke in upon the mongrels, arrested Morton, and subsequently deported him, hoping that this would be the end of it. But the man would come back again and again to the scene of his transgression and his triumph. And again and again the English authorities would hound and persecute him. Throughout the New England winter of 1644–45 they kept the old man in a drafty jail, in irons and without charges, and when at last they let him loose he was broken. When he died "old and crazy," the English were satisfied that this threat no longer existed.

But they could not lay to rest that larger threat of which Morton had been but a particular carrier, for it kept appearing. The wilderness that had spawned it would recede by the year and mile, and with it the Indians, but all along the gnawing frontier where contact was still to be had there was the profoundly disturbing and puzzling phenomenon of "Indianization." On the other hand, there were but few examples of Indians who had volunteered to go white and who had remained so. And even the missionaries, to say nothing of their lay captors, seemed ashamed of the pathetic show these converts made.

Observing a prisoner exchange between the Iroquois and the French in upper New York in 1699, Cadwallader Colden is blunt: ". . . notwithstanding the French Commissioners took all the Pains possible to carry Home the French, that were Prisoners with the Five Nations, and they had full Liberty from the Indians, few of them could be persuaded to return." Nor, he has to admit, is this merely a reflection on the quality of French colonial life, "for the English had as much Difficulty" in persuading their redeemed to come home, despite what Colden would claim were the obvious superiority of English ways:

> No Arguments, no Intreaties, nor Tears of their Friends and Relations, could persuade many of them to leave their new Indian Friends and Acquaintance; several of them that were by the Caressings of their Relations persuaded to come Home, in a little Time grew tired of our Manner of living, and run away again to the Indians, and ended their Days with them. On the other Hand, Indian Children have been carefully educated among the English, cloathed and taught, yet, I think, there is not one Instance, that any of these, after they had Liberty to go among their own People, and were come to Age, would remain with the English, but returned to their own Nations, and became as fond of the Indian Manner of Life as those that knew nothing of a civilized Manner of Living.

And, he concludes, what he says of this particular prisoner exchange "has been found true on many other Occasions."

Benjamin Franklin was even more pointed: When an Indian child is raised in white civilization, he remarks, the civilizing somehow does not stick, and at the first opportunity he will go back to his red relations, from whence there is no hope whatever of redeeming him. But

> when white persons of either sex have been taken prisoners young by the Indians, and have lived a while among them, tho' ransomed by their Friends, and treated with all imaginable tenderness to prevail with them to stay among the English, yet in a Short time they become disgusted with our manner of life, and the care and pains that are necessary to support it, and take the first good Opportunity of escaping again into the Woods, from whence there is no reclaiming them.

Colden's New York neighbor Crèvecoeur, for all his subsequently celebrated prating about this new person who was an American, almost unwittingly reveals in the latter portion of his *Letters From an American Farmer* that the only *really* new persons are those who have forsaken white civilization for the tribes. "As long as we keep ourselves busy tilling the earth," he writes, "there is no fear of any of us becoming wild." And yet, conditions being what they then were, it was not that simple. It was not always possible to keep one's head looking down at the soil shearing away from the bright plow blade. There was always the great woods, and the life to be lived within it was, Crèvecoeur admits, "singularly captivating," perhaps even superior to that so boasted of by the transplanted Europeans. For, as many knew to their rueful amazement, "thousands of Europeans are Indians, and we have no examples of even one of those aborigines having from choice become Europeans!"

So there were these infamous and embarrassing scenes such as at the big prisoner exchange on the Muskingum River in 1764, where the victorious whites were cheated of the feel of victory since the Indian prisoners went back to their defeated relations with great signs of joy while many of the redeemed white captives had to be brought bound hand and foot by their red captors lest they try to break away from this new captivity and run back to the old one.

And there were those who could not be redeemed at any cost, or who when redeemed by force simply wasted away in longing for that other life their fellows feared so much. Such a one was Mary Jemison, whose narrative was popular enough in the nineteenth century to reveal in its several editions the changing uses to which such documents could be put. But to actually read Jemison herself between the glosses and patent editorial intrusions with their standard laments for this poor, defenseless woman, forced to depend through the long and dreary years on the "tender mercies of the Indians!", is to find that she decidedly preferred her Indian existence to that offered by the solicitous whites.

Captured with her family in Pennsylvania in 1755 by a party of Shawnee and French, Jemison was subsequently sold to the Seneca with whom she was to live for more than seventy years. Hers was no "forest idyll" by any means. She lived through the Iroquois part of the Revolutionary War and with the rest of that proud people suffered their drastically altered status after peace had been concluded on American terms. Here she refused redemption as she had before, choosing her lot with the defeated Six Nations. Twice widowed, she endured now the heartbreaking dissolution of Iroquois culture, witnessing as her personal portion of it the destruction of her two sons. And yet when her story was taken down in 1823, Mary Jemison could still speak feelingly of the virtues of the Seneca in the prereservation period and lament their present condition. *That* lament, at least, is a genuine one. Above all, she stressed the love that bound the traditional Seneca family together, an emotional bond she was clearly unwilling to relinquish for the proffered comforts of white civilization. So when the Seneca sold their remaining lands on the Genesee River and removed to the northwest corner of New York State, Jemison did not hesitate long to follow after them, though she could have stayed on her fertile river flats, surrounded by white neighbors. She died a Seneca, and though her editors have tried to present her story as a tragic one, the real tragedy in it is the view it gives us of what white civilization did to the Iroquois.

Mary Jemison's story is paralleled in many respects by that of Frances Slocum, captured by the Delaware in Pennsylvania's Wyoming Valley in 1778. Like Jemison's mother, Frances Slocum's mother anguishes over the fact that she did not witness the murder of her little girl, preferring that instantaneous destruction to years of fruitless speculations on a worse fate. But when a traveling trader accidentally discovered this "Lost Sister of Wyoming" living with the Miami Indians sixty years later, he found that she in no way considered herself lost; that was an imputation that would have to await publication of her "tragic story."

When her whereabouts became known, her two surviving brothers hastened to Indiana Territory to redeem her. Here they gazed with horrified incredulity upon the spectacle of this old Indian woman moving serenely about her cabin amidst her dark and blanket-wrapped relations. And one brother cried out, "Oh God! Is that my sister?" Macon-a-quah felt none of this, only a suspicion that these whites had come to take her away from her people, as indeed they had. She refused to go, saying to them:

> . . . I have always lived with the Indians; they have always used me very kindly; I am used to them. The Great Spirit has allowed me to live with them, and I wish to live and die with them. Your wah-puh-mone [looking-glass] may be larger than mine, but this is my home. I do not want to live any better, or anywhere else, and I think the Great Spirit has permitted me to live so long because I have always lived with the Indians.

But the brothers could not leave it at that. They came back again several years later to meet with the same refusal, by which time Frances Slocum had become a minor national cause. In 1840 her tribe had been coerced into signing away its remaining lands in Indiana and removing west of the Mississippi. But the Slocums' congressional representative was able to push through Congress a special resolution exempting this solitary white woman from the fate of her tribal relations, and John Quincy Adams was one of those who argued eloquently for its passage.

When the Miami finally did remove, they left Ma-con-a-quah behind on a 640-acre tract of rich land, now to be surrounded by the surging whites, who immediately desecrated the aboriginal graves. Finding the old woman defenseless and a white Indian at that, they boldly encroached on her property, stealing livestock and ponies. In this situation Ma-con-a-quah lingered a year and then lapsed into a resigned illness in which she refused all white medicine, saying that as all her people had gone away, she wished now to do so.

In death she was betrayed by the Christian burial her family had to provide even though they knew better than any that she had remained unconverted. Nor was this the end of it: more than seventy years after Ma-con-a-quah's death, her family was still trying to bleach that aboriginal stain out of her. At a meeting of the American Association for the Advancement of Science, Charles E. Slocum told the Anthropology Section that Frances's Christian Quaker upbringing had protected her from Indianization throughout those long years of bondage. It was like some impenetrable stockade inside of which she had always been white.

Perhaps the most absorbing of such stories is that recorded in 1819 by Edwin James who took down and translated the reminiscences of John Tanner who had been captured in 1789 from his father's wilderness clearing at the mouth of the Big Miami River. John Tanner Sr. had been a preacher from Virginia who, like thousands of other whites, had moved into what was still very much contested territory. A band of Ottawa, watching their opportunity from the woods about the small field in which the father and his slaves dropped corn, kidnapped the boy and took him up into the old Northwest Territory where he was to replace the dead son of an Ottawa woman.

For thirty years Tanner shared the hard life of the Ottawa and Ojibwa in these woods and waters, moving with them in their seasonal migrations, following the game trails, harvesting fish and wild rice from the cold northern lakes and streams. And like Mary Jemison's, his was no romance of the woods. He suffered the abuse of a cruel stepfather who regularly beat him, who rubbed his childish face in a mass of steaming dung, and once tried to kill him for having fallen asleep on a hunting stand. Like the others, he endured the incidents of aboriginal existence, the seasonal starvation, the frostbite, the wounds of hunting accidents. And like them too he suffered the new penalties imposed by the encroaching white presence: horrible bouts of drunkenness when the Indians, fouled on traders' rum, would sell all they had so painfully gathered for a few more kegs of the stuff and then fall into days and nights of insane behavior, stabbing, shooting, and robbing one another, biting off fingers and noses, until the last bit of poison had drained out with their blood; smallpox; diphtheria; and the inevitable and rapid depletion of the ancient hunting grounds.

Still, he moved on with them, out of the hearing of English or even French tongues, refusing his chances to come out, until there was nothing but the sounds of campfires, insects, bird songs, and—in winter—the agonized clacking of the straitened trees and the crunch of snowshoes. At last he became so fully possessed by this life that it entered his dreams.

His second and last stepmother, a remarkable old chieftainess named Net-no-kwa, had taught him the power of dreaming as she had taught him that only slightly lesser power of aboriginal mother love (beside which many with the opportunity of comparison found white mother love a pale substitute). On several occasions Tanner remembered that Net-no-kwa had saved her little band from starvation through prayers and dreams of deliverance, once sending him out into the snow-bound woods to find in a meadow the hole of a bear she had dreamed. And Tanner had gone, his musket charged for big game, and in an unknown place had come into a small clearing where he had fallen partway into a bear's hole. Scraping the snow away, he exposed that shaggy, sleeping head and then killed the dream bear, digging afterward into its eyes and wound with a stick to make sure of it.

Now he dreamed his own dreams. Late one fall Tanner and his band became seriously reduced through hunger. The weather had turned very cold, but no snow had fallen and the tracks of the game were all but invisible. When they could find some, the animals would hear their approach over the iron ground and its dry, brittle leaves. In this situation the white Indian resorted to a medicine hunt:

Half the night I sung and prayed and then lay down to sleep. I saw, in my dream, a beautiful young man come down through the hole in the top of my lodge, and he stood directly before me. "What," said he, "is this noise and crying that I hear? Do I not know when you are hungry and in distress? I look down upon you at all times, and it is not necessary you should call me with such loud cries." Then pointing directly toward the sun's setting, he said, "do you see those tracks?" "Yes," I answered, "they are the tracks of two moose." "I give you those two moose to eat." Then pointing in an opposite direction, towards the place of the sun's rising, he showed me a bear track, and said, "that I also give you." He then went out at the door of my lodge, and as he raised the blanket, I saw that the snow was falling rapidly.

At the very crack of dawn Tanner started from his lodge through the heavily falling snow, following the dream course. Before noon he came on the tracks of two moose and soon killed them both, "a male and a female, and extremely fat."

The dreams and their successful conclusions came out of the profoundest depths of the aboriginal religion, nurtured by that not-wilderness they knew so well, and Tanner came to feel this connection as much as any of them, as once when a company of canoes had started across one of those inland lakes whose waters can become so dangerous in the instant of a storm. Well launched upon the waters, the old chief in charge stopped their passage and standing in his canoe addressed a prayer to the Great Spirit. "You," he said, "have made this lake, and you have made us, your children; you can now cause that the water shall remain smooth, while we pass over in safety." Then, singing, he dropped his offering of tobacco upon the surface.

Here it was: the aborigines, exposed in their frail crafts, in the midst of an untamed power they could only hope to supplicate and draw upon. They passed over in safety, and Tanner passed on with them and would doubtless have escaped recording in the white man's history—except for an entry in a trader's journal—had not that history been so relentlessly transforming the world he had come to love. For it was the whites who drove Tanner from the deep woods.

The agency of transformation in the Northwest Territory was the rise of Indian prophets who came to the tribes early in the nineteenth century, bearing their messages of reform aimed against that increasing and increasingly disastrous dependency of the tribes on the whites. These were the charismatic or would-be charismatic figures who in that ageless pattern had meditated on the problems of their people and had come forth with programs for regeneration. The first of these to affect Tanner's people was the Shawnee prophet Tenskwatawa, brother to the great Tecumseh. A reformed alcoholic, like Handsome Lake of the Seneca, Tenskwatawa began about 1805 to preach a return to precon-

tact aboriginal ways, an avoidance of whites, and a discarding of white
goods upon which the tribes had come to depend: flints, steel, and
rum. In the way of such matters this message, and kindred others that
followed it with telling frequency as the tribes felt the white threat
more keenly, could not fail to jeopardize a white man's standing among
Indians. And since Tanner had a whisper left in him of the white man's
rational skepticism, he made the mistake of ridiculing the prophetic
commandments.

Eventually a prophet appeared in the very midst of Tanner's band, a
man of poor reputation and poorer means named Ais-kaw-ba-wis,
whom Tanner openly laughed at for his pretended intercourse with the
Great Spirit. But so altered and desperate was the condition of the people that this man, widely rumored to have eaten his own wife to fend
off starvation, was soon given solemn credence. And there came a day
when he called the band together in a large lodge for the purpose of
passing on the latest communication from the Great Spirit. He gave the
white Indian an ominous prophecy. This long line, said he, drawing
one in the dirt of the lodge floor, represents the life of the Indians. But
this short and crooked one here represents your life, which shall not be
one half of theirs.

For John Tanner as a white Indian it was to prove a dark prophecy.
Not long afterward a relative tomahawked him from behind as he sat
among his family in his lodge, and the blow would have been a fatal
one had it not been blunted by the thick moose-hide capote he was
wearing. On another occasion, as he lay convalescing in his lodge, his
mother-in-law came in from the field and beat him badly about the
head with a hoe. The prophet had told the woman that Tanner had
wished the white man's disease (measles) on her child that had died.

It was time to come out, but not quite time enough: on his way to
the settlements with his three children and wife, the latter set him up
for assassination. Paddling hard against a river current and in close to
the bank, Tanner heard a shot, saw the paddle drop from his deadened
hand and his assailant fleeing behind a blanket of musket smoke. He
would not die, but he was clearly through as an Indian. Even his dream
visitors had bade him farewell.

Forced out into the settlements, he became for a time a trader,
though because he would not use rum he was an ineffective one, and
then an interpreter for the famous Indian scholar Henry Rowe Schoolcraft at Sault Ste. Marie. At last he was an outcast, the white savage of
his settlement. In the woods he had hesitated to come out, knowing

that however precarious his situation was becoming, it would be even more so among the whites of frontier Michigan. He knew too well the hatred in such fringe areas of those like himself with his braided hair and barbaric tongue. Now he felt it all.

At first, he seems to have gotten along passably with Schoolcraft, but soon enough the scholar embraced community sentiment and turned Tanner out. We have an image from Schoolcraft of his moody employee, a misanthrope of the woods, following the scholar into a boathouse in a walking altercation and there menacing Schoolcraft until driven away with a heavy walking stick. Thereafter Schoolcraft observed that "This being, in his strange manners and opinions, at least appears to offer a realization of Shakespeare's idea of Caliban. . . ." Literary allusion, like history, had moved this far inland.

Now the white savage lived by himself in a shack on the edge of the settlement, frozen between two ways of life with access to neither. In his reminiscences he had told his editor a story of an Indian who had died and had gone up to the gates of heaven seeking admission there but was turned away because he was an Indian. So he had turned then toward the heaven of his fathers, but when he arrived there he was turned away also. "You have been ashamed of us while you lived," the head chief told him. "You have chosen to worship the white man's God. Go now to his village, and let him provide for you." Tanner had never repudiated his Indian past and so, ironically, he shared a similar fate.

For the community of Sault Ste. Marie the summer of 1846 began with a mysterious flash fire that burned the Schoolcraft house to the ground. Then on July 6, James Schoolcraft, Henry's popular younger brother, was shot and killed from ambush. On that same day Tanner's shack also burned and the white savage disappeared. Suspicions instantly focused on Tanner, and a Lieutenant Tilden from nearby Fort Brady led a posse into the bush in search of him. They never found him. Only the next spring did a French trapper report his discovery of a skeleton in the woods, beside it a rusted musket and some coins partly melted by a forest fire that had burned through in the intervening year. A Sault Ste. Marie storekeeper identified the coins as those he had given Tanner in trade two days before the murder.

So apparently the story ended. Except that Lieutenant Tilden, who had led that posse so many years ago, reportedly made a deathbed confession to the murders of both young Schoolcraft and Tanner. His motive for the first had been rivalry over a woman of the settlement, and as for Tanner, the officer had accurately calculated the way in which he might use local opinion to cover his tracks. Perhaps an earlier confession would not have mattered much to the frontier community, anyway, for Tanner's crime of cultural preference encompassed all others.

But for us who have come after, facing choices apparently more complex in a country without borders and borderline figures like Tanner, there is only the rude monument of his narrative: rich, dense, fierce, like that life he led with the tribes. In it some may see faint old trails, game tracks over our hard and frozen ground, leading in other directions than to shacks on the edges of settlements where scapegoats crouch in mortal limbo.

But what of these faint trails and their directions? For these we must go back to the beginning of the captivity narrative in our country and follow old Cabeza de Vaca and his three companions as they stumbled and searched their way from disaster to a curious kind of salvation.

For the better part of six years Vaca and the others served as slaves for various bands on the coast of what is now Texas. Thorns tore their flesh as they dragged bundles of firewood, and blazing stones blistered their feet as they gathered plant food or hauled water. Vaca says that they shed their skins twice a year like snakes and were often beaten and threatened with death by their captors. How, then, did they survive? How did they find their way back through thousands of miles of unknown territory, moving through the ranges of mutually hostile groups whose languages and customs they could not know?

The answer is by no means clear, nor susceptible to rational understanding. But we do know that sometime prior to 1534 one or more of these survivors had been pressed into additional service as a shaman or medicine man and that the curings were successful. After the fall of 1534 when the four slaves united and escaped the Gulf Coast bands to the north, their reputations as shamans opened a few entrances into the broken country. Their neighbors had heard something of these whites in their new capacity, and at first this was enough.

Vaca tells us that at first they were reluctant to trade on this reputation with those to whom they had come, partly perhaps for fear of failure, and partly out of historic Christian injunctions against thaumaturgy. But their hosts pressed them to perform, and so, tentatively, against their wills, they succumbed to the customs of the country and began to practice; first, Alonso Maldonado del Castillo and then the others—Andres Dorantes, the black, Estevánico, and Vaca. At one place Vaca was brought into the presence of a man who appeared dead. The white man noted his blank eyes, the silence in his wrists. Nevertheless, the Indians insisted on prayer, and Vaca mixed his Christian prayers with traditional native practices. That night the Indians reported that the dead man had arisen, had talked and eaten, and was to all appearances healthy.

There were other, less spectacular cures as the men moved northward and then turned toward the west, but the most remarkable one of all was the slow cure that worked on the men themselves, transforming them unwittingly from conquistadores into faith healers. From bringers of death, even feeders on death, they had become enhancers of life. Forced to participate in the native beliefs, they had discovered the vast, untapped spiritual reservoir the New World had always constituted. And discovering this, what had been a fatal, sealed wilderness to them magically opened. Carrying native gourd rattles as badges of their office, they were themselves carried by large throngs of natives in their triumphant passage from one territory to the next.

Everywhere they went they breathed blessings, touched bodies and food in gestures of sanctification. A genuine spiritual power seemed to spark from their fingertips and quicken the crowds that danced and sang their arrivals at the villages. At one village the inhabitants presented them with six hundred opened deer hearts, which the shamans divided among the multitude that now traveled with them. "Throughout all these countries the people who were at war immediately made friends," Vaca writes, "that they might come to meet us, and bring what they possessed. In this way we left all the land at peace. . . ."

All of it, that is, that yet lay beyond the reaches of the marauding Christians, for as they neared the borders of civilized activity, the miraculous spiritual procession began to encounter sure and ominous signs, and among the huddled, fugitive people they now met, the dreaded watchword was, "The Christians are coming." Coming to enslave, to kill, to despoil. So as once again the four survivors stood in the presence of their countrymen, they encountered, as Haniel Long so well says, vestiges of themselves as they had once been. And after the first long stare of mutual disbelief there could only be anger and misunderstanding, for Diego de Alcaraz, the captain of the slaving expedition they had come upon, wanted immediately to make captives of those who followed the white shamans.

Vaca refused, and Alcaraz then attempted to destroy the shamans' authority with the Indians by telling the latter that these four miserable men were not the lords of the land but were instead the subjects of those they now confronted, mounted, armed, and with their coffle of captives. But the Indians, says Vaca,

> cared little or nothing for what was told them; and conversing among themselves said the Christians lied: that we had come whence the sun rises, and they whence it goes down; we healed the sick, they killed the sound; that we had come naked and barefoot, while they arrived in clothing and on horses with lances; that we were not covetous of anything, but all that was given to us we directly turned to give, remaining with nothing; that the others had the only purpose to rob whomsoever they found, bestowing nothing on any one.

And, he concludes, "I could not convince the Indians that we were of the Christians."

Here, then, is a captivity narrative that does not merely impersonate the mythic pattern of heroic quest, for in its interior there is true initiation. Hesitant, fearful, ignorant as we all must be in the presence of such matters, Cabeza de Vaca and his companions were granted a brief but vivid glimpse of that spiritual bond between humans that always lies awaiting acknowledgment. By themselves they could not change the course of their civilization, and we have noted the wholly inappropriate uses to which Soto and Coronado put Vaca's story. Vaca himself went on to hunt for the fabulous riches of Manoa along the Paraguay, and Estevánico was killed by Indians while serving as an advance man for Coronado. But in this narrative we may follow a trail that leads to a more genuine treasure than those so tragically sought in so many places under so many names. White Indians, these. . . .

# The Vanishing
# New World

L iving in a land so transformed by the burdens we brought, we find it almost incredible that North America was not long ago an unscarred place of overpowering beauty and fecundity, where the hand of humankind was dark, aboriginal, and mythologically bound, so that it brushed but lightly its environment. Traveling the cancerous stretch of the eastern seaboard, from the English outposts in Massachusetts, down through the Outer Banks of North Carolina, and on into the Island of Flowers, Florida, where Soto and Narvaez trudged off on their nightmare errands, we have some trouble imagining either beauty or fecundity. The old chronicles that tell of these seem traveler's tales to us now, though we have no doubt that we are in the presence of tremendous industry and power. Nothing less could have changed this landscape so totally from what it once was that even with our narrative guides we are at a loss to find vestiges of that old New World.

For most of us even more unsuspected and irrecoverable are the forces that nerved and sinewed our industry and that still drive us toward the acquisition and consumption of more and more power. For what underlay our clearing of the continent were the ancient fears and divisions that we brought to the New World along with the primitive precursors of the technology that would assist in transforming the continent. Haunted by these fears, driven by our divisions, we slashed and hacked at the wilderness we saw so that within three centuries of Cortés's penetration of the mainland a world millions of years in the making vanished into the voracious, insatiable maw of an alien civilization. Musing on this time scale, one begins to sense the enormity of what we brought to our entrance here. And one begins to sense also that it was here in America that Western man became loosed into a strange, ungovernable freedom so that what we now live amidst is the culminating artifact of the civilization of the West. Most of us take our landscape for granted, assuming that this is the way it naturally had to be. Many of us assent to the oft-stated fact that no nation in history ever plundered its own resources so rapidly as we have. Yet even this assent seems inadequate to our current situation in what remains to us of the New World.

To those who followed Columbus and Cortés the New World truly seemed incredible, not only because of what civilization had made of the Old World but because of the natural endowments of the one they now began to enter. The land often announced itself with a heavy scent miles out into the ocean, and the coasting whites with their nostrils full of salt and the sour odors of confinement recorded their delight with the odor of forests and verges in bloom. Giovanni di Verrazano in 1524 smelled the cedars of the East Coast a hundred leagues out. Raleigh's colonists scented what they thought a garden, though they would soon enough make it something else. The men of Henry Hudson's *Half Moon,* already disposed to hate and fear the natives, were temporarily disarmed by the fragrance of the New Jersey shore, while ships running farther up the coast occasionally swam through large beds of floating flowers.

Wherever they came inland they found that these announcements had been in no way false: the land, wilderness though it was, was a rich riot of color and sound, of game and luxuriant vegetation. Even if some of the most glowing descriptions of the New World were in fact real-estate advertisements, given then as now to calculated falsehood, still the theme of beauty in abundance is so pervasive that it transcends any scheme, insisting its truth upon the reluctant and hesitant pens of the white observers. Had they been other than they were, they might have written a new mythology here. As it was, they took inventory, around the margins of which one feels the spectacular presence of America.

Waterfowl took flight under their advances with thunderous wings, and deer in unconcerned droves browsed lush meadowlands. Squirrels and huge turkeys barked and gobbled in the endless forests that stretched all the way from the coast to the huge river that Soto had crossed and recrossed and been buried in. Elsewhere ground fruits lavished themselves on the land: scarlet blankets of strawberries painted the bellies of the horses and the legs of the horsemen who rode through them, and swollen clusters of grapes bowered the streams and rivers.

When the whites penetrated the western watercourses they found the life there as abundant as it had been along the eastern seaboard where sturgeon, giant lobsters, and shad were so plentiful that settlers grew nauseated on them. Out west, Pierre Radisson in the middle of the seventeenth century found otters so numerous in the streams that they hindered the progress of the little expedition's canoes. Gigantic catfish thumped ominously against the frail crafts of Jesuit fathers and voyageurs, while overhead flocks of passenger pigeons traveled the skies in such numbers that for hours at a time the sunlight would be obscured.

When Daniel Boone and the Long Hunters crept through the Cumberlands into Kentucky, they discovered newer variations on this theme of abundance in a land of canebrakes, clover, bluegrass, wild grains, and salt licks where a thousand animals might be glimpsed in a single lucky moment. They saw the buffalo whose enormous presence Vaca and Coronado had earlier reported and whose relatives, the woods buffalo, were to be found in considerable numbers as far east as Pennsylvania and upper New York. Here these few whites were on the very margin of the prairies that stretched they knew not how far toward sunset, prairies that in spring glinted like an ocean running under wind—sunflower, golden alexander, prairie lily, silphium, blazing star, golden rod, sky blue aster, purple gentian, big bluestem.

All of it seemed so lavish and exhaustless that it tempted the whites to tales of exaggeration, some imported from the Old World, some locally grown. There was, for example, the story about the Fortunate Hunter: charged simultaneously by a bear and a moose, he sent his only shot into a rock squarely between them, the bullet splitting and each half killing its beast. The fragments of the rock killed a squirrel in a nearby tree, while the recoil of the hunter's rifle knocked him into a stream from which he emerged with his pockets brimming with fish.

But while the old chronicles tell us their tales of abundance and show us occasionally individuals like William Bartram who actually *paused* to muse on this magnificence, caught up by it into aboriginal-like meditations, they tell us something else as well. They tell us that this was a world the whites wanted not as it was but only as they might remake it. For the other great theme of the narratives is that of waste, destruction, and frantic spoliation. As Peter Matthiessen notes, within a century of settlement the whites in the East had broken the wilderness from the coast well into Pennsylvania, a staggering achievement in which they had cut their way through vast stands of hardwood and white pine, cut them up into houses, boat ribs, ship masts, and items that would enable them to continue moving on—gun stocks, ax handles, singletrees, and wagon hubs. These trees, like the Indians and wild life they sheltered, stood in the way, not only of Progress, which was obvious, but of deeper notions of order and light. Occasionally there seems to have been a brief pause when some ancient woods giant crashed earthward and white men stood still in the dust and sunlight of its clearing to count its annual rings. So John Bakeless records that two oaks felled in Pennsylvania in 1833 were 460 and 390 years old, meaning that they had been in their maturity when Columbus had made his landfall. William Penn's Treaty Elm in Philadelphia had begun to grow when Cabeza de Vaca had embarked with Narvaez. And when at last the whites would attack the giant sequoias of the far West they would find trees that predated Christ.

With the forests went the wild life and the tribes, the former killed off and the latter dispossessed. What lands the tribes still held east of the Mississippi prior to the winning of the Revolution were thereafter claimed by the thirteen states—huge, hoggish claims these, utterly insupportable by any moral or international code and indicative in their greed and ignorance of the future treatment of the lands. In the Louisiana Purchase of 1803 the United States claimed another 756,961,280 acres of tribal holdings, though it would not be until late in the nineteenth century, when the process of land grabbing was over, that someone would point out that neither the defeated British nor the impetuous French had ever had the right to convey to the United States more than they themselves had possessed. This was Helen Hunt Jackson, who showed in A Century of Dishonor that the lands so cavalierly transferred by white hands had actually belonged to the tribes.

Few were considering this vexing matter in the early years of the Republic. Precedents had long been established in which neither law nor morality could protect the tribal possessions against the savage hunger of the transplanted Europeans, and though the infant American government from time to time made noises and gestures of protection, it was powerless against its citizens who meant to grab lands and then clear them of all vestiges of nature. For a time it was suggested that someplace well west of the line of settlement might be given the tribes, but that line was so ragged and advancing so rapidly that this idea was seen as impractical and unenforceable. Then in the aftermath of Major Stephen Long's expedition into the great West in 1819, it was suggested that since this region was an uninhabitable desert it would be suitable for the tribes of the East. Eventually this plan was adopted as the Removal Act, which Andrew Jackson signed into operation in 1830. It provided for the deportation of the tribes east of the Mississippi River into the "desert" west of it, where they were somehow to make their ways into strange lands already claimed by resident tribes. But even this was not enough, not expeditious enough for the white settlers who were now pouring into the regions of the old Southwest and the Mississippi Valley, taking their own sort of possession and ruthlessly clearing out the Indians in open defiance of the governmental timetable. The Seminole War, the Cherokee removal, and the Blackhawk War were three especially cruel results of this freestyle policy that by 1840 achieved its end. By this time Congress had designated the Great Plains as "permanent" Indian territory and had created an official bureau to oversee Indian affairs and to regulate commerce with the tribes.

As for the wildlife of the East, the turkey, prairie chicken, wolf, and elk were all driven out. So were the cougar and the woods buffalo. Strangely appropriate killer heroes now appeared in white folklore, such as the Pennsylvanian Aaron Hall who decimated the local panther population and in winter would festoon the premises of his cabin with their frozen carcasses. Another hunter of the same region was credited with having killed two thousand buffalo. The last of these creatures reported east of the Mississippi was a solitary vagrant shot in West Virginia in 1825, by which time the wilderness of the East was a faded and vaguely disquieting memory, perhaps most evidenced by its negative reminders: vanished forests, erosion, opened, parched lands, and small pockets of aboriginal slums where the demoralized fugitives who had somehow escaped deportation hung on surrounded by fields of white hatred.

As the inheritors rushed into the Mississippi Valley, the rage to clear and claim reached new levels. Into what Mark Twain with unconscious irony had called the "body of the nation" the whites now plunged their axes, plows, and other weapons, girdling and burning stands of oak and cypress at a rate that reached twenty-five million forest acres a year, and exterminating the wildlife in massacres. In Indiana Territory John Audubon witnessed one of these bloody spectacles, the victims of which were the passenger pigeons soon to be wholly exterminated. At sunset of a random day hundreds of settlers gathered in a woods known to be the roosting place of a huge flock. Armed with poles, torches, sulphur pots, firearms, and domestic hogs for the fattening, they waited for the winging in. Then the fearsome cry went up, "Here they come!"

and the massacre commenced. In the glaring light of the torches the work went forward well past midnight as the heaps of dead birds mounted, and the din of the firing was so intense that Audubon could not distinguish the reports of even the nearest guns. In 1821 at New Orleans, Audubon witnessed a similar slaughter of golden plover.

In Hinckley Township, Ohio, on the day before Christmas 1818 five hundred men and boys from neighboring areas gathered to clear out pests. They marched forward into the forest blasting away, and when they came together at last they accounted for the lives of three hundred deer, twenty-one bears, seventeen wolves, and uncounted numbers of foxes and other small game.

Early on the original violater, Boone, had become aghast at the destruction that had been unleashed. A failed land speculator himself who had refused his own opportunity to become a white Indian, he lamented to Audubon the almost instantaneous depletion of the vast stores of game and the replacement of them by noisome herds of settlers and speculators more successful than he.

Of this last breed, there now appeared to be almost as many as passenger pigeons, only they were impossible to quarry. With a telling consistency the federal government was unable to frame land acts or ordinances that would favor the individual settler over the speculator, for in truth the government itself was but a manifestation of a culture long committed to speculation in unknown lands. And while Roy Robbins has shown that the history of the American public domain has been that of a protracted battle between the settler/squatter and the speculator, in fact there seems never to have been such a class as *permanent* settlers since most of them thought of themselves as at least potential realtors. Like the old conquistadores who would carve up the plundered lands of the islands and New Spain and then move on, few of these Americans would condescend to settle and grow roots. Instead, they would drive restlessly, relentlessly on, nameless, petty, misguided Coronados searching for their Cibolas ever westward.

Attracted by the fertility of the soil and the prospects of instant gain, hundreds of thousands left the East for the Mississippi Valley. "Ohio Fever" threatened to depopulate Connecticut, once the much coveted land of the Mohican, and large sections of western Massachusetts and upper New York became deserted, the stone walls of their fields spilling into the weeds, the houses, barns, and outbuildings crumbling down to their sills, and the pruned-back woodlands recommencing their slow and ironic advances. Speculators in the newer region gobbled up huge portions of land, holding them out to feverish hordes, driving up the prices, and selling ever smaller individual lots to those who could afford no larger. Metropolises in woods and swamps were mapped out on the grid plan that, as Lewis Mumford notes, had the preeminent advantage of facilitating the sale and resale of lots. And sold and resold they were, each individual buying up as much as he possibly could and far more than he ever intended to use, betting against tomorrow that newcomers would purchase from him at double his original investment. Almost none of these metropolises ever materialized, of course, though some survived as county seats presided over by hulking courthouses, outsized, stranded monuments to baseless ambition and expectation.

In Michigan the land hunger and speculation amounted to a craze, and fraud was rampant. One eastern migrant, Caroline Kirkland, has left us an insightful memoir of this time and climate. Her husband had been seized with the fever in upper New York and had hauled his family west to Michigan in 1835 to settle in what is now Livingston County. Like many others, he was duped by a speculator, but the family stayed on for seven years in the midst of its swampy clearing, during which period Mrs. Kirkland raised her family, tended the stock, and took notes on the blasted, messy landscape, the raw little outposts of the new order, the stump-speech politics, and the manners of the nameless drifters who settled the backwoods for a few months, ran up debts, and then cleared out for the West with their neighbors' pigs and chickens in tow. Above all, she noted the speculation craze, a developer behind every stump, ribbons and strings everywhere demarcating the site of the next city. This entire region, Roy Robbins says, "was regarded as a lottery office, to which individuals from all over the world might resort to accumulate wealth, under the favors of the capricious and blind goddess Chance."

Before the spectacle of what was happening to America, both foreign observers and Americans alike were in awe. The movement of so many people, the mad pace of settlement—chopping, clearing, buying, and quickly selling out—had there ever been anything like this in the whole history of the Western world? And what did it mean? Most Americans in their haste seemed not to care, while foreigners traversing middle America wondered as they took astonished notes on the rapid despoilment of the wilderness. They marveled at the vicious shoddiness of the settlements with their taverns, their gouging and biting contests, their fugitives and hermits and robber gangs; the frenzied ways of the settlers, bolting their gluttonous piles of food, draining jugs of whiskey and high wine, horseracing, knife-throwing, and endlessly speculating. Occasionally these foreign travelers might stumble across ancient Indian earthworks or other jumbled artifacts of the natives now kicked out of the way and produce standard sentiments on the fate of empires.

In 1837 the new nation's insatiable lust for land finally produced a financial panic that spread back from the western frontier to the population centers of the East. The previous year had been the biggest yet in land sales, and an extent of territory the size of New England had passed into the control of the speculators. Now there was a crash that revealed not merely the enormous economic inflation of the times but their spiritual bankruptcy as well. Yet, surveying this scene, Horace Greeley could produce a remarkably characteristic solution: Move On: Go West. And those who could go did so, leaving those who could not stranded on their farms and in their little towns with outsized grid plans that would remain unrealized. Succeeding generations in this region would reap the crop of bitterness, frustration, and xenophobia that lurks there still—people, as the Midwesterner Glenway Wescott observed, born where they do not like to live.

The others went on. They bridged the great river and entered the grasslands that stretched all the way to the big mountains: tall grass prairies to the east, short grass plains westward, and beyond that the tough bunch grasses of the semidesert. This was the territory that had been described by Major Stephen Long as an uninhabitable desert. But closer experience revealed that the same blue-stem sod grass that grew in the rich lands of Illinois was to be found stretching west across the Mississippi. Henry Nash Smith has charted with clarity and grace the course of American thinking about this trans-Mississippi region, and what is most striking is the evidence of the sudden realization that there were *no limits to what could be done to America:* that the advancing, driven people need not put up with any permanent barriers to their civilization or remain forever uneasy about the specter of the trackless West with its wild people. All of this too could be claimed for civilization. Within a very few years of the appearance on the charts of the "Great American Desert," men were planning the penetration of it by railroads.

In 1830 there were only twenty-three miles of rails in the entire country. In 1850, when attention turned in earnest to the spanning of the continent, there were nine thousand miles of railroad, and a decade later that figure had tripled. In 1856 and 1857 Congress made grants to forty-five western railroads, which began to spur out across the stretches like cactus, breeding the thorns of bribery, political corruption, manipulation of public opinion through the newspapers, land speculation, discriminatory rates, and hatred. The railroads also bred towns along their routes, towns projected as cities of the near future, full of what Robert Louis Stevenson saw as "gold and lust and death," towns that shriveled into dead husks when the rail routes changed. Here again was that hunger beyond consumption. Robert Hine quotes an associate of railroad magnate Colis P. Huntington on the nature of the man's "ravenous pursuit." Huntington was always "on the scent, incapable of fatigue, delighting in his strength and the use of it, and the full love of combat. . . . If the Great Wall of China were put in his path, he would attack it with his nails." As it was, there was no wall, only half a continent; but there were Chinese aplenty, some of them appearing as smudges at the edges of photographs of rail construction.

Following the thrusting rail routes, and often in advance of them, breaking the ground on foot or wagon, were the restless migrants who moved in such numbers that even the inflation-warped journalists of the 1840s and 50s sometimes seemed alarmed at the rate and scale of the phenomenon. Rumors of richer lands and vaster fortunes to be made westward caused settlers to leave decent homes in the Mississippi Valley almost as soon as they had established them. Horace Greeley noted in 1857 that the men "who are building up the villages of last year's origin on the incipient railroads of Iowa, were last year doing the like in Illinois, and three years since in Ohio. He who is doing well in the newest settlement is looking sharply around for a chance to do better in Nebraska or along the lines of the railroads leading thither which are soon to be constructed." He predicted another inflation-caused panic, which indeed occurred in this very year.

Sometimes the rumors of western wealth seemed substantiated, as happened with the California gold strike at the end of the 1840s. These rumors quickened to fantastic life the old golden dream of the New World. As they ebbed, they left the West littered with haunted, stubble-cheeked prospectors, "eternals" seeking out lost lodes, rumors of Moctezuma's treasure, fabulous veins overlooked in earlier rushes, or caches of stolen jewels taken from explorers and settlers by marauding Indians. Pathetic scavengers, these, moping through a gigantic, glittering landscape and seeing nothing of it except the next horizon.

Most often, the rumors arose out of nothing more substantial than those recondite forces that drove the whites to fill up the continent's spaces with their presence. The ghastly story of the Donner Party should be understood as a parable of this, for here was a group like thousands of similar groups, traveling across barely charted spaces toward the vaguest rumors of More. According to their historian, C. F. McGlashan, many of them had been solid citizens of Ohio, Tennessee, Illinois, and Missouri, and yet here they were, crawling across the Great Plains, across a desert, and on into the high Sierras with scarcely a notion of their destination. McGlashan, interviewing the survivors, records that many had joined the procession without even knowing that it was going to California, only that it was going somewhere. Winter caught them in the mountains, imprisoning them in their miserable hovels until, perishing one after another, the survivors ate the frozen and emaciated corpses of the dead. Two Indian guides were revolted at such hunger until they themselves were shot and consumed as the desperate stragglers went over the pass and down into rescue.

Facing east from California's shores, as now we can, we have a clear view of this gigantic process, especially the portion of it that occurred west of the Mississippi. For out there the camera caught what we are pleased to call the "Winning of the West." Here are the track crews, shadow-faced, slouch-hatted men with their mules, laying track at two miles a day. Lonely figures are posed on bark-covered ties that stretch off into blank horizons. Rail tickets scream like circus posters, advertising transportation to "ALL POINTS IN THE MINING DISTRICTS." "Ho! for the GOLD MINES!" "1865! 1865!"

Here are the mining towns blasted out of the mountains with their tin roofs glinting in bleak contrast to the new wood of their walls and to the muddy streets and cluttered creeks. A jungle of advertising shingles hangs above the porches of the stores—Dentist, Wholesale Liquor Dealer, Bank—and beneath slouch the miners, shaggy, unkempt, hopeful. One sits on a crate in a Black Hills camp, a rifle across his knees to guard his claim. The names of the towns are Deadwood, Gold Hill, Montezuma's Works, Sugar Loaf.

Here are the timber miners who came out with the railroads after cutting as much as four million board-feet a year in the North Woods. Shattered hulks move along the skid roads and bull teams slog along while skid greasers pour from their rancid cans; locomotives three and four abreast haul redwood sections. Dwarfish figures are posed astride the ruins of primeval trees or beside heaps of slaughtered game or atop mountain crags that offer perspectives almost none of them could grasp.

Here are the silenced, solemn faces of the "hostiles" who vainly opposed all this: Sitting Bull, Satanta in his soldier's uniform with epaulets, Lone Wolf and Dull Knife in a photographer's studio.

Here are the soldiers, white and black, who fought the tribes for the possession of Indian territory: overstuffed generals in beards, buttons, and braid; lounging officers at Fort Ellis, Montana Territory, their coats open, their trousers saggy and boots dirty, the obligatory hound at the bottom of the steps of their quarters.

And here are the hostiles and soldiers together under the wide flaps of the treaty tent at Fort Laramie in 1868, the Indians in blankets and buffalo robes, their braids wrapped in weasel fur; the soldiers on their camp stools with William Tecumseh Sherman in their midst, his burning eyes fixed on those he had determined to destroy, treaty or no.

Here is a photograph that appears to epitomize the whole process, for it is of a literal land race: high noon, April 22, 1889, and the blurred forms of 10,000 whites racing off the starting line and into another section of "permanent Indian territory." This is Oklahoma, the devastated soil of which would in a mere three decades be visible on the East Coast in dense red dust clouds that rolled out into the Atlantic.

Here near the end of the westward rush is a photograph of a little man in a yard, surrounded by what appears to be chips and flakes of an indeterminate nature; they are actually buffalo bones, and the well-known and much-lamented destruction of this animal is as concise a way of understanding what was done to America as we are likely to find, for the dates and numbers of this destruction are at once finite and suggestive.

Well into the eighteenth century the woods buffalo was to be found in good numbers inland from the eastern coastline, but as we have seen, by the middle of that century it had become scarce and was finally extirpated from the East shortly after the turn of the century with Pennsylvania hunters doing most of the final work during the winter of 1799–1800.

Few in those days knew of the reports of Vaca and Coronado that described the herds of the West, but Daniel Boone and his friends had seen buffalo in great numbers in Kentucky and had an inkling of what might be found farther on. Then Lewis and Clark made a reality of the guesswork by returning from the great West with descriptions of buffalo browsing in herds that stretched farther than a man could see. They estimated one herd at twenty thousand, and this figure was more than confirmed by subsequent travelers who told of riding whole days at a time through a single herd. They told of plains that looked dark and dense with the shaggy beasts cropping slowly windward, of herds as huge as 100,000.

At first none could find long-range uses for this bounty. The first whites in the West used the buffalo approximately as the Indians had for centuries: they ate the flesh, wrapped themselves in the robes, and fueled their fires with the dried dung. Gradually, however, a trade in buffalo robes grew up, especially as other fur-bearing animals were hunted out. By 1835 more buffalo robes were being shipped down the Missouri than beaver pelts—almost fifty thousand of them annually. A little more than twenty years later this total had mounted to seventy-five thousand robes a year, and by this time indiscriminate slaughterings in the name of sport were increasing the carcass count. Wealthy gunners from the East and from the Old World journeyed out to the Great Plains to burn their powder against the huge and obliging targets. David Dary recounts one lengthy expedition by an Irish nobleman that bagged more than 2,000 buffalo—as well as 105 bears and 1,600 elk and deer.

By the 1860s the railroads and the military had begun their westward advances, and meat was needed. Both organizations hired professional buffalo hunters to supply it, and one of these was William F. Cody, not many years away from a national herohood based in good measure on his prowess in this killing. For five hundred dollars a month Cody killed buffalo, and in the year and a half that he shot for the Kansas Pacific he dropped an estimated 4,280. Now he began to enjoy a local fame, not unlike that of celebrated killer heroes of the East, and in a contest to see who could kill more buffalo in a single day he bested one Billy Comstock, sixty-nine to forty-six.

The growing reputation of Cody as a buffalo killer together with the slow rumor that there might not be an inexhaustible supply of this big game prompted more expeditions from the East as well as from places like Chicago, St. Louis, and Kansas City, the jocund hunters rolling out now in the comfort of railroad palace cars. Cody was once guide to an ornate excursion from New York City in 1871 that stopped off first at Chicago's Lincoln Park Zoo to get an idea of what a real live buffalo looked like. In Nebraska they detrained and moved off into the plains in a convoy of army ambulances, garnished with greyhounds, a French chef, waiters in full dress, the finest silverware and china, and three hundred of the 5th Cavalry's finest to prevent untoward encounters with real live Indians.

At this same time American, British, and German tanners arrived at the conclusion that the buffalo was worth a great deal more than the ostentatious trophies aristocrats took, more even than robes for carriage and sleigh rides or for frontier coats. They had developed processes by which buffalo hides could be made into high-grade leather. A new way had thus been found to use up the New World, and the word quickly spread that dealers would buy hides at any time of the year, with the long seasonal hair or shorn of it. Newspapers and advertising circulars blasted the news, and by the summer of 1872 a new sort of miner moved in great quantity over the Plains, searching out the herds. In October, November, and December of that year the Atchison, Topeka & Santa Fe carried 43,029 hides to eastern markets. Between that year and 1874 an estimated 3,158,730 buffalo were killed by the white hide hunters.

Now a remarkably sophisticated technology sprang into being to expedite the new industry. Remington and Sharps arms companies both developed big high-powered rifles, the latter being the weapon of choice among the hunters. Looking at them now as they lie scoured and embalmed in the museum cases of the West, one gets another view of that fearsome ingenuity at work in America, for these were truly lethal monsters whose barrels were but a few inches less than three feet long and packing enough muzzle force to drop a buffalo in its tracks. Their big, heavy, soft bullets caused terrific damage as they flattened into and slammed through the innards of the animals. Equipped with telescopic sights and mounted on portable shooting stands, they were more than adequate tools for the trade.

So too were the men who crouched behind them, lining up their prey in the cross hairs. A good buffalo runner and his crew of skinners might take up to sixty buffalo a day, and early in the the 1870s they perfected the method of the still hunt, a more economical, efficient way to kill large numbers without having to chase after them. A carefully concealed hunter would pick out a likely target and hit it about midlength in the body, a mortal wound that caused the animal intense agony before it died. In its death antics it would attract others of this curious species who would gather around to watch. So the hunter would hit another one in the same place and so on until he had killed all who remained on the spot or until his ammunition ran out. One Thomas C. Dixon is reported by David Dary as having killed 120 buffalo in forty minutes using the still-hunt method.

Naturally, the complaints of the Plains tribes about this carnage were brushed aside by government representatives as well as by the military, which was by now pursuing a de facto policy of Indian extermination and rightly saw the extermination of the buffalo as a shortcut to this end. Whites almost always pointed out to the complaining chiefs that numbers of their own people were trading in hides too, adding that fears of the disappearance of the buffalo were foolish. When buffalo runners from Dodge City asked the commanding officer at Fort Dodge whether he thought they ought to go into Indian territory after a herd, they were told, "Boys, if I were a buffalo hunter, I would hunt where the buffalo are."

This same officer, Colonel Richard I. Dodge, has left us this telling first-hand account of the rapid extermination of the southern herds:

> In 1872 I was stationed at Fort Dodge, on the Arkansas, and was out on many hunting excursions. Except that one or two would be shot, as occasion required, for beef, no attention whatever was paid to buffalo (though our march lay through countless throngs), unless there were strangers with us. In the fall of that year three English gentlemen went out with me for a short hunt, and in their excitement bagged more buffalo than would have supplied a brigade. From within a few miles of the post our pleasure was actually marred by their numbers, as they interfered with our pursuit of other game.
>
> In the fall of 1873 I went with some of the same gentlemen over the same ground. Where there were myriads of buffalo the year before, there were now myriads of carcasses. The air was foul with sickening stench, and the vast plain . . . was a dead, solitary, putrid desert.
>
> In October, 1874, I was on a short trip to the buffalo region south of Sidney barracks. A few buffalo were encountered, but there seemed to be more hunters than buffalo.

Dodge estimated that in the three years he writes of here, "at least five millions of buffalo were slaughtered for their hides."

By 1880 the hunt was over on the southern plains. Astonishingly, the buffalo were all gone. But in the north, on the other side of the railroad line, there were still large herds in Montana, Wyoming, and the western portions of Dakota Territory; the hunters went where the buffalo were. Now, as if realizing an end in sight, the killing reached a general, sustained roar: 200,000 hides were shipped out in 1882. And that about did it. In 1883 only a fourth as many were shipped, and the year after that a pitiful three hundred.

The hunters could not believe this at first, but rather like buffalo in a still hunt they hung about the littered and rotting plains waiting for the return of the herds; surely they must have browsed northward into

Canada and would return. They never did, and so, sadly, the runners
and their skinners packed up and moved on, to be succeeded by the last
humans who would live off the buffalo: the bone hunters who gathered
up the bleached refuse of the grand orgy and shipped it off to glue factories, sugar refineries, and fertilizer plants. The buffalo had taken its
place in a lengthening list of exterminated or endangered New World
species that includes the great auk, the green turtle, the jaguar, the
masked bobwhite, the passenger pigeon, the lake sturgeon, the sand-
shoal duck, the sea-mink, and several varieties of bighorn sheep—to
name but a few. Except for a few pathetic relics preserved in Yellow-
stone Park by conservationists, the last vestiges of the great herds were
these piles of bones that waited patiently on Western railroad sidings in
the early 1880s. By the end of the decade even these had vanished,
and it was only occasionally that a sod buster would turn up a fragment
with his plow and hurl a last curse at the vanished wilderness.

During this century of the final clearing of America there were certain individuals and even a few groups who regarded the spectacle with
sadness and even disgust. Were it not for these solitary lights it would
be more difficult for us to comprehend what had been done and why,
for in that gap between what they stood for and what the majority was
practicing, the issues and motives stand out in relief. I think here of
Ralph Waldo Emerson, of John Wesley Powell, and of John Muir who
quite literally jerked himself out of the stale bed of his father's Presby-
terianism to become a kind of mystic and an effective voice of conservation. I think mostly of Thoreau whose steps always tended westward
and who took Emerson to heart, even took him further afield and
deeper into nature than the master himself had ever wished to go.

It is probably significant that Emerson gained his audience in 1837,
the year of the Panic. For several years he had been asking irritating
questions about the nature of Christian conduct and ritual and about
Americans' existence in their lands. To him his contemporaries seemed
peculiarly incurious about these lands, dead to the natural life they still
harbored. Their condition of alienation jarred with the current cock-
crowing nationalism. When he asked why Americans should not seek
an "original" relationship with the universe, the query had a more trou-
blesome depth to it than most (perhaps Emerson included) could have
acknowledged. This was because from his Concord study, musing on
the news of the day, the din of limitless growth and exploitation,
Emerson sensed that the story of the New World was going to be the

story of the Old World again; that there would be no fresh beginning here but only a brutal monument to meretriciousness and emptiness of spirit.

It was the younger man, Thoreau, who best intuited the depth and urgency of Emerson's question. It eventually sent him to Walden Pond and to meditating there on the savage life: whether the departed aborigines whose scattered relics were all around him in the woods might not have once enjoyed that original relationship of which Emerson had spoken and which now seemed so unavailable to nineteenth-century whites. In his hut on the edge of the jewel-like pond, surrounded by the gentle wooded bowl of its hills, he thought about the meanings of savagery, even of spending a day as the animals might. And throughout the book of his experiences there one finds evidence of his intuitive drive to dig, to burrow, to go earthward toward that original relationship. He tells us of digging his cellar where a woodchuck had formerly dug his burrow, down through the roots of sumach and blackberry "and the lowest stain of vegetation. . . ." In his bean field he writes that in his hoeing he "disturbed the ashes of unchronicled nations who in primeval years lived under these heavens," bringing to the light of his scrutiny their implements of war and hunting and husbandry. In winter we find him cutting a hole in the pond ice and gazing through it into the silent, secret world beneath where magical pickerel held themselves still as shadows. And in a more didactic passage, he urges us to settle ourselves, not like the restless settlers of his day, but as a people who would grow roots in the earth, wedging our feet downward through the slush of opinion and appearance "till we come to a hard bottom and rocks in place, which we can call *reality*. . . ."

For Thoreau, such a reality seemed more and more to be somehow connected to the Indian, though in his Walden days he only ruminates at the edges of the connection, remarking here and there that mythology might be the record of a primitive, original relationship with nature. In his journal he writes, "I do not know where to find in any literature, whether ancient or modern, any adequate account of that nature with which I am acquainted. Mythology comes nearest to it of any." But perhaps myth could only arise from contact with the unfathomed, unfathomable wilderness, and with his fellow countrymen all about him laying feverish waste to the wilderness, where might such a profound confrontation take place? We need the tonic of the wilderness, he writes, we need to witness "our own limits transgressed and some life pasturing freely where we never wander." Well to the west of where

Thoreau sat, Americans were determining that there should be no limits, no wilderness left unfathomed.

Seeking the ground of such a confrontation, Thoreau left Concord, tame enough territory after all, and journeyed up into Maine in 1846, again in 1853, and for the last time in 1857. During this period the purity and profundity of his quest was revealed, as Philip Gura has pointed out, for in Maine Thoreau found that the speculators, the fur traders, and timber miners had long preceded him and that any confrontation with the old New World was difficult to come by. In *The Maine Woods* he records his dismay at the messy, chewed-up landscape he encountered, and his even greater dismay at the degraded aboriginals he observed living a fringe existence in their cheerless slum dwellings. On his second trip he went deeper in with a Penobscot guide who whistled a marching song of the frontier instead of chanting the images of mythology. But he went back yet again, convinced his quest could be realized somewhere in the Maine interior, if only he could penetrate it far enough. By now he was sure that his earlier intimations had been leading in the right direction, that what he had been seeking was an *ab*original relationship to the universe.

This time his guide was the remarkable Penobscot Joe Polis. Thoreau heard him talk to muskrats, saw the strange phosphorescence of moose-wood in a campfire, and finally achieved his meeting with utter wildness atop Mountain Ktaadn. Here at last was that reality he had heroically sought, that confrontation with the Other that had always been the great, unsuspected treasure of the New World. Here was Nature, savage, awful, and beautiful too:

> I looked with awe at the ground I trod on, to see what the Powers had made there, the form and fashion and material of their work. This was that Earth of which we have heard, made out of Chaos and Old Night. Here was no man's garden, but the unhandselled globe. It was not lawn, nor pasture, nor mead, nor woodland, nor lea, nor arable, nor waste-land. It was the fresh and natural surface of the planet Earth, as it was made forever and ever. . . .

He felt its terrific presence then, as the old and vanished inhabitants had felt it everywhere, the presence of

> a force not bound to be kind to man. It was a place for heathenism and superstitious rites,—to be inhabited by men nearer of kin to the rocks and to wild animals than we.

He felt possessed and unafraid of that possession:

> What is this Titan that has possession of me? Talk of mysteries!—Think of our life in nature,—daily to be shown matter, to come in contact with it,—rocks, trees, wind on our cheeks! the *solid* earth! the *actual* world! the *common sense! Contact! Contact! Who* are we? *where* are we?

As Thoreau well knew, these are the questions answered by all mythologies, and in this moment at least they were answered for him. Nature in such an untouched state, he reflected, must have made a thousand kindred revelations to the Indians that it had never made to the conquering white man. I think the key word here is "revelations"—to feel as the aborigines had that messages can come at any moment from divinity, that divinity has not been sealed off by canon and dogma and empty ritual, that miracles can happen *for us*.

Thoreau sickened and died before he could go deeper in his quest, his last words revealing the profundity of his commitment to it: "Moose . . . Indian. . . ." But he left us enough evidence to suggest that he might have gone on to learn to write some words in that divine picture language that is myth, so to supersede Emerson who had said, "We too must write bibles." We too are writing them, Thoreau might have said: I have had intelligence with the earth.

# A Dance of the Dispossessed

Of the conquest of the New World no aspect is better known and less understood than the very last, brief episode glimpsed in the preceding pages. Variously disguised as the "Winning of the West," or "Cowboys and Indians," this ragged end of a giant swath of Western history is as popular in Latin America (where wilderness and natives continue to be affected by the old motivations) as in Paris's Latin Quarter where one may purchase ten-gallon hats, lariats, and synthetic cowboy boots from the Orient. Partly we owe this popularity to a rich and various documentation, for in the American West it was not only the explorers, traders, and casual travelers who made first-hand accounts. In these last years of wilderness America newspaper journalists and journalistic artists, painters, itinerant photographers, and finally publicists, anthropologists, and historians all made their tours of the region generally understood to be doomed and they recorded it for an ever-expanding audience. Today in America the audience continues to grow and remains as credulous as ever for fictions of the victory and the defeated adversaries. Few understand or care that this final episode is in reality a commentary on Western Civilization that is as ironic as it is tragic. For in their response to spiritual dispossession, the defeated Plains tribes recapitulated the origins of that now bankrupt religion to which their dispossessors paid empty homage, even to laying out the last red victims in a reservation church and burying them behind it in the sanctified graveyard.

Maybe it was the sheer space that seemed so fascinating to the whites. Or maybe it was the style of the Plains culture with its nomadism, its horse herds, hunters, warfare, and apparently aimless, leisurely movement over the land. Or maybe it was the immensely colorful costumes of these people who seemed to play at life with a sort of fierce artfulness. At bottom, maybe it was the sense that was to be gained over and again that the tribes loved this life, thought themselves the privileged and elect of all creation to be permitted the gusty, wide-skied joys of their world. This made them seem arrogant to the whites, but it also compelled a kind of admiration that yet lingers in our culture.

For the tribes themselves, of course, it was all of these things and far more profoundly than for those who encountered them, dispossessed them, and then memorialized a fictitious version of this. Space, as Charles Olson says, comes big here, big and beautiful and merciless. In comparison, the lands east of the Mississippi seem tangled and dark, their vistas difficult to come to, and even then somehow enclosed, as in Cherokee country, for example, where you might tear your way through a "laurel hell" or mount steadily through huge, straight stands of gum and sycamore where the sunlight is filtered and gloomy, up to a mountain peak where you gazed out over more of the same and knew what it would cost you to attain the next ridge. Here on the Plains vistas were endless, even in the mountains, and the skies were commensurate, seeming to stretch both taut and depthless over it all. You could see endlessly upward at night, as when the tribal children might stretch themselves on the buffalo grass and gaze up into a night sky filled with stars and stories. You could see out endlessly across plains to ridges that appeared available in the atmospheric purity that invisibly intervened, subtly painting the distances in alluring shades of gray, blue, and mauve.

Closer in, where the nomads might be camping, there were the greens of the lush life of the watercourses; the cottonwoods that drowsed, shedding in the breeze. Fallen, the cottonwood made a greasy, smarting smoke for the lodge fires; better to use the aspen that twirled its leaves aloft on limber spindles that twinkled like a child's toy. Or the ponderosas that bunched dark in the distances of their slopes, making green the shadows of their needles underneath. Through this spaciousness moved game in great plenty: the gigantic herds of buffalo; antelope, fleet and musky on their sage diet; deer, bear, bighorn sheep, quail, and fish.

But it was merciless, too, in its stretches, in its seasonal moods. As for the tribes of the woodlands, there was always the specter of seasonal starvation, of snows and winds that would pile them into eye-searing drifts, of scarce game, and scarcely adequate shelter in the lodges among the winter-stripped groves; frozen rivers and streams; temperatures so fathomless that hunters might occasionally come upon a small bunch of animals standing frozen and upright in the curling snow, the hair of their hides pointing away from the direction of their destruction. In that case the hunters would be lucky, but often, their garments like iron, snow-blinded and empty, they would return to the pinched lodges for a thin broth or nothing. For these reasons they marked their histories in Winter Counts. And so, in spring, as the winds became constant breezes and the snows sank into the ground and ran in the rivers, the people would come out of the lodges to lie upon the tepid earth; the dogs that had survived winter's desperate pots would roll in fat places, and the long-haired, starveling pony herds would begin to assault the first green shoots. They had come through.

These extremes encouraged a gorgeousness of dress and deportment, especially among the males, their clothing tasseled, fringed, and feathered, their faces painted with animal fats and the colors of the earth. Even the favored ponies were painted and feathered. Here the tribes seemed to accept with a joyous and creative fervor all items of trade that might come to them: shell from both coasts, mirrors, brass, German silver, soldier paraphernalia, calico, copper. When they got hold of rifles they decorated the stocks with brass studs to make them their own.

So when the early white artists penetrated beyond the Mississippi and up the Missouri, they found a cultural panoply that almost beggared their craft. Sometimes the display so outstripped their capacities that they *bought* the costumes they could not otherwise capture, only to find that somehow their purchases would not travel, the feathers wilting as if they had been plucked from live birds, the skins stiffening or mildewing as they were brought back in their confining trunks. Still, before the camera or yet the cinema these artists—Catlin, Bodmer, Miller, Bierstadt—got down the true, vibrant variety of the Plains culture before it was reduced to the caricature of the war bonnet and the blanket. In their paintings you can sometimes come close to the frontal shock of first impact, to the smells of the Indian camp or village, the herds of horses, the strutting headmen and warriors, and the often handsome women who did so much to make the entire display possible. Sometimes, as in the case of George Catlin, you can also get the sense that in depicting this life the artist was nervously edging close to his anonymous predecessors of the Old World's caves, and like them was taking the outline of a mythic world.

As the whites claimed, and with increasing pointedness as the nineteenth century unearthed the realities of the western lands, these Indians were mostly nomads. Except for a few groups like the Mandans, who planted extensively and moved between seasonal villages, they followed the game, the seasons, the watercourses, and the faint but traditionally grooved trails that led them through their cycle. In the days before the horse they had been even more dependent on the seasons and the availability of game, having to await and surround the buffalo like Ice Age hunters after mastodon when it was the work of many people to bring one huge beast to the earth and the fire. Or else they would construct the cliff-edge devices we glimpsed in the Blackfoot narrative of the woman made captive by the buffalo bull. When they were successful with these there would be meat aplenty for a time and feast-giving among the lodges, and the women were expert in preserving the surplus in various forms such as pemmican—flat cakes of mashed chokecherries or sarvis berries mixed with buffalo tallow and pulverized meat. Pemmican could keep a long time in its rawhide storage pouches, but there were times when even it would be gone, and then the people became expert at sucking in their bellies.

With the coming of the horse to the Great Plains in the years following the Pueblo Revolt against the Spanish (1690) matters changed. The people learned the use of the horse and became expert riders. No longer did they have to await the coming of the game; they could go after it. A single bowman, mounted on his trained buffalo pony, could kill several buffalo on the run and then pack the meat back to camp. The tribes had thus gained a leverage, a margin against the spaces they moved through. Mounted, they became aristocratic nomads, subject still to the moods of the region and indeed deeply impressed by these. But with the addition of the horse they were freed significantly from the grip of their powerful environment, and they ranged through it with an elan that lifted the heart.

For this reason the horse became a status symbol, and outsized herds grazed at the edges of the camps. They were held as a kind of money, used in marriage payments, given as gifts, wagered in gambling, raced, preened—and stolen. George Bent, a mixed-blood Cheyenne, said that horses were to the Plains tribes what gold was to the whites, an inexact analogy but good enough to suggest the multifarious ways beyond immediate, practical use that the horse figured in Plains culture. Little boys among the Crow, for example, were trained in the art of horse stealing by teachers who would send them sneaking through their own village to pilfer racks of drying meat from the lodges; the meat was symbolic of the horses they would one day steal from the Arapaho, and of course the trick was to take the meat without being detected by the sharp-eyed women who guarded the product of their industry. Here the penalty for detection would be a thrashing and a scolding by the women and ridicule from their peers. Later, in manhood, it might be death.

The horse became the central symbolic item in the intricate and mortal game of ritual warfare the Plains tribes were playing when encountered by whites. They fought each other over other matters, it is true— territorial boundary disputes, insults, revenge, inherited traditions of enmity—but most of the endemic skirmishing that gave a special cast to Plains life was connected in one way or another to horse-stealing raids. And in these no one stole horses because he was poor in them. He stole to gain honor, to test and display his powers, and to humiliate the enemy. The real prizes in this male game were honor, pride, and courage, and for these the warriors were always willing to die. But to kill and scalp was not the ultimate intent of the game. "I looked forward to the warpath," recalled the Lakota, Standing Bear, "not as a calling nor for the purpose of slaying my fellowman, but solely to prove my worth to myself and my people." So it is not surprising that the greatest honor in the game went to the warrior who could contemptuously spare his enemy in battle and merely strike him with a symbolic "coup" stick.

Still less was the intent of the game the subjugation or annihilation of the enemy. As much as the Lakota affected scorn for the Crow and proudly wore Crow hair as fringe for their shirts, they never conceived the idea of dispossessing the Crow of their lands and all their horses, of reducing them to a subject people or killing them all off. They would joyously steal Crow ponies, kill, scalp, and mutilate resisting Crow warriors, and go out on the warpath to take bloody vengeance for tribesmen killed on these raids. But when that score had been settled, the game could begin all over again.

So they played on, living on the buffalo, skirmishing, moving so lightly over the Plains that to the whites they seemed to leave no least traces of tenure on their lands. There were, of course, the deepest sorts of traces, spiritual ones, but the mistake was natural to a people who themselves had no roots anywhere despite the often pathetic protestations everywhere erected. Though the Plains tribes were nomads, every aspect of Plains life was sacred to them: stones, earth, plants,

animals, and the lives of the tribes. It was all "holy," as the Oglala
Lakota shaman Black Elk said, with "us two-leggeds sharing it with the
four-leggeds and the wings of the air and all green things; for these are
children of one mother and their father is one Spirit." Each tribe believed its territory had been donated to it by the Great Spirit for its
special use, and that use must reflect the sacredness of the donation and
the donor. The whole landscape was thus infused with spirit life, a
mental fact that the child learned very early. Soon, Standing Bear recalled,

> the child began to realize that wisdom was all about and everywhere and that
> there were many things to know. There was no such thing as emptiness in
> the world. Even in the sky there were no vacant places. Everywhere there
> was life, visible and invisible, and every object possessed something that
> would be good for us to have also—even the stones.

This sense of sacred attachment to the lands thus amounted to a different kind of possession than the whites were prepared to understand
as they looked about these spaces and found them empty of visible
marks of tenancy. They were beyond understanding that the tribes here
lived by traditions that sprang out of these lands like grass or rock formations. At the eastern entrance to the Black Hills, for example, the
whites found no permanent habitations nor even a trace that there had
ever been any at the place they would call "Buffalo Gap." But the land
was sanctified as the location of the Great Race of Cheyenne mythology
in which the Great Spirit had arranged a contest to determine whether
the buffalo should continue to eat the people or the other way around.
During the race the animals had bled from their mouths and "the
ground there is still red with this blood." Everywhere there were such
marks of divinity and of tribal participation in the divine plan. One old
Crow warrior said to another, "You are sunk in this ground here up to
your armpits." So were they all, and even deeper than that, for the
lands were the very stuff of their dreams and visions.

In dreams and in visions the lands manifested the presence and abid-
ing concern of the Great Spirit. So the people actively courted such
states as ways of remaining attuned to the divinity that brooded in the
wind and clouds and grasses, that vivified both human and animal existence. This was their kind of possession, and they celebrated it in their
medicine bundles, their songs, stories, and communal ceremonies,

down through the years of the Winter Counts into a season harsher than any of them could imagine.

All of this began to change with the first trickle of whites into the region. Solitary and ineffectual-seeming, they were in reality the advance personnel of a trade nexus that had already leeched the tribes of the East, pathetic remnants of which were now to be found scattered here and there in the West.

In the East, as throughout aboriginal America, the tribes had always carried on extensive trade among themselves. But with the coming of the Europeans, trading took on a new and sinister velocity, for the new party to the bargaining had no relationship whatever to the lands other than what could be realized economically from them. To them the lands were satanic rather than sacred, and the traders and their employees could tolerate the wilderness only in the hope that eventually they could make enough money to leave it behind and return to civilization to live like humans. So they would grimly push out into the woods beyond the farthest reach of civilization—but not so far that they could not be supplied with trade goods. Here they would establish a post and make it known that they stood ready to supply the needs of the resident tribes in return for pelts taken in trapping and hunting.

But in order for such a scheme to work, the Indian hunters, trappers, and dressers had to be made to want the imported items that the trader had to offer. Here again we encounter the clash between history and myth, with the whites, driven to enormous technological ingenuity, producing a vast array of seductive items for peoples of the globe whose spiritual contentments had kept their own technologies at comparatively simple levels. Regarding this phenomenon, enacted everywhere whites invaded the wildernesses, we know now that there has been no people on earth capable of resisting this seduction, for none has been able to see the hidden and devious byways that lead inevitably from the consumption of the new luxuries to the destruction of the myths that give life its meaning. From the acceptance of guns, powder, shot, flints, metal traps, woolen blankets, capes, and metal cookware to deportation, the reservation, and cultural extermination is an unforeseeable way. All that is known is all that can be: and this is that these new luxuries make daily living easier, and myths or no, all humans have wanted some relief, some margin against the beloved lands. None have known how much margin would be too much.

Only here and there, the inspired ones, the crazed and dangerous, the visionary prophets, have been capable of divining the dead end of such trade, and this is why they have always been the targets of white hatred—and often the ignorant scorn of their own people. In the East the messages of the Delaware Prophet, the Shawnee Prophet, and Tecumseh were all attempts to turn the people away from trading and back to their aboriginal ways. They were too late, for not only had the tribes become dependent on the trade items that had supplanted their traditional crafts, but many of the people had become physiologically dependent as well through the calculated introduction of alcohol into the trade.

As an inducement to the sale of pelts at the lowest possible price, traders began the generous custom of tossing in a drink or two of some alcoholic beverage for the chiefs and headmen. Swiftly the custom became an institution as the traders noted its lubricious effects. Now it became standard practice to get the Indians helplessly drunk at the outset of negotiations and then buy all they had for a few drinks more. When the Indians at last sobered up, aware in the blinding light of day that they were destitute, they would have to beg the traders for credit to get the necessary supplies to go out hunting and trapping again, and so the cycle of exploitation and dependency would deepen. It is not too much to say that white traders intentionally created large numbers of alcoholics among the tribes in furtherance of their exploitation of the wilderness, and they viciously fought any attempts to outlaw the use of this poison they dispensed in exchange. Thus these pushers would push on: a locality would be hunted out and the trader would decamp westward, leaving in his wake a devastated landscape and a mendicant bunch of aboriginals to be despised by the white settlers moving in behind.

The stain of this trade had already begun to seep westward before Lewis and Clark, but it encountered resistance since the tribes of the Great Plains were living well enough to be unimpressed at first with the luxuries white civilization could offer them. Pierre-Antoine Tabeau, who was on the upper Missouri before Lewis and Clark as an agent for a St. Louis fur trader, remarked that the buffalo furnished these tribes

> not only everything of absolute necessity but also much that is useful and even superfluous. The flesh is a substantial food and one to their liking; the skin serves for lodging them well and for clothing them in every season; the horns and bones give them implements and necessary tools; the sinews give them thread; the paunches make their vessels; and the spun wool yields the

women ornaments and other superfluities. Finally, there is the head which serves them for household gods.

It is evident from such a catalog, he concludes, "that with the bow and arrow the Savage of the Upper Missouri can do easily without our trade, which becomes necessary to them only after it has created the needs." Tabeau's experienced vision of the trade nexus for which he now served as a scout in the new territory also produced this characterization of the three stages of intercourse with the "Savages": "the age of gold, that of the first meeting; the age of iron, that of the beginning of their insight; and that of brass when a very long intercourse has mitigated their ferocity a little and our trade has become indispensable to them."

And quickly it did become indispensable, the Plains tribes taking to the rifle and its associated items; to alcohol, coffee, sugar, blankets, and various trinkets of adornment. Chief Plenty-Coups of the Crow remembered that it had been some time before breech-loading rifles had reached his people, but that when they had finally come

> I did not rest until I owned one, giving ten finely dressed robes for it. Such a gun could be loaded on a running horse, and I laid my bow away forever. But some of the older men stuck tight to their familiar weapon. I could understand why they did so before the cartridge gun came, but after that the bow seemed only a plaything.

More than two centuries earlier William Bradford had noted this same revolution among the tribes of Massachusetts, and of course it was hardly limited to major items like firearms. Asked if he had ever made fire with two sticks, Plenty-Coups replied that he had but that it required strong hands and much time. "The match," he said, "is a wonderful thing. I never light my pipe or a fire with a match, but that I remember when flint and steel was the only way we knew; and even that came to us from the white man. Before that we made fire by hard work." The Cheyenne George Bent remarked that with the horse and the gun his people had been as proud and independent as any that had ever lived. But he did not understand that they could not maintain that fragile independence under the relentless pressure of the white traders with white trade goods and white designs on the lands.

After the middle of the nineteenth century these pressures incrementally increased so that by the 1860s the Plains tribes had become angry, hostile, but also baffled in their new dependency on this inexplicable people who supplied them with highly desired items but were never satisfied with the simple terms of the exchange. Only when it was all over would they know that the whites had not only wanted beaver, buffalo, rights-of-way, parcels of land for expanding settlements, and military posts to protect the settlers. They had wanted all of it. Everything. And even that would not be enough.

The end of the Plains culture had actually been forecast as early as 1862 in an area somewhat removed from the Plains tribes themselves. In that year the easternmost nation of the people the whites called the "Sioux" rose up in the midst of white settlement and thereby called attention to the lingering menace of the natives of America. The Santee Sioux had been victims of that process of trade, dependency, land cessions, and reservation existence already sketched. In 1862 they hung on in Minnesota, surrounded by white settlement and almost entirely dependent on a government dole. Yet the old days of freedom, of the hunt and the warpath, were still very much in memory, and the Santee knew also that just west of them their Plains relations yet lived those old ways. There was wide and deep resentment among the people, and it was kindled into a hopeless rage when in the late summer their dole was late in issue and they had missed their annual buffalo hunt in waiting for it. So now they took up arms and with them cleared a goodly portion of Minnesota of whites, killing an estimated 700 settlers before superior forces could flank and crush them.

But some Santee escaped westward into the Plains and others north into Canada, and while the government could do little about the latter group, the sanctuary of the Plains drew malefic attention. By this time the political voice of the western frontier was a powerful one that could not cry unheeded. And the cry was for Indian blood—and not merely Santee Sioux blood. The massacre in Minnesota had raised the old, clotted specter of the Savage, and this could hardly be laid to rest by the hanging of a few Sioux at Mankato, Minnesota, the day after Christmas, 1862. On the frontier there was talk that America could not abide the presence of the Sioux or any savages, that the Plains, as the last stronghold of the savages, would have to be cleared before settlers could feel safe and before the nation could achieve its special destiny.

The cry was heeded, and after the Civil War the heavy guns were deliberately swiveled west under the direction of General William Tecumseh Sherman who, surveying the situation of the Plains, wrote to his brother, "The more we can kill this year, the less will have to be killed the next war, for the more I see of these Indians the more convinced I am that all have to be killed or maintained as a species of pauper. Their attempts at civilization are simply ridiculous."

Sherman, his subalterns, and successors did their best to bring about the condition the general here contemplated. Those Indians who could be efficiently killed would be; those who provoked the wrath of the white frontier with their "outrages" would also be killed, whatever the cost; and the remainder would be maintained as paupers on their steadily shrinking reservations until at some future point nature would take its course and the last bit of aboriginal red would be expunged from America. Only then would the gates to the economic millennium be flung wide and the rush into it commenced.

In the life of a traditional culture the critical issue is the orderly transmission of custom from the older generations to the younger, but in this swiftly altering world of the Plains the process was disrupted among the tribes since the external conditions from which the customs had sprung were no longer what they had been. Increasingly the customs did not fit. Increasingly there was a disconcerting disharmony between traditional tribal life and the history of which the tribes were now unwillingly a part. All felt this, but its effects on the children were most significant. They grew up feeling the ominous presence of the whites and sensing in the behavior of the adults a baffled inability to combat that presence. The great warriors fought the whites but could win nothing decisive. Even victories seemed to end always in further land cessions. Once-respected leaders inexplicably became "peace chiefs" or drifted off into moody, uncommunicative alcoholism. Clan and kinship structures began to disintegrate through disease, death, and dislocation. Some bands debased themselves in the eyes of the rest of their people by taking up residence in the shadows of the soldier forts or trading posts and sponging shamelessly, while other groups joined the whites outright and rode against other Indians in battles that had no vestige of ritual about them. And some tribes like the Blackfoot appeared to be cooperating in their own destruction by slaughtering the buffalo in the great hide hunt.

One of the children who experienced all this has left a remarkably vivid reminiscence of it. This was Black Elk, the Oglala Lakota, who in late years dictated his story to the white poet John G. Neihardt. A half-blind relic, a museum piece really, waiting on the Pine Ridge reservation of our Dust Bowl 1930s to be called to the Grandfathers, Black Elk told how it was to grow up in a world coming apart, the fragile fabric of his culture fraying and spindling even as he tried to learn it. Born in the Moon of the Popping Trees in the Winter When the Four Crows Were Killed (December 1863), Black Elk was in a line of visionaries and dreamers. When he was three his father was wounded in the famous Fetterman Fight that prompted Sherman to another call for the extermination of the Sioux. When he was five his people were forced to make extensive land cessions to the whites. He was nine in the year the white tanners perfected the method that would precipitate the great hide hunt, and in this same year he had a vision of his own. Lying ill in his parents' lodge, the child was taken in dream to the sky world on the traditional journey of the hero who visits far and dangerous places and then returns with power for his people. Here the boy witnessed a magnificent display of the wild creatures of the Plains—horses, buffalo, birds—all dancing together and in a cloud lodge, the Six Grandfathers representing the Powers of the Earth. All this, they told him, is to teach you so that you may go back to your people and remind them that they may flourish only in remembering that all of life is holy.

It was an overwhelming vision, permanently affecting, and also tragic in its implications, for the boy who became a man brooding on it had seen in the center of his great vision the end of the Plains way of life. He had seen the wild creatures and his own nation walking reluctantly the black road of war, death, and destruction that led west to a cataclysmic sundown. Around them surged the winds of discord and confusion; each creature seemed on its own errand and none followed the common way. His dream was a dream of the end of dreaming, of an end to the days when people lived through the power of their dreams, and even as he dreamed this they were all awakening to the nightmare of another people's history. So he was unable to translate his vision into the power for which the Grandfathers had designed it, becoming instead of the hero of myth just another witness to history.

Neither as a hunter, nor as a warrior who took a scalp in the Custer battle, nor as a shaman, nor even as a sojourner in the white world whose ways he tried to fathom—in none of these roles was Black Elk able to save the people. And all the while he saw around him the truth of his dream of the end: the creatures of the Plains were indeed walking that black road toward sundown. In the decade following his vision it had all come true, and he had been powerless to prevent it, the hero's quest a failure, the vision not one of life but of suicidal despair. The big victory of the Greasy Grass (Little Big Horn) had really signaled the end of tribal resistance; Sitting Bull had fled to Canada; Crazy Horse had been assassinated; the Cheyenne had surrendered and been deported southward to a desert where the water was bad and the people daily died; the buffalo had disappeared; and the Lakota had given up a third of their remaining territory, including the sacred Black Hills. By 1882 the tribes were on their reservations, their main staple of life was gone, and those who survived had indeed been reduced to a species of pauper. Somehow General Sherman's vision had been the stronger.

At first, the people could not believe all this. Their world had begun to spin so rapidly in the 1860s that a centrifugal force had been created that threw fragments of their cultures off the edges. Through the 1870s that spinning had made them all dizzy. They tottered, fell, or sought refuge in strange ways, fighting against traditional friends, policing their own people in the uniforms of the enemy. The old men of the Blackfoot, bewildered by the disappearance of the buffalo, had recourse to the tatterdemalion remnants of myth, for they remembered that in the ancient days the malicious Raven had hidden the buffalo in a cave until they were released by the culture hero, Old Man, who thus saved the people from starving. Perhaps now some evil person had hidden them again. But who could come to the rescue?

And who for the Arapaho? Who for the Kiowa with Satanta a suicide in the white man's prison? Who for the Cheyenne with Roman Nose killed on Beecher's Island and Little Wolf, the great chief, turned drunken murderer of a fellow tribesman in a trader's store? Who for the Sioux with Red Cloud wrapped in silence, Spotted Tail a peace chief and then he too murdered by a fellow tribesman; and Crazy Horse stabbed in the back at Fort Robinson? As Black Elk wondered, if these great men can do nothing, what can I do?

In Grandmother's Land, Canada, the lone great figure of Sitting Bull clung to his bands and herds and rifles. Interviewed there by an enterprising New York reporter, the shaman scoffed at the notion that it had really been the Indians who had destroyed themselves by killing off the buffalo. The buffalo have disappeared in America, he said, because the country there is "poisoned with blood. . . ." It is strange, he remarked,

> that the Americans should complain that the Indians kill buffaloes. We kill buffalo as we kill other animals, for food and clothing, and to make our lodges warm. They kill buffaloes—for what? Go through your country. See the thousands of carcasses rotting on the Plains. Your young men shoot for pleasure. All they take from a dead buffalo is his tail, or his head, or his horns, perhaps, to show they have killed a buffalo. What is this?

A good question, to which the reporter had no answer.

At last, even Sitting Bull became resigned to the new West and came in to the Standing Rock reservation to rejoin his people in that land "poisoned with blood." And the spectacle was heartbreaking, even to some of the whites who witnessed it. For here, as throughout the Plains, were the former aristocrats of the wide spaces, penned up, unhorsed, and subsisting meekly on the loathsome trash that passed in official documents as food. Colonel Richard I. Dodge, hardly a friend of the Indian, noted on the reservations the scavengings of the slop buckets and dump piles, the voracious attacks on carrion, the ways in which decorum, custom, and self-respect all gave way to "the cravings of an empty stomach." An army surgeon reported the Santee Sioux standing in line on their Crow Creek reservation for helpings of a glutinous refuse ladled into their pails from a sawmill vat. The surgeon reported numerous deaths from starvation and a variety of illnesses arising from malnutrition.

Meanwhile, no opportunity was overlooked to destroy whatever of the native cultures might have survived. It was as if even now that the West had been won and the Indians so reduced that their bitterest enemies were generally satisfied, the whites could not rest easy with the least whisper or rag of the old "wild" ways. Within the eroded little enclosures of the reservations the rage for white order, the rage to clear and burn and purge the continent, continued unabated.

The Dawes Severalty Act of 1887 was a signal manifestation of this rage and the fear that drove it. The act provided for the dismemberment of the tribally held lands into individual allotments, the ostensible idea being to thereby make the Indians into responsible landholders with individual stakes in the free enterprise system. Citizenship, it was called. Cynics concluded then and subsequently that the real purpose of the act was to facilitate the transfer of Indian lands to whites since the allotments could eventually be sold, whereas the tribally held lands could not so easily be pried loose. And in a way they were right: by 1906 seventy-five million acres had passed into the control of whites. But in another way this cynical observation did not reach deep enough, for the real thrust of the act was at the very concept of the tribe itself, that communal, nonprogressive, aboriginal way of living that still seemed so threatening. Thus the Secretary of the Interior, witnessing the work of the act in 1892, was able to smugly report that the "continuance of tribal relations [has] been broken to such a degree that what remains of these obstacles to the Indian's progress is light and easily removed."

There were other strikes at the concept of tribe. Standing Bear recalled that Indians were prohibited from giving to those of their tribal relations who were in greater need—unless the givers were wealthy themselves. Few were, so the old way of sharing was broken, for it might, even in these wretched circumstances, sustain some small portion of that indefinable contentment the white captors sensed had always been there. They themselves had lacked it so long that the ache had been made a virtue, the very engine of Western progress. Robert Berkhofer quotes an extraordinary statement by a prominent "friend of the Indian" that quite artlessly reveals this. In order to bring our savage brother to the light of civilization, this philanthropist noted, we must make him selfish:

We need to *awaken in him wants*. In his dull savagery he must be touched by the wings of the divine angel of discontent. . . . Discontent with the tepee and the starving rations of the Indian camp in winter is needed to get the Indian out of the blanket and into trousers—and trousers with a pocket in them, and with a *pocket that aches to be filled with dollars!*

Perhaps most distressing to the tribes were the calculated efforts to turn their children from the traditional ways. These efforts were called "educating the children for citizenship," and doubtless some of them were sincerely motivated. But almost all of them were ignorant of tribal realities or else contemptuous of them, and the effect on the adults of witnessing their children taken away to virtual prisons and shorn there of their hair, traditional clothing, language, and religion was both depressing and angering. They knew too well that if the younger generations were ripped from the tribal ways, soon enough there would be no tribes, and some, like Sitting Bull, had no doubt that this was the ultimate goal of the whites. Sitting Bull at least could keep his young son, Crow Foot, close to his side, instructing him daily in the old ways. Others could not do so, and their children were forcibly taken to far places where many sickened and died, or committed suicide, or ran away; places where their pride was cruelly broken on inflexible rules and alien standards. Standing Bear never forgot his education-by-humiliation at the Carlisle School. One day in a reading class the teacher conceived the idea of trying or testing the strength of the pupils in the class. A paragraph in the reading book was selected for the experiment. A pupil was asked to rise and read the paragraph while the rest listened and corrected any mistakes. Even if no mistakes were made the teacher, it seems, wanted the pupils to state that they were sure that they had made no errors in reading. One after another the pupils read as called upon and each one in turn sat down bewildered and discouraged. My time came and I made no errors. However, upon the teacher's question, "Are you sure that you have made no error?" I, of course, tried again, reading just as I had the first time. But again she said, "Are you sure?" So the third and fourth times I read, receiving no comment from her. For the fifth time I stood and read. Even for the sixth and seventh times I read. I began to tremble and I could not see my words plainly. I was terribly hurt and mystified. But for the eighth and ninth times I read. It was growing more terrible. Still, the teacher gave no sign of approval, so I read for the tenth time! I started on the paragraph for the eleventh time, but before I was through, everything went black and I sat down thoroughly cowed and humiliated *for the first time in my life.* . . .

To this litany of moral and cultural dispossession must be added a cluster of natural calamities that befell the tribes, and especially the Sioux, at the end of the 1880s. Throughout the Plains in the last years of that decade there were extensive crop failures and livestock diseases that caused many of the white settlers to give up in disgust and move on to the West Coast or back to the Mississippi Valley. The tribes, clamped on their reservations, had no choice but to stay and starve. There were human epidemics also that caused great fatality among the malnourished young: measles, flu, whooping cough, and tuberculosis.

For the Sioux this time was even more critical, for in 1889 they were coerced by whites who coveted their better lands into ceding another eleven million acres, about half of what they had left. Sitting Bull campaigned strenuously against this cession, but the whites now had enough influential Sioux in their pockets, and it was approved even though the tribe had not been fully paid for its cessions under the 1876 agreement. Nor was this all: after cynically raising the tribal rations while the latest cession was under consideration, the agencies then dropped them to new and unlawful lows. So for the Sioux, as for the Cheyenne, Kiowa, Arapaho, and Crow, 1889 was as bad a year as they had ever experienced, one blacker than any vision had ever shown them, and it was widely feared that the time was upon them when they would cease to exist as a people. Incredibly but surely, they were themselves going the way of the buffalo.

But in this terminally depressing winter of 1889, when they were resentful but bereft even of anger, "psychotic" as we might term them, word came from over the mountains to the west. The word was of one with great powers, a prophet who had talked with God. More word in the summer from tribes westward that this being, a messiah as some were now calling him, had promised to bring back the buffalo and the dead ancestors and to cause the disappearance of the mysterious oppressors, the whites. Thus in the fall of the year a delegation of Sioux, Cheyenne, and Arapaho, under the guidance of some Shoshoni emissaries, went west into Nevada to seek an audience with the man himself. The word they brought back the following spring went like a great wind through the tribes. It was said the man was Jesus himself, the Son of God; that they had seen the marks of his crucifixion, and that after his rejection by the whites, he had turned to his Indian children; that he would exterminate the whites in a giant earthquake; that he had removed his hat and within its crown they had seen the whole earth. Best of all, some had seen their dead relations, and on the way back to the Plains had seen buffalo and actually killed one, sacrificing it to the messiah in the prescribed fashion.

The returned delegates told of the messiah's creed, a pacific one built around a kind of cut-down decalogue and emphasizing that the tribes must not cause trouble with their new religion. In due time, perhaps in the spring of the next year, perhaps in the summer near the time when they had formerly held their Sun Dances, the new earth would come, sliding green and laden with game and ghosts, down over the old worn-out and poisoned one the whites had made. The whites would be swallowed up, but the believing Indians would be uplifted, like eagles, like crows. Meanwhile, they must remain peaceful, practicing the Ghost Dance, purified, painted with a sacred red paint the delegates brought back, their hands joined as they sang to the Father:

> *He!* When I met him approaching—
> *He!* When I met him approaching—
> My children, my children—
> I then saw the multitude plainly.
> I then saw the multitude plainly.

So sang the Arapaho of the vision of a dancer who, supine upon the earth, had been enabled to dream once again and to see in that dream the messiah advancing at the head of the ghostly multitude. And they sang too of the end of the oppressors:

*I'yehe'*! my children—*Uhi'yehe'heye'*!
*I'yehe'*! my children—*Uhi'yehe'heye'*!
*I'yehe'*! we have rendered them desolate—*Eye'ãe'yuhe'yu*!
*I'yehe'*! we have rendered them desolate—*Eye'ãe'yuhe'yu*!

Into the chant, rising now in volume from all over the lands the whites had thought ineradicably theirs, came the voices of the Cheyenne:

Our father has come,
Our father has come,
The earth has come,
The earth has come,
It is rising—*Eye'ye'*!
It is rising—*Eye'ye'*!
It is humming—*Ahe'e'ye'*!
It is humming—*Ahe' e' ye'*!

And the voices of the Sioux:

The whole world is coming,
A nation is coming, a nation is coming,
The Eagle has brought the message to the tribe.
The father says so, the father says so.
Over the whole earth they are coming.
The buffalo are coming, the buffalo are coming,
The Crow has brought the message to the tribe,
The father says so, the father says so.

And the voices of the Assiniboin, Gros Ventre, Arikara, and Mandan. Tellingly, some of the most active disseminators of the new gospel were returned students from the white man's world who used his methods, his writing system and mail service, to spread a message utterly contrary to white teachings. Even old Red Cloud, silent, half-blind, and huddled within himself these many years, declared himself a believer. So did Sitting Bull at Standing Rock, and these two figures from the great old days were the crucially necessary disciples of the movement. Whites cynically observed that these two old men had attached themselves to the movement as a way of revitalizing nothing more than their own political positions. But as always in these matters, their cynicism was founded on their own spiritual condition and their ignorance of the deeper issues involved. Sitting Bull answered them best when he posed the cognate question: What difference does it make if this new religion is not true? What else is there we can hope for? What else indeed, and desperate as this hope might now seem to us, it was truly all the tribes had.

Thus the failed hero, Black Elk, who had for many years carried his great vision within him like a thing inert, a stone, now heard of the messiah and of the Ghost Dancers. But for a time he stayed away, believing it was only despair that made the people believe, "just as a man who is starving may dream of plenty of everything good to eat." But such was the contagion of belief that like other doubters he became a dancer, and in his dance he had a vision of a

> beautiful land where many, many people were camping in a great circle. I could see that they were happy and had plenty. Everywhere there were drying racks full of meat. The air was clear and beautiful with a living light that was everywhere. All around the circle, feeding on the green, green grass were fat and happy horses; and animals of all kinds were scattered all over the green hills, and singing hunters were returning with their meat.

When he came out of his vision he was surrounded by dancers clamoring to know what he had seen: such was the need to be possessed once again. Subsequently, Black Elk had another dream and in this one saw sacred shirts and dresses for the dancers to wear. He began making them and dispensing them among the Sioux as the religion began to proliferate.

The spectacle made the whites bristle in fear. Here at the end of the Century of Progress when few could doubt that the nation was verging on the millennium was this ugly and threatening atavism of the old, forgotten New World that had seemed so safely buried beneath the towns, cities, and railroads. Civilization had so thoroughly established its imperative order throughout what had been the "howling wilderness" that it had lately begun to indulge in a harmless sort of nostalgia, lamenting the "Vanishing American," stamping his granitic features on seals and coins, casting his defeated yet heroic figure in bronze or carving it in marble. It seemed safe enough now. And then for the wilderness to howl again with the chants of these dancers, for "Poor Lo" to pull back from that abyss of cultural extermination over which the white artists had shown him poised! And not only on the Plains but west of the Rockies, where the Paiute, the Bannock, Shoshoni, Gosiute, and Ute were dancing. And it was known that disciples had been into the Southwest and perhaps into the East spreading this thing. Who could tell where it might end? Were they to give it all back to the Indians, to make mockery of all the sacrifices, the spilled blood, the terrors and sweats and humped backs of pioneers who had cut and blasted and heaped up stones to make clearings where there had been darkness? Ever since they had been driven to encounter the native populations of the globe, the whites had deeply feared the drums of rebellion, and so here, even buffered by a ludicrous numerical and technological advantage, settlers fell into a strange hysteria, writing newspapers and badgering congressional representatives and the Department of the Interior for protection.

The hysterical reaction focused on the Sioux, and there were predictions of a "general Indian uprising" emanating from this people. Attempts were made on the Sioux reservations to stop the dancing, and leaders were briefly detained, but the Sioux dancers confirmed the whites' deepest fears by telling the agents that they would fight and die rather than desist. The agent at Pine Ridge (South Dakota) daily sent out frenzied telegrams to his superiors informing them that the situation was out of hand and could be saved only by a massive show of military force. The President responded by directing the Secretary of War to take charge in the event of the "general uprising." And the Secretary of War indeed took charge by dispatching three thousand troops into Sioux country in the middle of November, 1890. Only here and there less fearful individuals took the time to point out that a comparative handful of half-starved prisoners, mostly unarmed, surrounded by white power, and seeking merely to recover a portion of their lost religious freedom, could pose no actual threat to the nation. Such voices had to be ignored, for the threat was ultimately not a military one.

And so once again commenced the Hunt for the Wild Man, the shaman, the ultimate doer of this evil. Red Cloud was too old for the role, and his attachment to the movement was mostly ceremonial, but there was Sitting Bull, and he would fit handsomely. After a few ineffectual efforts to talk Sitting Bull out of his adherence to the movement—including the charade of sending his old friend Buffalo Bill out to talk with him—plans were laid to remove this person in whom so much of the demonism of the continent seemed concentrated. So at daybreak of December 15 a detachment of Indian police surrounded the old shaman's cabin with ominously vague orders that he should neither escape nor be rescued by the Ghost Dancers.

But the barking of the camp dogs had alerted the Ghost Dancers, and when the policemen pushed the old man out through his cabin door into the gray of the winter dawn, they found themselves in a crowd of angry and excited men and women. We have been expecting this, the dancers said: you shall not have him. In the inevitable fierce little fight that ensued, the police made sure that they did have Sitting Bull, once in the body and again in the face. Then they reentered the cabin and executed the son, Crow Foot, as he begged for his life. *That* branch, anyway, had been nipped, and their white employers should be satisfied.

The murder of Sitting Bull was what led with terrific inexorability to the savage conclusion of this business, for it confirmed the fears of the Sioux Ghost Dancers that the troops had been brought in to murder them all. Badly frightened but defiant Sioux huddled under Chief Big Foot in the Badlands off the reservations where refugees from Standing Rock came bearing news of the murder. While they danced and awaited further news the troops surrounded them and began a kind of negotiation that ended with the dancers being prodded toward the Pine Ridge reservation on December 27. On the 28th the troops camped about twenty miles from the agency and plans were made to take from the dancers whatever arms they might have.

Another American dawn, but singular in its way, for what it broke upon was the end of something that vastly predated the corporeal realities of those who now faced each other across a tiny patch of clay-dry Indian soil. On the slopes of the little hillocks stood 470 white men muffled and capped against the weather, their rifles greased, glinting, and government-issued, and their four big rapid-fire cannons trained downhill. Among them, grim-faced, expectant members of the U.S. 7th Cavalry—Custer's outfit—sat their big horses and remembered the Sioux at Little Big Horn. And on the flats below some 340 Ghost Dancers camped in canvas lodges. The Indians were ordered to surrender their arms. They looked at the surrounding guns and into their lodges full of women and children and then delivered up an obviously fallacious cache. So now the close-buttoned, anonymous-faced soldiers moved in, poking and kicking through the lodges while the women called to their men and the children cried. A shaman kept blowing an eagle-bone whistle. Some lodges were overturned and their contents scattered. The sun rose.

A pitful treasure of rifles, some of purely talismanic value, was thus discovered, but when a soldier attempted to search the body of a young man, the latter drew a concealed gun and fired. In an instant the Indian threat, such as it had been, was exterminated as the gunners and riflemen dropped almost all of the men where they stood at the entrances to the lodges. But now the real battle began, inevitable and terrific. A whole civilization—Chicago and St. Louis and the older outposts eastward; fortresses, soldiers, and slave castles beyond the edges of the Old World; driven kings, commanders, prelates, and nameless spear carriers—all its gathered force poured down the greased barrels into the screaming women and children who fled westward along winter's dry gulch toward no refuge. The soldiers pursued them, mile after mile, while behind them a sullen smoke drifted up from the smoldering canvases of the lodges.

Black Elk was there, having ridden desperately out with others from Pine Ridge when the sounds of the heavy guns announced sure annihilation. Armed with his vision and nothing more, he galloped his lathered pony through the smoking destruction and out along the gulch,

> and what we saw was terrible. Dead and wounded women and children and little babies . . . scattered all along there where they had been trying to run away. The soldiers had followed along the gulch, as they ran, and murdered them in there. Sometimes they were in heaps because they had huddled together, and some were scattered all along. Sometimes bunches of them had been killed and torn to pieces where the wagon guns hit them. I saw a little baby trying to suck its mother, but she was bloody and dead.

Numbers here are meaningless, but General Nelson A. Miles, in charge of operations, spoke truly when he laconically reported, "I think very few Indians have escaped."

Very few had, and when a blizzard swooped down on the countryside, those who lay out wounded and exposed died silently under the winds and snow. Burial details on New Year's Day, 1891, discovered a few who had survived under the corpses or blankets or jumbled ruck of the lodges, but the rest were frozen in the impossible poses of violent death. The soldiers and the civilian vultures who accompanied them stripped the corpses of their effects, the most prized trophies being those associated with the Ghost Dance. Then the bodies were heaped on wagons and hauled to the hillock from which the cannons had poured death, and here they were tumbled into a common trench hacked out of the frozen clay. In the bright and bitter sunlight members of the burial detail posed beside the stacked remains, the arms and legs of the naked corpses looking as if they were still struggling to dance. In after years some of the stolen ceremonial paraphernalia found its way into various museums, and so it is possible for us to pause now before glass cases and gaze within at the moldering rags of Ghost shirts, torn with bullets and fire, and stained so faintly with an archaic blood.

Among the voices raised in the aftermath was that of Buffalo Bill. He had responded to an urgent appeal from General Miles and had left a lucrative European tour with his Wild West show to come out to Standing Rock for a parley with Sitting Bull. Miles knew that Cody was one of the few whites Sitting Bull would talk with and had reasoned that Cody might be able to avert major trouble by talking the shaman out of his sponsorship of the dancing. But Miles had been overruled, and Cody had been recalled without ever having seen Sitting Bull. Cody now felt he had been sent on a fool's errand, and he was right. Disgruntled, he announced that he was returning to his European tour.

But eventually Buffalo Bill was recompensed for his aborted service: he was allotted a hundred ex-Ghost Dancers to tour with his show, including No-Neck, a child who had been found alive beneath the body of his father on the field at Wounded Knee. These additions did their bit to enhance the already fabulous popularity of the show, both in America and in the Old World where great throngs turned out everywhere to see the heroic scout and his entourage of feathered exotics, just as their ancestors had centuries earlier when Columbus had put his captive Arawaks through their paces. In these settings there was still something strangely fascinating about the savages and the old, wild life they represented, and though Cody was no thinker, he did possess a sense of that fascination and an ability to exploit it.

That sense had been derived from his own life, for as a child Cody too had felt the fascination of the great West, and from the record, so had his family, which emigrated from Ireland to Massachusetts, and from there to Iowa, where Cody was born in 1846. There is a legend that tells us that the spot near Davenport where Cody was born was the site of the last war dance performed by the Sac and Fox tribes before they went on the warpath in the doomed attempt to resist removal that is known as the Blackhawk War. Whatever the historical truth of this, there was plenty of frontier in Cody's young life as his family moved on with the line of settlement, out of Iowa and into Kansas, to Leavenworth, a raw town on the bluffs of the Missouri. Here wagon trains made up for the army posts in Colorado, Utah, and New Mexico, and here little Will Cody greedily absorbed the exciting sense of the Wild West that always tantalizingly began just a bit beyond wherever one was. Here he hung about the mule skinners and bullwhackers and sutlers, and here he saw the scouts and listened for hours to their talk. A bit later he was actually to see the legendary Jim Bridger and Kit Carson and follow them about to note indelibly their least mannerisms. He made a childish vow to become such a heroic figure himself when opportunity would permit.

As we know, he amply fulfilled that vow. As a pony-express rider and teamster, he served an apprenticeship to the herohood that came to him in the days of the Plains campaigns when he scouted for George Armstrong Custer and General Phil Sheridan, becoming eventually a much-lionized Chief of Scouts. And there was, too, his fame as a buffalo hunter. All of this led to a career as a showman, in the long course of which he assumed proportions that dwarfed his boyhood models, Bridger and Carson.

At first the tool of the wily picaro E. Z. C. Judson (a.k.a. Ned Buntline), who wrote melodramatic farces and dime novels using him as a kind of lay figure, Cody quickly outgrew his creator, whose vision he found too confining. Cody's mind had been permanently shaped by the amplitude of the West, and he wanted to bring that amplitude to the stage where it could be experienced by the millions who could never themselves know the grandeur of a vanished world. What he began dreaming about was a giant spectacle with horses, Indians, buffalo, and riders, something that could be packaged and moved on tour. And this is what he eventually put together, adding to the vital nucleus the trappings of fancy riding, trick shooting, and a stage holdup. For more than three decades Buffalo Bill brought the Wild West, stuffed and theatrically preserved, to loving audiences of the Western world.

The secret of Cody's public success was that he himself had been a heroic killer of that old, wild life he now displayed the remains of, for his fame rested securely on the reports that he had killed Indians and buffalo in untold numbers. In this he was symbolic of that process of taming the wilderness that had seemed so imperative to whites even while the wilderness had been so perilously seductive. In his figure this ambivalence was delightfully resolved, the heroic killer displaying to his audiences the trophies of his triumphs. Inside the ropes and canvases of hundreds of arenas the wilderness lived on in the only way whites could accept it: tamed, scaled down, and under the direction of the white hero/magician. Knowing who Buffalo Bill was, what his eye, hand, and rifle had done, and then seeing him ride in at the head of his company of savages and buffalo, was a dramatic illustration of what civilization was all about.

As the tours and the applause went on and as the real Wild West steadily receded into memory, there came a change in the mind of the hero. He began to be haunted by the suspicion that he and his kind had killed the thing he had most loved. For Cody knew, better than any of his fans or even those responsible for the promotion of his show, that he had been one of a singular breed of white men who had destroyed a whole way of life in the name of civilization. In his later days, therefore, he became increasingly uneasy about the fact that his reputation rested on his prowess as a killer. He no longer would answer questions as to how many Indians he had killed, preferring to emphasize his more "positive" contributions. He was a real-estate developer now, rather like what Daniel Boone had wished to become, and he had a scheme to irrigate the Shoshone River Valley. He wanted to be known not as a killer but as "the man who opened up Wyoming to the best of civilization."

But even this would not do for, however obscurely, Cody knew that civilization had meant the death of the Plains way of life, and he realized that that way of life had been loved intensely by others besides himself—those who had lived it as he had or in other ways. He had known the great chiefs Red Cloud, Sitting Bull, and American Horse, and also the many tribespeople who toured with him, and he could not have escaped knowing what these people had lost to him and his civilization. He could also see in the aging companions of his youth a sense of loss beyond mere nostalgia. America had been deprived of something indefinable but precious with the destruction of the buffalo, the tribes, and the free lands. Cody came to feel himself less hero than romantic mortician, less master of ceremonies than relic of vanished times.

As if to combat this uneasiness, he and his associates went in for greater and ever more exotic spectacles. He still retained the riders, shooters, and stage coach. He still staged the Custer massacre and his own famous duel with the Cheyenne chief Yellow Hand in which he had taken the first scalp to avenge Custer. But now there appeared incongruous German grenadiers, the "Potsdamer Reds," in their fire-engine red uniforms. And French soldiers in sky blue, English hussars, Mexicans, Cossacks, Arabs, and Cubans in response to the Spanish-American War (in the taking of San Juan Hill, Cody played Teddy

Roosevelt). There were also Hawaiians, Filipinos, Japanese magicians, a snake enchantress, midgets, a boy giant, Venetian glassblowers, a sword swallower, and even mind readers.

The more unwieldy and incoherent the show became, the more un-cannily it revealed the inner state of its hero, who was himself becom-ing unwieldy and incoherent, both with drink and with the suspicions that haunted him. His personal life was coming apart like the moth-eaten props of thirty years' packing and unfolding. He sued his wife, Louisa, for divorce, claiming that on four occasions she had tried to poison him. Louisa, perhaps herself deranged by a lifetime of loneliness, stashed conveniently at North Platte, Nebraska, while her husband toured the world and learned about women, filed a countersuit in which there were crazy rumors of Queen Victoria's involvement. The judge threw the embarrassing mess out of court.

At last the show and its hero came asunder in bankruptcy, but still the old scout went on—what was left to him but to act out to the end what had become his private nightmare? Now he was owned by other shows: propped up in his saddle astride a very sedate old plug, he would enter the arena slowly and give a sweeping bow and a wave of his Stetson in "Buffalo Bill's Farewell." He did the "Farewell" over and again as if mumbling to himself a series of old regrets. Once his wig came off with his Stetson, exposing to the cruel light of high wattage his all-but-naked skull with its wisps of white hair like a tattered scalp-lock.

He developed a great fear of dying in the arena before hundreds of anonymous faces, but he did not. He survived as an increasingly eccen-tric drunk who was available for whatever guest performances were oc-casionally offered. He hoped to recoup his fortunes by making a movie of the Wild West, and a company was organized to reenact the massacre of the Ghost Dancers at Wounded Knee. They traveled out to Pine Ridge where an Indian village was reconstructed along Wounded Knee Creek, but when the Sioux learned that the shooting was to take place on the precise site of the massacre, their mood turned ugly. The old scout had to call on the services of No-Neck once again, this time to avert real and not staged violence, and the film was made. It did little to benefit Cody's financial situation, and to the end he remained the prop-erty of others.

In the late fall of 1916 Cody entered his terminal illness, and like his theatrical farewells it went on and on, into January 1917, with the newspapers reporting his slippages, small recoveries, and deeper declines. Through the last days Cody's life flickered inside his head in a jumble of hallucinatory images: he raved on about Indians, buffalo, and plans to go on the road this season with an even bigger, more extravagant show. At the very end his mind cleared briefly, and he directed that his body be shipped back from Denver where he lay dying to the Big Horn Basin above his town of Cody, Wyoming. But he was buried at the crest of Lookout Mountain above Denver, his widow apparently revenging herself in this decision, which had been suggested to her by a circus promoter.

And a circus affair it was, the Sells-Floto troup parading with the cortege up the muddy slopes of the mountain to its summit where the old showman who had so much feared exposure in death was uncovered for final viewing. From this lofty spot, so a reporter for the Denver *Post* intoned, Buffalo Bill Cody could look eastward toward the ever-rising dawn that each morning would reveal anew the concrete realization of his dream: a great city humming where once there had been nothing but a wild New World.

# Notes on Sources

PART ONE   Loomings

ESTRANGEMENT

There are hundreds of books and uncounted essays on various aspects of exploration. So far as I know, there is only one that attempts to treat it as a phenomenon accomplished by a single coherent civilization. But Walter Prescott
Webb's emphases in *The Great Frontier* (Boston, 1952), are almost exclusively
economic, social, and political. And he almost totally assumes the wilderness
and its life in assessing the impact of its exploration on various Western institutions. The case for the pre-Columbian exploration of the Americans by non-
aboriginals is so conclusively proven, I think, that it is not worthwhile to take
the space to defend it here. But it is essential to know that others did go out and
return again or, if they stayed, disappeared into the aboriginal landscape. This
fact makes the European invasion of the Americas, Africa, and the Far East the
more striking in contrast. Cyrus Gordon's *Before Columbus: Links Between the
Old World and Ancient America* (New York, 1971) reintroduced an old subject to
new readers in authoritative fashion when it appeared. Barry Fell, *America B.C.:
Ancient Settlers in the New World* (New York, 1976), gives a far more comprehensive view of pre-Columbian activity and powerfully suggests its scope;
doubters should have a look at his book. Henriette Mertz's *Pale Ink: Two Ancient Records of Chinese Explorations in America* (Chicago, 1972) explores the
probability that the Chinese made regular visits to the Americas beginning
about 2250 B.C.

Lewis Mumford. *Technics and Civilization.* New York, Burlingame, 1963.
Howard Mumford Jones. *O Strange New World; American Culture: The Formative
    Years.* New York, 1964.
———. "The Greatness of the Nineteenth Century," *History and the Contempo-
    rary: Essays in Nineteenth-Century Literature.* Madison, Wis., 1964.

## THE NECESSITY OF MYTH

It is often difficult to demarcate the specific origins of our most cherished convictions since, after a time, they come to seem so wholly ours. As for my attitude toward myth, however, I can still recall the sense of joyous discovery that attended my first reading of Joseph Campbell's *The Hero with a Thousand Faces* (New York, 1956). At that time I was a graduate student in folklore and finding out what an enormous mountain of rubbish had been printed on this subject. Campbell's book was the first I had encountered that even approximated the poetry and profundity of its subject. To me it remains the finest, and I am indebted to it throughout my own essay. After Campbell's work I found Robert Duncan's *The Truth and Life of Myth: An Essay in Essential Autobiography* (Fremont, Mich., 1968) wonderfully rich and allusive. Earl W. Count, "Myth as World-View: A Biosocial Synthesis," in Stanley Diamond (ed.), *Culture in History: Essays in Honor of Paul Radin* (New York, 1960), supplied a view of myth as a vital human function that I have found of continuing value. More recent observations on the relationship of the archaic brain to the neocortex and on the continued operations of the former are found in Carl Sagan's *The Dragons of Eden: Speculations on the Evolution of Human Intelligence* (New York, 1977).

John N. Bleibtreu. *The Parable of the Beast*. Toronto, 1968.

Joseph Campbell. *The Masks of God: Primitive Mythology*. New York, 1970.

Grahame Clark. *The Stone Age Hunters*. New York, 1967.

Frank Hamilton Cushing. *Zuñi Breadstuff*. New York, 1920.

James G. Frazer. *The (New) Golden Bough*, rev. and ed. Theodore Gaster. New York, 1964.

Sigmund Freud. *Civilization and Its Discontents*, trans. Joan Riviere. Garden City, N.Y., 1958.

P. V. Glob. *The Bog People: Iron Age Man Preserved*. London, 1969.

Carl G. Jung. *Memories, Dreams, and Reflections*, ed. Aniela Jaffe and trans. Richard and Clara Winston. New York, 1963.

————. *Modern Man in Search of a Soul*, trans. W. S. Dell and Cary F. Baynes. New York, 1933.

————. "On Approaching the Unconscious," Jung et al., *Man and His Symbols*. New York, 1968.

Andreas Lommel. *Prehistoric and Primitive Man*. Middlesex, England, 1968.

Konrad Lorenz. *King Solomon's Ring*, trans. Marjorie K. Wilson. London, 1965.

Arthur O. Lovejoy and George Boas. *Primitivism and Related Ideas in Antiquity*. New York, 1965.

Eugene Marais. *The Soul of the Ape*. New York, 1969.

Henry A. Murray, ed. *Myth and Myth-Making*. New York, 1960.

Lewis Thomas. *The Lives of a Cell: Notes of a Biology Watcher*. New York, 1974.

Peter J. Ucko and Andrew Rosenfeld. *Paleolithic Cave Art.* New York, Toronto, 1973.

Olivia Vlahos. *Human Beginnings.* New York, 1972.

A. C. Vroman. "The Enchanted Mesa," in Ruth I. Mahood, ed., *Photographer of the Southwest: Adam Clark Vroman, 1856–1916.* N.p., 1974.

## BEARINGS FROM THE ANCIENT NEAR EAST

The sense I try to impart of the struggles and achievements of the peoples of the ancient Near East was most impressively suggested to me by two works of Lewis Mumford: *Technics and Civilization* and *The City in History: Its Origins, Its Transformations, and Its Prospects* (New York, 1961); by Paul Shepard's brilliant evocation of that struggle in *Man and the Landscape: A Historic View of the Esthetics of Nature* (New York, 1972); and by Henri Frankfort's panoramic *The Birth of Civilization in the Near East* (Bloomington, Ind., 1951). Edward Hyams, *Soil and Civilization* (London, New York, 1952), and V. Gordon Childe, *Man Makes Himself* (New York, 1963), were both instructive on the reciprocal effects of humans and their specific environments. As to the spiritual consequences of this struggle, I have drawn on Theodore Gaster's fine collection of texts from the region, *The Oldest Stories in the World* (Boston, 1958), which has the additional merit of indicating the widely shared narrative motifs occasioned by the areal cultural traits and history; Samuel Noah Kramer's collection of studies of the areal mythologies in Kramer (ed.), *Mythologies of the Ancient World* (Garden City, N.Y., 1961); S. H. Hooke, *Middle Eastern Mythology* (Middlesex, England, 1975); and Thorkild Jacobsen's study of the evolution of Near Eastern religious forms, *The Treasures of Darkness: A History of Mesopotamian Religion* (New Haven, London, 1976). These last two works contain generous amounts of translations of the old sacred texts. Here as elsewhere, of course, the conclusions I draw are my own and in some instances conflict with those drawn by these authorities.

Bruno Bettelheim. *Symbolic Wounds: Puberty Rites and the Envious Male.* Glencoe, Ill., 1954.

Joseph Campbell. *The Masks of God: Occidental Mythology.* New York, 1970.

Donald P. Cole. *Nomads of the Nomads: The Āl Murrah Bedouin of the Empty Quarter.* Chicago, 1975.

Carleton S. Coon. *Caravan: The Story of the Middle East.* New York, 1958.

James G. Frazer. *The (New) Golden Bough,* rev. and ed. Theodore Gaster. New York, 1964.

Sabatino Moscati. *Ancient Semitic Civilizations.* New York, 1960.

Erich Neumann. *The Great Mother: An Analysis of the Archetype,* trans. Ralph Mannheim. Princeton, N.J., 1972.

Amaury de Reincourt. *Sex and Power in History.* New York, 1974.

## THE PEOPLE OF THE BOOK

My primary experience here has been with the established texts of the scriptures themselves, both in the King James version of the Old Testament and in Herbert G. May and Bruce Metzger (eds.), *The Oxford Annotated Bible* (New York, 1962). Beyond these, I have learned much and found heartening confirmations in three significant studies of ancient Israelite culture and history: Salo Wittmayer Baron, *A Social and Religious History of the Jews* (New York, 1952), Vol. I; Johannes Pedersen, *Israel: Its Life and Culture,* trans. H. Milford (Copenhagen, London, 1959), 2 vols.; and Max Weber, *Ancient Judaism,* trans. Hans Guth and Don Martindale (New York, 1967). There were also helpful hints on the cultural significations of the texts in Julius A. Bewer, *The Literature of the Old Testament,* rev. Emil G. Kraeling (New York, London, 1962), and Cornelius Loew, *Myth, Sacred History, and Philosophy: The Pre-Christian Heritage of the West* (New York, 1967). George H. Williams has some richly suggestive remarks about the role of wilderness in Israelite thought in *Wilderness and Paradise in Christian Thought* (New York, 1962).

Donald P. Cole. *Nomads of the Nomads.* Chicago, 1975.

Carleton S. Coon. *Caravan.* New York, 1958.

Henri Frankfort. *The Birth of Civilization in the Near East.* Bloomington, Ind., 1951.

Sigmund Freud. *Moses and Monotheism,* trans. Katherine Jones. New York, 1967.

————. *Totem and Taboo: Resemblances Between the Psychic Lives of Savages and Neurotics,* trans. A. H. Brill. New York, 1946.

S. H. Hooke. *Middle Eastern Mythology.* Middlesex, England, 1975.

Carl O. Sauer. *Agricultural Origins and Dispersals.* New York, 1952.

Paul Shepard. *Man in the Landscape.* New York, 1972.

W. W. Swidler. "Adaptive Processes Regulating Nomad-Sedentary Interaction in the Middle East," in Cynthia Nelson, ed., *The Desert and the Sown: Nomads in the Wider Society,* Berkeley, Cal., 1973.

## A CRISIS CULT

For the cultural background within which Christianity arose and made its appeal I have relied on Michael Grant, *The Climax of Rome: The Final Achievements of the Ancient World, A.D. 161–337* (New York, Toronto, London, 1968), which is both comprehensive and incisive; E. R. Dodds, *Pagan and Christian in An Age of Anxiety; Some Aspects of Religious Experience from Marcus Aurelius to Constantine* (New York, 1970); Frederick Grant, *Ancient Roman Religion* (Indianapolis, 1957), which has a very informative introductory essay and a generous selection of texts illustrating the evolution of Roman religion to the Christian era; and George C. Brauer's *The Age of the Soldier Emperors: Imperial Rome A.D. 244–284* (Park Ridge, N.J., 1975). As I have indicated in my text, my debts are great to Weston La Barre, *The Ghost Dance: The Origins of Religion* (New York, 1972), and Anthony F. C. Wallace, "Revitalization Movements," *American Anthropologist* (April 1956) and *The Death and Rebirth of the Seneca Nation* (New York, 1970). The latter was my teacher in graduate school many years ago and his approaches to the study of human cultures were decisive in my life and work. Max Weber's study of charismatic movements in *The Theory of Social and Economic Organization*, ed. Talcott Parsons, trans. A. M. Henderson, Parsons (New York, 1947), has to some extent been superseded by these later scholars, but his conceptualizations remained useful to me. For my view of the early Church I have, of course, spent time with the texts of the New Testament. Henry Chadwick, *The Early Church* (Baltimore, Harmondsworth, England, 1967), and Burnett Hillman Streeter, *The Primitive Church* (New York, 1929), gave informative general views of early Christian struggles and resolutions. So did A. D. Nock's classic *Conversion: The Old and the New in Religion from Alexander the Great to Augustine of Hippo* (London, Oxford, New York, 1972). These last three studies made it possible for me to better understand and appreciate that monument of the formative years, Augustine's *The City of God*, trans. Marcus Dods (New York, 1950). Whatever one thinks of Christianity and the history it has made, reading this work in its appropriate context is a deeply moving experience.

(No Authors). *The Lost Books of the Bible and the Forgotten Books of Eden*. New York, 1974.

Norman O. Brown. *Life Against Death: The Psychoanalytic Meaning of History*. New York, 1959.

———. *Love's Body*. New York, 1966.

Mircea Eliade. *The Myth of the Eternal Return; or, Cosmos and History*, trans. Willard Trask. Princeton, N.J., 1971.

James G. Frazer. *The (New) Golden Bough*, rev. and ed. Theodore Gaster. New York, 1964.

Carl G. Jung. *Memories, Dreams, and Reflections*, ed. Aniela Jaffe and trans. Richard and Clara Winston. New York, 1963.

Felix M. Keesing. *Culture Change: An Analysis and Bibliography of Anthropological Sources to 1952*. Palo Alto, London, 1953.

Henry Charles Lea. *Chapters from the Religious History of Spain Connected with the Inquisition*. Philadelphia, 1890.

Arthur Cushman McGiffert. *A History of Christian Thought*. New York, London, 1954, Vol. I.

Walter Nigg. *The Heretics*, ed., trans. Richard and Clara Winston. New York, 1962.

Martin P. Nilsson. *A History of Greek Religion*. New York, 1964.

Amaury de Reincourt. *Sex and Power in History*. New York, 1974.

E. C. Rust. *Towards a Theological Understanding of History*. New York, 1963.

Evelyn Underhill. *Mysticism: A Study in the Nature and Development of Man's Spiritual Consciousness*. New York, 1961.

Anthony F. C. Wallace. *Culture and Personality*. New York, 1961.

Hans Zinsser. *Rats, Lice, and History*. New York, 1971.

## HECATOMBS

Norman Cohn's *The Pursuit of the Millennium; Revolutionary Millennarians and Mystical Anarchists of the Middle Ages* (New York, 1970) was an eye-opener into a little known and highly revealing chapter in the history of Christian civilization; reading it is like witnessing a vivisection on the body of the Church. Leon Festinger et al., *When Prophecy Fails: A Social and Psychological Study of a Modern Group that Predicted the End of the World* (New York, 1964), can be used to update and supplement Cohn. The great work on the Inquisition is of course that of Henry Charles Lea, *A History of the Inquisition of the Middle Ages* (New York, 1888), 3 vols. In its massiveness and density it takes one back into a fearful time; one does not have to share Lea's biases to read it for instruction. As for that old mythic world of pre-Christian Europe, I know of no work that more powerfully evokes it than that of Marija Gimbutas, *The Gods and Goddesses of Old Europe, 7000–3500 B.C.: Myths, Legends, and Cult Images* (Berkeley, Los Angeles, 1974). Even though the period she surveys is far earlier than that considered here, and its geographical range is restricted to southern Europe, still, to read of the many findings of cultic objects and to gaze upon their rude vigor in photographs is to get a vivid, shocking sense of that spiritual residue that underlay Christian civilization and that must have lived on in one way or another through the centuries of Christian history. For as the great students of history have told us, nothing so powerful can ever be truly forgotten.

T. S. R. Boase. *Death in the Middle Ages: Mortality, Judgment, and Remembrance*. New York, 1972.

Frederick Duncalf. "The Councils of Piacenza and Clermont," in Marshall W. Baldwin, ed., *A History of the Crusades*, Vol I, *The First Hundred Years*. Madison, Wis., London, 1969.

A. D. J. Macfarlane. *Witchcraft in Tudor and Stuart England*. New York, Evanton, N.J., 1972.

John T. McNeill and Helena Garner, eds. *Medieval Handbooks of Penance*. New York, 1938.

Margaret Murray. *The Witch-Cult in Western Europe*. Oxford, 1921.

Frederick Bliss Luquiens, trans. *The Song of Roland*. London, 1966.

Sidney Painter. "Western Europe on the Eve of the Crusades," in Baldwin, ed., *A History of the Crusades*, Vol. I.

Jean Seznec. *The Survival of the Pagan Gods: The Mythological Tradition and Its Place in Renaissance Humanism and Art,* trans. Barbara F. Sessions. Princeton, N.J., 1972.

Keith Thomas. *Religion and the Decline of Magic.* New York, 1971.

J. M. Wallace-Hadrill. *The Barbarian West, A.D. 400–1000.* New York, 1962.

Helen Wadell. *The Desert Fathers.* London, 1936.

William, Archbishop of Tyre. *A History of Deeds Done Beyond the Sea,* ed. and trans. Emily Atwater Babcock, A. C. Krey. New York, 1943, 2 vols.

George H. Williams. *Wilderness and Paradise in Christian Thought.* New York, 1962.

## LOOMINGS

C. Raymond Beazley's *The Dawn of Modern Geography* (Oxford, 1906), 3 vols., is a mine of information on the relationship between the Christian religion and geographical science. For the concept of mental maps I am again indebted to Wallace; a discussion of the concept is to be found in his *Culture and Personality.* Peter Gould and Rodney White, *Mental Maps* (Baltimore, Harmondsworth, England, 1974), have worked out the concept somewhat more elaborately. The sense I try to convey here of the self-inclosed worlds of myth-bound peoples has never been more impressively conveyed to me than in Peter Matthiessen's *Under the Mountain Wall: A Chronicle of Two Seasons in the Stone Age* (New York, 1962).

Leo Bagrow. *History of Cartography,* rev. and ed. R. A. Skelton and trans. D. L. Paisey. Cambridge, 1964.

Manuel Komroff, ed. *The Travels of Marco Polo.* Garden City, N.Y., 1930.

Henry Charles Lea. *A History of the Inquisition of the Middle Ages.* New York, 1888.

Lewis Mumford. *Technics and Civilization.* New York, Burlingame, 1963.

B. Traven. *The Treasure of the Sierra Madre.* New York, 1935.

## PART TWO Rites of Passage

## MYTHIC ZONES

Each of the narratives I use here has been adapted from a translation of ethnographic material. I have tried for a strict fidelity to the content of each narrative, but in the main the language is my own. No doubt this is not an improvement over the narratives as they were traditionally recited in their appropriate settings, but in a few cases it may be a slight gain over the more formal diction of the translations on which I have relied. For the sketch of the mythology of the Arawak who migrated to the Antilles, I have pieced together from two sources what little is known of these gentle folk who were so quickly destroyed by the Spaniards. Columbus's bastard son, Ferdinand, includes a primitive ethnographic report of them in his *Life of the Admiral Christopher Columbus By His Son Ferdinand,* ed. and trans. Benjamin Keen (New Brunswick, 1959). A much more sympathetic but necessarily more speculative reconstruction of Arawakan beliefs is found in Fred Olsen, *On the Trail of the Arawaks* (Norman, Okla., 1974). The narrative of the origin of disease and death is taken from Walter E. Roth, "An Inquiry into the Animism and Folk-Lore of the Guiana Indians," *Bureau of American Ethnology, 30th Annual Report, 1908–1909* (Washington, 1915). The heroic protectors of the Amazon tribes, the brothers Orlando Villas Boas and Claudio Villas Boas, collected the myth of the journey to the battlefield of the dead in *Xingu: The Indians, Their Myths* (New York, 1970). Frank Hamilton Cushing's *Zuñi Breadstuff* has been a classic since it was first printed more than half a century ago; I drew on it for the story of Há-wi-k'uh. George Bird Grinnell made crucial collections of traditions of the Cheyenne, Pawnee, and Blackfoot just as their cultures were coming apart. The one I use is from *Blackfoot Lodge Tales* (Lincoln, Neb., 1962). The "Story-Telling Stone" is a justly famous myth, first printed by Jeremiah Curtin in *Seneca Indian Myths* (New York, 1923). Joseph Campbell's paradigm of the hero story is in *The Hero with a Thousand Faces.*

## DEFLORATION

Though much of the primary material relating to the voyages of Columbus has been lost, there is still enough to give us a fairly accurate picture of his activities. J. M. Cohen, ed. and trans., *The Four Voyages of Christopher Columbus* (Baltimore, Harmondworth, England, 1969), puts together a fine synoptic collection of relevant documents. Benjamin Keen's edition of *The Life of the Admiral Christopher Columbus By His Son Ferdinand* contains a version of some of that lost primary material. For some years Samuel Eliot Morison's *Admiral of the Ocean Sea: A Life of Christopher Columbus* (Boston, 1942) has been considered the standard biography; it was most useful to me in detailing the Admiral's navigational achievements. As for the background of Columbus's voyages, I have relied on two works by C. Raymond Beazley: *The Dawn of Modern Geography,* Vol. III; and *Prince Henry the Navigator: The Hero of Modern Portugal and of Modern Discovery* . . . (London, New York, 1923). Andrée M. Collard's edition of Las Casas's *History of the Indies* (New York, 1971) puts the impassioned words of this hero of history in the hands of the contemporary reader. And though the accuracy of every assertion of Las Casas is open to question, this should not be used—as it so often has been—to discredit him. In addition to the brutal fact that the island Arawak have been extinct for more than two centuries, there are too many contemporary corroborations of Spanish practices in the islands to dismiss Las Casas as a raging ideologue or his words as merely a stick the English employed to pummel their colonialist rivals. To cite but two such corroborations, both the anonymous chronicler of the Soto expedition, the "Knight of Elvas," and Augustín de Zárate in his account of the sack of Peru remark on the incredibly rapid disappearance of the island natives. Zárate's remark is the more striking in that it comes in the midst of a criticism of Las Casas and his ilk: "Indeed the native population had been decreasing at such a rate that soon none would have been left in New Spain or Peru or the other countries where they still survived. In Santo Domingo, Cuba, Puerto Rico, Jamaica and other islands there was already no memory of any natives at all." Zárate wrote this sometime after his visit to the New World in 1544. See J. M. Cohen, ed. and trans., *The Discovery and Conquest of Peru* (Baltimore, Harmondsworth, England, 1968). For the Knight of Elvas's description of conditions on Cuba, see Edward G. Bourne, ed., and Buckingham Smith, trans., *Narratives of the Career of Hernando de Soto . . .* (New York, 1922), Vol. I. Finally, Lewis Hanke, *Aristotle and the American Indian: A Study of Race Prejudice in the Modern World* (Bloomington, Indiana; London, 1975), has a moving account of Las Casas in his own time.

Pierre Bontier and Jean le Verrier. *The Canarian; or, Book of the Conquest and Conversion of the Canarians . . .*, ed. and trans. Richard Henry Major. London, 1872.

Ernle Bradford. *Christopher Columbus.* New York, 1973.

Carl J. Jung. *Modern Man in Search of a Soul,* trans. W. S. Dell and Gary F. Baynes. New York, 1933.

Woodbury Lowerey. *The Spanish Settlements Within the Present Limits of the United States.* New York, London, 1911. 2 vols.

Richard Henry Major. *The Life of Prince Henry of Portugal . . .* London, 1868.

Fred Olsen. *On the Trail of the Arawaks.* Norman, Okla., 1974.

Darcy Ribeiro. *The Americas and Civilization,* trans. Linton Thomas Barrett and Marie McDavid Barrett. New York, 1971.

George R. Stewart. *Names on the Land: A Historical Account of Place-Naming in the United States.* Boston, 1958.

## PENETRATION

For my view of the conquest of the Aztec empire I have relied heavily, as have so many before me, on the first-hand account of Bernal Díaz del Castillo, *The Discovery and Conquest of Mexico,* ed. and trans., A. P. Maudslay (London, 1928); and on Cortés's dispatches to his king in the Irwin R. Blacker and Harry M. Rosen edition, *Conquest: Dispatches of Cortés From the New World* (New York, 1962). Miguel Leon-Portilla, ed., *The Broken Spears: The Aztec Account of the Conquest of Mexico,* trans. Maria Garibay K. and Lysander Kemp (Boston,

1962), tells the story from the native point of view. I have also relied on Leon-Portilla for my sense of the Aztec's preconquest history, as well as on Laurette Séjourné, *Burning Water: Thought and Religion in Ancient Mexico* (New York, 1960). The history of the Valley of Mexico before the coming of the Europeans is remarkably complex—much more so than my treatment of it here can afford to suggest. Yet I believe I have been faithful to its main outlines as these are presently understood. Two works that do fine jobs of tracing the interlaced lines of migration and development in the region are Eric Wolf, *Sons of the Shaking Earth: The People of Mexico and Guatemala* . . . (Chicago, London, 1959), and Ignacio Bernal, *Mexico Before Cortez: Art, History and Legend* (Garden City, N.Y., 1975). Alfonso Caso, *The Aztecs: People of the Sun*, trans. Lowell Dunham (Norman, Okla., 1970), contains a lucid description of the Aztec religious pantheon and has the additional merit of magnificent illustrations of the Aztec gods by Miguel Corvarrubias.

Daniel G. Brinton. *American Hero Myths.* Philadelphia, 1882.

Diego Durán. *Book of the Gods and Rites and the Ancient Calendar,* trans., ed. Fernando Horcasitas and Doris Heyden. Norman, Okla., 1971.

Francisco Lopez de Gómara. *Cortés: The Life of the Conqueror by His Secretary,* ed. and trans., Leslie Byrd Simpson. Berkeley, Los Angeles, 1966.

Bartolomé de Las Casas. *History of the Indies,* trans. Andreé M. Collard. New York, 1971.

Claude Lévi-Strauss. *Tristes Tropiques: An Anthropological Study of the Primitive Societies in Brazil,* trans. John Russell. New York, 1972.

Salvador de Madariaga. *Hernán Cortés: Conqueror of Mexico.* London, Sydney, Toronto, 1968.

Edmundo O'Gorman. *The Invention of America: An Inquiry into the Historical Nature of the New World and the Meaning of Its History.* Bloomington, Ind., 1961.

## THE LOST COLONY

Anyone studying the Lost Colony must begin and end with the scrupulous collections of David Beers Quinn for the Hakluyt Society, *The Voyages and Colonizing Enterprises of Sir Humphrey Gilbert* (London, 1940), 2 vols.; and *The Roanoke Voyages, 1584–1590* (London, 1955), 2 vols. The latter is especially crucial and contains the accounts of Peckham, Barlowe, Lane, Harriot, and White upon which I draw for my construction of the Roanoke venture. There is also a fine facsimile edition of Harriot's *Briefe and true report of the new found land of Virginia* as the Flemish historian/iconographer Theodore de Bry published it in 1590, including his copies of John White's paintings (New York, 1972). The largest uncited debt here (and elsewhere) is to Roy Harvey Pearce for his studies of the impact of the New World on the European mind, especially the classic *Savagism and Civilization: A Study of the Indian and the American Mind* (Baltimore, 1967). Other literature on the conjectured fate of the Lost Colony is to be found in the references to the chapter "Possession."

Francis Bacon. *The Physical and Metaphysical Works of Lord Bacon . . .* , ed. Josephy Devey. London, 1860.

Leo Bagrow. *History of Cartography*, ed. R. A. Skelton and trans. D. L. Paisey. Cambridge, 1964.

Marie Boas. *The Scientific Renaissance, 1450–1630*. New York, 1966.

Paul Green. *The Lost Colony: A Symphonic Drama of American History*. Chapel Hill, N.C., 1954.

A. R. Hall. *The Scientific Revolution, 1500–1800: The Formation of the Modern Scientific Attitude*. Boston, 1966.

V. T. Harlow, ed. *Raleigh's Last Voyage*. New York, 1971.

Francis Jennings. *The Invasion of America: Indians, Colonialism, and the Cant of Conquest*. Chapel Hill, N.C., 1975.

Alvin Josephy. *The Patriot Chiefs*. New York, 1961.

Woodbury Lowerey. *The Spanish Settlements Within the Present Limits of the United States*. New York, London, 1911. 2 vols.

Perry Miller. "The Marrow of Puritan Divinity," *Errand into the Wilderness*. New York, 1964.

Lewis Mumford. *Technics and Civilization*. New York, Burlingame, 1963.

Darcy Ribeiro. *The Americas and Civilization*, trans. Linton Thomas Barrett and Marie McDavid Barrett. New York, 1971.

E. M. W. Tillyard. *The Elizabethan World Picture: A Study of the Idea of Order in the Age of Shakespeare, Donne and Milton*. New York, n.d.

Max Weber. *The Protestant Ethic and the Spirit of Capitalism: The Relationships Between the Economic and Social Life of Modern Culture*, trans. Talcott Parsons. New York, 1930.

## THINGS OF DARKNESS

The literature on the English colonial experience in New England is formidably vast. My view of that experience has been most powerfully shaped by the narrative of Bradford in Samuel E. Morison's admirable edition, *Of Plymouth Plantation* (New York, 1970); by Cotton Mather's *Magnalia Christi Americana; or the Ecclesiastical History of New England* (London, 1702); and by the Reverend William Hubbard, *A Narrative of the Troubles with the Indians . . .*, ed. Samuel G. Drake (Roxbury, Mass., 1865), 2 vols. I am indebted for longer and more comprehensive views to the scholarship of Perry Miller, *The New England Mind: The Seventeenth Century* (Boston, 1961); *The New England Mind: From Colony to Province* (Boston, 1961); and especially his *Errand into the Wilderness*, (New York, 1964) which is rich in hints and suggestions. Larzer Ziff's *Puritanism in America: New Culture in a New World* (New York, 1974) seems to me informed and intelligent, though as his subtitle indicates, I have differed considerably about what was new in the Puritan experience. Francis Jennings' *The Invasion of America* (Chapel Hill, 1975) and Richard Slotkin's *Regeneration Through Violence* (Middletown, Connecticut, 1973) have been more than merely helpful; they encouraged me to believe that despite the mountain of serious scholarship on the subject there was much more to be said. Kai T. Erikson, *Wayward Puritans: A Study in the Sociology of Deviance* (New York, London, and Sydney, 1966), was both informative on New England dissenters and provocative in his analysis of their position in relation to orthodoxy. The scholar of early American culture, Mason I. Lowance, was most helpful to me in understanding the significance of Puritan typology. Behind all of this, like Augustine behind the pulpits of the Puritan preachers, looms William Carlos Williams' *In the American Grain* (New York, 1925). If some of what I say here has a familiar ring, it is a tribute to the poet's profound understanding of the roots of American culture and not, as Edward Dahlberg would say, an instance of "high-born stealth."

Augustine. *The City of God,* trans. Marcus Dods. New York, 1950.

Sacvan Bercovitch, ed. *Typology and Early American Literature.* Amherst, 1972.

Richard Bernheimer. *Wild Men of the Middle Ages: A Study in Art, Sentiment, and Demonology.* New York, 1970.

George Lincoln Burr, ed. *Narratives of the Witchcraft Cases, 1648–1706* in *Original Narratives of Early American History.* New York, 1972.

Norman Cohn. *The Pursuit of the Millennium.* New York, 1970.

John Dillenberger, ed.. *John Calvin: Selections from His Writings.* New York, 1971.

Samuel G. Drake, ed. *The Old Indian Chronicle; Being a Collection of Exceeding Rare Tracts, Written and Published in the Time of King Philip's War.* Boston, 1867.

Joseph Haroutunian and Louise Pettibone Smith, eds. and trans. *Calvin's Commentaries* in *The Library of Christian Classics.* London, 1958. Vol. XXIII.

Dwight B. Heath, ed. *Mourt's Relation.* New York, 1963.

Cotton Mather. *The Wonders of the Invisible World: Being an Account of the Tryals of Several Witches Lately Executed in New England.* London, 1862.

Increase Mather. *A Relation of the Troubles which have hapned in New-England, By reason of the Indians there . . .* New York, 1972.

——. *Remarkable Providences.* London, 1890.

Samuel Mather. *The Figures and Types of the Old Testament . . .,* ed. Mason I. Lowance, Jr. New York, London, 1969.

Johannes Pedersen. *Israel: Its Life and Culture,* trans. H. Milford. Copenhagen, London, 1959. 2 vols.

Samuel Purchas. *Hakluytus Posthumus; or, Purchas His Pilgrims.* Glasgow, 1906. Vol. XIX.

Mary Rowlandson. "Narrative of the Captivity of Mrs. Mary Rowlandson, 1682 . . ." in Charles H. Lincoln ed., *Narratives of Early American History*. New York, 1941.

William Shakespeare. *The Tempest,* ed. Robert Langbaum. New York, 1964.

Herbert M. Sylvester. *Indian Wars of New England*. Boston, 1910. Vol. II.

E. M. W. Tillyard. *The Elizabethan World Picture*. New York, n.d.

Benjamin Warfield. *Calvin and Augustine,* ed. Samuel G. Craig. Philadelphia, 1956.

Wilcomb Washburn. *Red Man's Land—White Man's Law: A Study of the Past and Present Status of the American Indian*. New York, 1971.

Hayden White. "The Forms of Wildness: Archeology of an Idea," in Edward Dudley and Maximillian E. Novak, eds., *The Wild Man Within*. Pittsburgh, 1972.

Roger Williams. *A Key into the Language of America,* ed. John J. Teunissen and Evelyn J. Hinz. Detroit, 1973.

PART THREE  Haunts

POSSESSION

The starting point for an interest in captivity narratives is now *The Garland Library of Narratives of North American Indian Captivities,* 311 titles in 111 volumes under the general editorship of the eminent Americanist Wilcomb Washburn. In reprinting these often very rare documents, Garland and Washburn have rendered an important service to Americanists, the effects of which will continue to be felt for years to come. A. Irving Hallowell's seminal article, "American Indians, White and Black: The Phenomenon of Transculturation," *Current Anthropology* (December 1963), has inspired a number of studies by calling intelligent and suggestive attention to the phenomenon. Among these are James Axtell, "The White Indians of Colonial America," *William and Mary Quarterly* (January 1975); and J. Norman Heard, *White into Red: A Study of the Assimilation of White Persons Captured by Indians* (Metuchen, N.J., 1973). All of these have been helpful to me in formulating my views in this chapter, but above all I have been impressed by the power and what I conceive to be the latent messages of the Vaca and Tanner narratives, and these have lain for some years within me awaiting proper expression. Haniel Long's wonderful little excursus on Vaca, *The Power Within Us* (New York, 1944), delighted me when I ran across it—as confirmations of intuitions always do. And Walter O'Meara, *The Last Portage* (Boston, 1962), was enlightening on Tanner's last years.

C. Raymond Beazley. *The Dawn of Modern Geography,* Vol. III. Oxford, 1906.

———. *Prince Henry the Navigator*. London, New York, 1923.

Erika Bourguignon. *Possession*. San Francisco, 1976.

Randolph G. Bourne, ed. *Narratives of the De Soto Expedition*, trans. Buckingham Smith. New York, 1922.

William Bradford. *Of Plymouth Plantation*, ed. Samuel E. Morison. New York, 1970.

Richard F. Burton, ed. *The Captivity of Hans Stade of Hesse in A.D. 1547–1555 Among the Wild Tribes of Eastern Brazil*, trans. Albert Tootal. London, 1874.

Cadwallader Colden. *The History of the Five Indian Nations; Depending on the Province of New-York in America*. Ithaca, London, 1964.

Joseph Conrad. *Heart of Darkness and The Secret Sharer*. New York, 1950.

———. *An Outcast of the Islands*. New York, 1964.

Cyclone Covey, ed. and trans. *Cabeza de Vaca's Adventures in the Unknown Interior of America*. New York, 1961.

Hector St. John de Crevecoeur. *Letters from an American Farmer; and Sketches of Eighteenth-Century America*. New York, 1963.

Adolph L. Dial and David K. Eliades. *The Only Land I Know: A History of the Lumbee Indians*. San Francisco, 1975.

Frederick W. Hodge, ed. *The Narrative of Alvar Nuñez Cabeça de Vaca* in *Spanish Explorers in the Southern United States, 1528–1543*. New York, 1907.

Edwin James, ed. *A Narrative of the Captivity and Adventures of John Tanner During Thirty Years Residence Among the Indians of North America*. New York, 1975.

Alvin M. Josephy, Jr. *The Patriot Chiefs*. New York, 1961.

Carl G. Jung. *Memories, Dreams, and Reflections*, ed. Aniela Jaffe and trans. Richard and Clara Winston. New York, 1963.

Weston La Barre. *The Ghost Dance*. New York, 1972.

Perry Miller. *Errand into the Wilderness*. New York, 1964.

Thomas Morton. *The New English Canaan*. Boston, 1883.

A. D. Nock. *Conversion*. London, Oxford, New York, 1972.

Martha Bennett Phelps. *Frances Slocum: The Lost Sister of Wyoming*. Wilkes Barre, Pa., 1916.

J. Buchan Telfer, ed. and trans. *The Bondage and Travels of Johann Schiltberger, a Native of Bavaria, in Europe, Asia, and Africa, 1396–1427*. London, 1879.

James E. Seaver, ed. *A Narrative of the Life of Mary Jemison, 1824*. New York, 1977.

Richard VanDerBeets, ed. "Captivity of Father Isaac Jogues, of the Society of Jesus, Among the Mohawks," trans. John G. Shea, *Held Captive by Indians: Selected Narratives, 1642–1836*. Knoxville, Ky., 1973.

———. "The Indian Captivity Narrative as Ritual," *American Literature* (January 1972).

Anthony F. C. Wallace. *The Death and Rebirth of the Seneca Nation*. New York, 1970.

Stephen B. Weeks. "The Lost Colony of Roanoke: Its Fate and Survival,"
Papers of the American Historical Association (October 1891).

## THE VANISHING NEW WORLD

Three works of exceptional scope and grace guided me in my sense of the clearing of America. These are: John Bakeless, The Eyes of Discovery: America as Seen by the First Explorers (New York, 1961); Peter Matthiessen, Wildlife in America (New York, 1959); and Roy M. Robbins, Our Landed Heritage: The Public Domain, 1776–1936 (Lincoln, Neb., 1962). The first two enjoy the great advantage of being based largely on first-hand accounts of that clearing while at the same time the authors have had personal experience with the things reported by their sources. Behind these works are two standards, Leo Marx's The Machine in the Garden: Technology and the Pastoral Ideal in America (New York, 1964); and Henry Nash Smith's Virgin Land: The American West as Symbol and Myth (Cambridge, Mass., 1950), which inevitably influence one's view of what happened in the nineteenth century to America's landscape. In the course of preparing an abortive collection of Mississippi Valley travel narratives some years ago, I read a good many such documents, but I list here only those I drew on directly for this chapter. My experience with these narratives makes me trust and admire the more the large observations of the authors listed above.

Maria Audubon and Elliott Coues, eds. Audubon and his Journals. London, 1898. 2 vols.

Thomas Ashe. Travels in America Performed in the Year 1806 . . . London, 1809.

William Bartram. The Travels of William Bartram, ed. Mark van Doren. New York, 1928.

Randolph G. Bourne, ed. Narratives of the De Soto Expedition, trans. Buckingham Smith, 2 vols. New York, 1922.

Mary Clavers (Caroline M. Kirkland), A New Home—Who'll Follow? New York and Boston, 1839.

Robert M. Coates. The Outlaw Years: The History of the Land Pirates of the Natchez Trace. New York, 1930.

Elliot Coues, ed. History of the Expedition under the command of Lewis and Clark. New York, 1965. 3 vols.

David A. Dary. The Buffalo Book; The Saga of an American Symbol. New York, 1974.

J. Frank Dobie. Apache Gold and Yaqui Silver. Albuquerque, N.M., 1976.

Richard I. Dodge. Our Wild Indians: Thirty-Three Years' Personal Experience Among the Red Men of the Great West . . . Hartford, 1883.

Ralph Waldo Emerson. The Complete Works of Ralph Waldo Emerson. Cambridge, 1903.

Philip F. Gura. "Thoreau's Maine Woods Indians: More Representative Men," *American Literature* (November 1977).

James D. Horan and Paul Sann. *Pictorial History of the Wild West* . . . New York, 1954.

Edward Hyams. *Soil and Civilization*. New York, 1976.

Robert V. Hine. *The American West: An Interpretive History*. Boston, 1973.

Helen Hunt Jackson. *A Century of Dishonor*. New York, 1965.

C. F. McGlashan. *History of the Donner Party; a Tragedy of the Sierra*. Stanford, Cal., 1974.

Lewis Mumford. *The Brown Decades*. New York, 1931.

———. *Sticks and Stones*. New York, 1924.

Roderick Nash. *Wilderness and the American Mind*. (New Haven, London, 1967).

Henry Blackman Sell and Victor Weybright. *Buffalo Bill and the Wild West*. New York, 1955.

Henry David Thoreau. *The Maine Woods*. Boston, 1873.

———. *Walden; or, Life in the Woods* and *On the Duty of Civil Disobedience*. New York, Toronto, 1960.

———. *Journals*, ed. Bradford Torrey and Francis H. Allen. Boston, 1949.

Walter Prescott Webb. *The Great Plains*. Boston, 1931.

Glenway Wescott. *Good-bye Wisconsin*. New York, 1928.

## A DANCE OF THE DISPOSSESSED

Until quite recently the Plains story was almost always understood in terms of the cowboys-and-Indians antagonism in which the only function of the latter was to serve as flat, featureless opponents against whom the former could test themselves and prove heroic in victory. Dee Brown's immensely popular *Bury My Heart at Wounded Knee* (New York, 1970) changed this, but for some time prior to this it had been possible to see things from a native perspective through the publication and reprinting of autobiographical documents of Plains Indians. Reading these, we are permitted a glimpse into the lodges of the tribes and so can feel something of that contentment, excitement, and loss. I have been especially impressed with the narratives of George Bent, Luther Standing Bear, Black Elk, and Plenty-Coups. (Because of their somewhat cumbersome citations these are listed below.) James Willard Schultz's *My Life as an Indian* (New York, 1907), though it is a white account of the Plains Indians, is a marvelous book, everywhere infused with a sense of loss made the more profound by the author's complicity in it. David Thomas's and Karen Ronnefeldt's skillful, beautiful edition of Prince Maxmillian's 1833 Missouri River expedition, *People of the First Man* (New York, 1976), gives the most vivid sense I know of the shock of first contact. Ralph Andrist's *The Long Death: The Last Days of the Plains Indians* (New York, 1964) remains the best general account of the Plains campaigns. The classic on the Ghost Dance movement is James Mooney's monograph, now reprinted as *The Ghost-Dance Religion and the Sioux Outbreak of 1890*, ed. A. F. C. Wallace (Chicago, 1965). Mooney was in the field within weeks of the massacre at Wounded Knee and collected records of the chants and dance patterns from survivors.

Annie Heloise Abel, ed. and trans. *Tabeau's Narrative of Loisel's Expedition to the Upper Missouri*. Norman, Okla., 1939.

Robert Berkhofer, Jr. *The White Man's Indian: Images of the American Indian from Columbus to the Present*. New York, 1976.

William Bradford. *Of Plymouth Plantation*, ed. Samuel E. Morison. New York, 1970.

George Catlin. *Letters and Notes on the Manners, Customs, and Conditions of the North American Indians* . . . New York, 1973. 2 vols.

Richard I. Dodge. *Our Wild Indians*. Hartford, Conn., 1884.

George Bird Grinnell. *Blackfoot Lodge Tales*. Lincoln, Neb., 1962.

——. *By Cheyenne Campfires*. New Haven, London, 1926.

——. *The Cheyenne Indians*. Lincoln, Neb., 1972. 2 vols.

George E. Hyde. *Life of George Bent Written from his Letters*, ed. Savoie Lottinville. Norman, Okla., 1968.

Weston La Barre. *The Ghost Dance*. New York, 1972.

Frank B. Linderman. *Plenty-Coups, Chief of the Crows*. Lincoln, Neb., 1962.

James McLaughlin. *My Friend the Indian*. Boston, New York, 1910.

Thomas Marquis. *Wooden Leg: A Warrior Who Fought Custer*. Lincoln, Neb., 1967.

Peter Nabokov. *Two Leggings: The Making of a Crow Warrior*. New York, 1967.

John G. Neihardt. *Black Elk Speaks; Being the Life of a Holy Man of the Oglala Sioux*. Lincoln, Neb., 1961.

Charles Olson. *Call Me Ishmael*. New York, 1947.

Henry Blackman Sell and Victor Weybright. *Buffalo Bill and the Wild West*. New York, 1955.

Michael A. Sievers. "The Historiography of . . . Wounded Knee," *South Dakota History* (Winter 1975).

Henry Nash Smith. *Virgin Land*. Cambridge, Mass., 1950.

Chief Luther Standing Bear. *Land of the Spotted Eagle*. Boston, New York, 1933.

John Stands in Timber and Margot Liberty. *Cheyenne Memories*. Lincoln, Neb., 1972.

Jerome Stillson. "Sitting Bull Talks," New York *Herald* (November 16, 1877).

Stanley Vestal. *New Sources of Indian History, 1850–1891; The Ghost Dance; The Prairie Sioux: A Miscellany*. Norman, Okla., 1934.

# Index